MOLECULAR THEORY OF CAPILLARITY

J. S. Rowlinson
Dr. Lee's Professor of Chemistry Emeritus
University of Oxford

and

B. Widom
Goldwin Smith Professor of Chemistry
Cornell University

DOVER PUBLICATIONS, INC.
Mineola, New York

Bibliographical Note

This Dover edition, first published in 2002, is an unabridged republication of the edition published by the Clarendon Press of Oxford University Press, Oxford and New York, in 1989. The work was originally published in 1982; reprints in 1984 and 1989 contained minor revisions and additional material.

Library of Congress Cataloging-in-Publication Data

Rowlinson, J. S. (John Shipley), 1926-
    Molecular theory of capillarity / by J.S. Rowlinson and B. Widom.
        p. cm.
    Originally published: Oxford, Oxfordshire: Clarendon Press, 1982.
    Includes bibliographical references and index.
    ISBN 0-486-42544-4 (pbk.)
        1. Capillarity. 2. Molecular theory. I. Widom, B. II. Title.

QC183 .R75 2002
541.3'3—dc21

2002074022

Manufactured in the United States of America
Dover Publications, Inc., 31 East 2nd Street, Mineola, N.Y. 11501

PREFACE

The tension at the surface of a liquid is one of the more striking manifestations of the forces that act between molecules, and attempts to explain it in these terms go back to the eighteenth century. The early attempts were in terms of crude mechanical models of a liquid, which we describe in Chapter 1. These pre-thermodynamic theories were abandoned in the nineteenth century to be replaced by more phenomenological or quasi-thermodynamic methods. We introduce these in the second and third chapters, and use them extensively in the last part of the book, since they are, when handled rightly, still powerful methods of treating complicated problems such as those of multi-phase equilibria and critical phenomena.

The re-establishment of a direct link between capillary phenomena and the intermolecular forces had to await the development of the statistical mechanics of systems that are highly inhomogeneous on the scale of length of the range of these forces. This topic is the subject of our long fourth chapter, the results of which provide a justification for and show the limitations of the earlier quasi-thermodynamic theories. The statistical results are applied in mean-field approximation to some tractable but artificial model systems in Chapter 5. More realistic models are treated both by computer simulation (Chapter 6) and by approximating the exact statistical equations (Chapter 7).

Our emphasis throughout is on the liquid–gas surface; we consider liquid–liquid surfaces in the last two chapters, but say little about the liquid–solid surface. It is in this restricted sense (for which we have the authority of both Gibbs and van der Waals) that we use the word capillarity. The statistical mechanics of the liquid–solid surface is an important subject, and one which is being pursued actively, but we believe it is best left for review at a later date, and by other authors.

We have adopted a narrative style in which some subjects may be visited several times as they are discussed, in turn, by mechanical, thermodynamic, quasi-thermodynamic, and statistical mechanical arguments. We ask, therefore, that those who look for information on any one topic be prepared to take several dips into the book. To help such readers we provide ample cross-references. This method of treatment is not dissimilar to that of Bakker in his classic book of 1928, *Kapillarität und Oberflächenspannung*, or to that of Ono and Kondo in their 1960 *Encyclopedia of Physics* article.

Our subject has a long history, an active present, and, we believe, an exciting future as it moves ever further into areas of great theoretical and practical interest such as the physics of membranes and colloids.

We are grateful to the following friends and colleagues for having read one or more sections: J. A. Barker, K. E. Gubbins, J. R. Henderson, P. Schofield, and J. C. Wheeler; and we thank Miss W. E. Nelson for typing the final version of the manuscript.

Oxford and Cornell   J. S. R.
August 1981   B. W.

PREFACE TO THE PAPERBACK EDITION

We have taken advantage of a further reprinting of this book to add new references, mainly to reviews, to cover some of the vast amount of work published in the last seven years. These references will be found at the end of each chapter except for those for Chapters 8 and 9 which follow Appendix 1, where there is more room for them.

Oxford and Cornell   J. S. R.
November 1988   B. W.

CONTENTS

PRINCIPAL SYMBOLS

The section numbers show where the symbol is first used or defined. Where more than one number is shown there is a significant change of use or definition in the later section(s).

A area § 2.2

a capillary length § 1.3; van der Waals constant §§ 1.4, 5.2; (subscript) specific component label § 5.4

b van der Waals excluded volume §§ 1.4, 9.1; (subscript) specific component label § 5.4

C coefficient of a curvature term § 2.4; a correlation function § 4.7; heat capacity § 9.3

c number of components § 2.2; lattice coordination number § 5.3; (superscript) critical § 1.4

$c(r_{12})$ direct correlation function § 4.2

$c^{(1)}(\mathbf{r})$ a one-body 'correlation function' § 4.2

D thickness of interface § 6.2

d range of intermolecular force § 1.2; dimensionality § 2.2; molecular diameter § 4.6

e (subscript) equimolar dividing surface § 2.4

$\mathbf{e}$ unit vector § 2.5

F a force § 1.2; Helmholtz free energy § 2.2

f an intermolecular force § 1.2; a distribution function § 4.2

g acceleration due to gravity § 1.3; pair distribution function § 4.2; (superscript) gas § 1.2

H Laplace's integral for the surface tension § 1.2; a correlation function § 4.7

$\mathcal{H}$ Hamiltonian § 4.2

h Planck's constant § 4.2; total correlation function § 4.2

i (subscript) generic component label § 2.3; generic molecular label § 4.2

J momentum density § 4.7

j a smooth function of density § 9.3; (subscript) generic molecular label § 4.2

K Laplace's integral for the internal pressure § 1.2; a 'kinetic energy' § 3.3; a constant § 9.4

k wetting coefficient § 1.3; Boltzmann's constant § 1.4

L a macroscopic length §§ 3.5, 4.9, 6.2, 8.6

l a microscopic length § 4.9; width of cell § 6.2; (superscript) liquid § 1.2; line or linear § 8.6

M a chemical potential § 3.1

m mass of a molecule § 1.3; second moment of $u(r)$ § 1.5, or of $c(r)$ § 4.6

N number of molecules § 1.4; (subscript) normal to the surface § 2.5

n molar amount of substance § 2.2; (superscript) order of a distribution function § 4.2

P a probability § 5.4

p (scalar) pressure § 1.2; order of a critical point § 9.3

(pp)   (preceding equation number) valid only for pair potentials § 4.2

$\mathbf{p}$ pressure tensor § 2.5

$\mathbf{p}$ momentum § 4.2

q wave number § 1.3; a probability § 5.4

R a macroscopic radius §§ 1.2, 1.3, 4.8, 8.6

r separation of two points or two molecules § 1.2; distance from centre of a drop § 4.8

$\mathbf{r}_i$ position of point or molecule i; $\mathbf{r}_{ij} = \mathbf{r}_j - \mathbf{r}_i$ § 4.2

S entropy § 2.2; coefficient of spreading § 8.2

s separation of two points or molecules § 1.2; dimensionality of a potential § 5.4; a chord length or thickness § 8.6; a field variable § 9.5; (subscript) surface of tension § 2.2; (superscript) solid § 1.3; surface excess § 2.3

$\mathbf{s}$ a horizontal vector § 4.7

T temperature § 1.3; (subscript) tangential to the surface § 2.5

t a field variable § 9.1

U internal energy § 1.2

$\mathscr{U}$ configurational energy § 4.2

u abbreviation of $u(\mathbf{r})$ § 4.2; (subscript) uniform § 4.2

$u(r)$ or
$u(r_{ij})$ intermolecular potential § 1.2

$u(\mathbf{r})$ a one-body potential § 4.2

$u(x)$ a function of density § 9.3

u_0, u_1 reference and perturbation potentials §§ 4.5, 7.2, 7.5

V volume § 1.2

$\mathscr{V}$ external potential § 4.2

v_0 a microscopic volume §§ 5.3, 5.5, 7.3

$v(\mathbf{r})$ external potential at $\mathbf{r}$ § 4.2

$v(x)$ a function of density § 9.3

W work § 2.2; a difference between two values of ψ § 3.1

$\mathscr{W}$ a covered volume § 5.5

w an energy § 8.4

x a horizontal distance § 4.3; a density §§ 5.4, 5.6, 8.3, 9.1; mole fraction § 8.5

y a horizontal distance § 4.3; a density §§ 5.4, 5.6, 8.3, 9.1

$y(r)$	a distribution function § 4.2
Z	phase integral § 4.2
z	height § 1.5
α	a coupling parameter § 4.5; contact angle § 8.2; a critical exponent § 9.3; (superscript) a phase label § 2.2
β	contact angle § 8.2; a critical exponent § 9.3; (superscript) a phase label § 2.2
Γ	adsorption § 2.3
γ	contact angle § 8.2; a critical exponent § 9.3; (superscript) a phase label § 8.2
Δ	difference between gas and liquid § 2.3; $= \frac{1}{2}(\rho^1 - \rho^g)$ § 5.5
δ	$= z_e - z_s$ § 2.4; a critical exponent § 9.3
ε	an interaction energy §§ 4.4, 5.3; an eigenfunction § 4.7; a parameter § 8.4; $= 4 - d$, where d is dimensionality § 9.7
ζ	height of a dividing surface § 4.9; a reduced activity §§ 5.1, 5.3
η	entropy density § 2.3; a reduced number density § 7.2; a critical exponent § 9.2
θ	angle of contact §§ 1.2, 2.2; a polar coordinate §§ 4.2, 8.6; $= \varepsilon/kT$ § 5.3
κ	isothermal compressibility §§ 4.2, 9.5
Λ	de Broglie wavelength § 4.2; linear adsorption § 8.6
λ	activity § 4.2; an eigenvalue § 4.7
μ	chemical potential § 2.2; a critical exponent § 2.3
ν	a thermodynamic field § 2.3; a critical exponent § 9.3
Ξ	grand partition function § 4.2
ξ	correlation length §§ 3.5, 8.3
π	$= pv_0/kT$ § 5.5
ρ	number density §§ 1.2, 5.3, 6.2
$\rho^{(n)}$	equilibrium distribution function § 4.2
$\hat{\rho}^{(n)}$	arbitrary distribution function § 4.5
σ	surface tension § 1.2
$\sigma(z)$	an effective density at height z § 5.5
τ	$= kT/\varepsilon$ § 6.2; line tension § 8.6
$\tau(z)$	an effective density at height z § 5.5
ϕ	energy density §§ 1.2, 5.5; an angle that describes 'distance' from a critical point § 9.5
$\phi^{(2)}$	energy density of a two-phase system § 5.5
Ψ	a difference of two values of ψ § 3.1
ψ	free-energy density § 2.3
Ω	grand potential § 2.2
ω	solid angle § 2.4; molecular orientation § 4.2; a probability § 5.1; a density § 9.6

Les Tubes Capillaires nous mettent entre les mains un indice palpable de la généralité de cette loi qui est la clef de la physique, le plus grand ressort de la nature, et le mobile universel de l'Univers.

J.-J. LeF. de Lalande (1768)

MOLECULAR THEORY OF CAPILLARITY

MECHANICAL MOLECULAR MODELS

1.1 Introduction

If a glass tube with a bore as small as the width of a hair (Latin: *capillus*) is dipped into water then the liquid rises in the tube to a height greater than that at which it stands outside. The effect is not small; the rise is about 3 cm in a tube with a bore of 1 mm. This apparent defiance of the laws of hydrostatics (which were an achievement of the seventeenth century) led to an increasing interest in capillary phenomena as the eighteenth century advanced. The interest was two-fold. The first was to see if one could characterize the surfaces of liquids and solids by some simple mechanical property, such as a state of tension, that could explain the observed phenomena. The things to be explained were, for example, why does water rise in a tube while mercury falls, why is the rise of water between parallel plates only a half of that in a tube with a diameter equal to the separation of the plates, and why is the rise inversely proportional to this diameter? The second cause of interest was the realization that here were effects which must arise from cohesive forces between the intimate particles of matter, and that the study of these effects should therefore tell something of those forces, and possibly of the particles themselves. In this book we follow the first question only sufficiently far to show that a satisfactory set of answers has been found; our interest lies, as did that of many of the best nineteenth-century physicists, in the second and more difficult question, or, more precisely, in its inverse—how are capillary phenomena to be explained in terms of intermolecular forces.

We could attempt an answer by summarizing the experimental results and then bringing to bear on them at once the whole armoury of modern thermodynamics and statistical mechanics. To do this, however, would be to throw away much of the insight that has been gained slowly over the last two centuries. Indeed the way we now look at capillary phenomena, and more generally at the properties of liquids, is conditioned by the history of the subject. In the opening chapters we follow the way the subject has developed, not with the aim of writing a strict history, but in order to trace the many strands of thought that have led to our present understanding.

In this first chapter we describe the early attempts to explain capillarity which were based on an inevitably inadequate understanding of the molecular structure and physics of fluids. Most of the equations of this chapter are therefore only crude approximations which are superseded by exact or, at least, more accurate equations in the later chapters.

1.2 Molecular mechanics

That matter was not indefinitely divisible but had an atomic or molecular structure was a working hypothesis for most scientists from the eighteenth century onwards. There was a minor reaction towards the end of the nineteenth century when a group of physicists who professed a positivist philosophy pointed out how indirect was the evidence for the existence of atoms, and their objections were not finally overcome until the early years of this century. If in retrospect, their doubts seem to us to be unreasonable we should, perhaps, remember that almost all those who then believed in atoms believed equally strongly in the material existence of an electromagnetic ether and, in the first half of the nineteenth century, often of a caloric fluid also. Nevertheless those who contributed most to the theories of gases and liquids did so with an assumption, usually explicit, of a discrete structure of matter. The units might be named *atoms* or *molecules* (e.g. Laplace) or merely *particles* (Young), but we will follow modern convention and use the word molecule for the constituent element of a gas, liquid, or solid.

The forces that might exist between molecules were as obscure as the particles themselves at the opening of the nineteenth century. The only force about which there was no doubt was Newtonian gravity. This acted between celestial bodies; it obviously acted between one such body (the Earth) and another of laboratory mass (e.g. an apple); Cavendish[1] had recently shown that it acted equally between two of laboratory mass, and so it was presumed to act also between molecules. In early work on liquids we find the masses of molecules and mass densities entering into equations where we should now write numbers of molecules and number densities. In a pure liquid all molecules have the same mass so the difference is unimportant. It was, however, clear before 1800 that gravitational forces were inadequate to explain capillary phenomena and other properties of liquids. The rise of a liquid in a glass tube is independent of the thickness of the glass;[2] thus only the forces from the molecules in the surface layer of the glass act on those in the liquid. Gravitational forces, however, fall off only as the inverse square of the distance and were known to act freely through intervening matter.

The nature of the intermolecular forces other than gravity was quite obscure, but speculation was not lacking. The Jesuit priest Roger Boscovich[4] believed that molecules repel at very short distances, attract at slightly larger separations and then show alternate repulsions and attractions of ever decreasing magnitude as the separation becomes ever larger. His ideas influenced both Faraday[5] and Kelvin[6] in the next century but were too elaborate to be directly useful to those who were to study the theory of capillarity. They wisely contented themselves with simpler hypotheses.

The cohesion of liquids and solids, the condensation of vapours to liquids, the wetting of solids by liquids and many other simple properties of matter all pointed to the presence of forces of attraction many times stronger than gravity but acting only at very short separations of the molecules. Laplace said that the only condition imposed on these forces by the phenomena were that they were 'insensible at sensible distances'. Little more could in fact be said until 1929.

The repulsive forces gave more trouble. Their presence could not be denied; they must balance the attractive forces and prevent the total collapse of matter, but their nature was quite obscure. Two misunderstandings complicated the issue. First, heat was often held to be the agent of repulsion[7] for, so the argument ran, if a liquid is heated it first expands and then boils, thus separating the molecules to much greater distances than in the solid. The second arose from a belief which went back to Newton that the observed pressure of a gas arose from static repulsions between the molecules, and not, as Daniel Bernoulli had argued in vain, from their collisions with the walls of the vessel.[8]

With this background it was natural that the first attempts to explain capillarity, or more generally the cohesion of liquids, were based on a static view of matter. Mechanics was the theoretical branch of science that was well understood; thermodynamics and kinetic theory lay still in the future. The key assumption in this mechanical treatment was that of strong but short-ranged attractive forces. Liquids at rest, whether in a capillary tube or not, are clearly at equilibrium, so these attractive forces must be balanced by repulsions. Since even less could be guessed about these than about the attractive forces, they were often passed over in silence, and, in Rayleigh's phrase, 'the attractive forces were left to perform the impossible feat of balancing themselves'.[9] Laplace[10-12] was the first to deal with the problem satisfactorily by supposing that the repulsive forces (of heat, as he supposed) could be replaced by an internal or intrinsic pressure that acted throughout an incompressible liquid. (This supposition leads to occasional uncertainty in nineteenth-century work as to exactly what is meant by the pressure in a liquid.) Our first task is to follow Laplace's calculation of the internal pressure. It must balance the cohesive force in the liquid and he identified this with the force per unit area that resists the pulling asunder of an infinite body of liquid into two semi-infinite bodies bounded by plane surfaces. Our derivation is closer to that of Maxwell[14] and Rayleigh[9] than to Laplace's original form, but there is no essential difference in the argument.

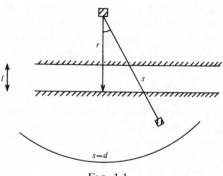

FIG. 1.1

Consider two semi-infinite bodies of liquid with sharp plane faces separated by vapour of negligible density of thickness l (Fig. 1.1), and let us take an element of volume in each. The first is in the upper body and is at height r above the plane surface of the lower body; its volume is $dx\,dy\,dr$. The second is in the lower body and its volume is $s^2 \sin \theta \, ds \, d\theta \, d\phi$, where the origin of the polar coordinates is the position of the first elementary volume. Let $f(s)$ be the force between two molecules at separation s, and let its range be d. Since the force is always attractive we have

$$f(s < d) < 0 \qquad f(s \geqslant d) = 0. \tag{1.1}$$

If the number density of molecules is ρ in both bodies then the vertical component of the force between the two elements of volume is

$$\rho \, dx \, dy \, dr \, \rho s^2 \sin \theta \, ds \, d\theta \, d\phi \, f(s) \cos \theta. \tag{1.2}$$

The total force of attraction per unit area (a positive quantity) is

$$F(l) = -\rho^2 \int_l^d dr \int_r^d ds \, s^2 f(s) \int_0^{\cos^{-1}(r/s)} d\theta \sin \theta \cos \theta \int_0^{2\pi} d\phi$$

$$= -\pi \rho^2 \int_l^d dr \int_r^d ds (s^2 - r^2) f(s). \tag{1.3}$$

Let $u(s)$ be the potential of the intermolecular force

$$u(s) = \int_s^d f(s') \, ds', \quad \text{or} \quad \frac{du}{ds} = -f(s) \quad \text{and} \quad u(s \geqslant d) = 0 \tag{1.4}$$

$$F(l) = \pi \rho^2 \int_l^d dr \int_r^d du(s) \, (s^2 - r^2)$$

$$= -2\pi \rho^2 \int_l^d dr \int_r^d ds \, su(s). \tag{1.5}$$

Integrate by parts again,

$$F(l) = -2\pi\rho^2 \int_l^d r(r-l)u(r)\,dr. \tag{1.6}$$

Laplace's internal pressure K is the attractive force per unit area between two planar surfaces in contact, that is, $F(0)$.

$$K \equiv F(0) = -\tfrac{1}{2}\rho^2 \int_0^d u(r)\,d\mathbf{r} \tag{1.7}$$

where $d\mathbf{r}$ is a volume element which may be written here $4\pi r^2\,dr$. Since $u(r)$ is everywhere negative or zero, by supposition, then K is positive. Laplace believed it to be large in relation to the atmospheric pressure, but it was left to Young (§ 1.6) to make the first realistic numerical estimate. The derivation above rests on the implicit assumption that the molecules are distributed uniformly with a density ρ, that is, that the liquid has no discernible structure on a scale of length measured by the range of the forces, d. Without this assumption we could not write (1.2) and (1.3) in these simple forms, but would have to ask how the presence of a molecule in the first volume element affected the probability of there being a molecule in the second. We return to this point in § 1.6.

By 1800 the concept of surface tension was commonplace; indeed it is almost irresistible to anyone who has tried to float a pin on water or perform similar experiments in childhood. What was lacking was a quantitative relation between this tension and the supposed intermolecular forces. A tension per unit length along an arbitrary line on the surface of a liquid must, in a coherent set of units, be equal to the work done in creating a unit area of free surface. This follows from the experiment of drawing out a film of liquid (Fig. 1.2). A measure of this work can be obtained at once[15] from the expression above for $F(l)$, (1.6). If we take the two semi-infinite bodies to be in contact, and then draw them apart

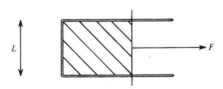

FIG. 1.2. A liquid film is held in a wire frame with the right-hand boundary fixed to a freely movable rider. The force F needed to balance the tension in the two-sided film is proportional to the length L. Let $F = 2\sigma L$. A displacement of the rider by a distance δx requires work $F\delta x = \sigma\delta A$, where δA is the increase in area. Thus the tension per unit length in a single surface, or *surface tension* σ, is numerically equal to the surface energy per unit area.

until their separation exceeds the range of the intermolecular forces, the work done per unit area is

$$H = \int_0^d F(l)\, dl = -\pi\rho^2 \int_0^d l^3 u(l)\, dl. \qquad (1.8)$$

The separation has produced two free surfaces and so the work done can be equated to twice the surface energy per unit area, which is equal to the surface tension σ; that is

$$\sigma = \tfrac{1}{2}H = -\tfrac{1}{8}\rho^2 \int_0^d r u(r)\, dr. \qquad (1.9)$$

Thus K is the integral of the intermolecular potential, or its zeroth moment, and H is its first moment. Whereas K is not directly accessible to experiment, H can be found if we can measure the surface tension. Before we turn to this point let us consider some further implications of the results so far obtained.

Let ϕ be the cohesive energy density at a point in the fluid, that is, the ratio $(\delta U/\delta V)$, where δU is the internal energy of a small sample of fluid δV which contains the point. For the molecular model we are using it is given by

$$2\phi = \rho^2 \int u(r)\, d\mathbf{r} \qquad (1.10)$$

where r is the distance from the point in question. Then, following Rayleigh,[9] we can identify Laplace's K as the difference of this potential 2ϕ between a point on the plane surface of the liquid, $2\phi_S$, and a point in the interior $2\phi_I$. At the surface the integration in (1.10) is restricted to a hemisphere of radius d, whilst in the interior it is taken over a complete sphere. Hence ϕ_S is half ϕ_I, or

$$2\phi_S - 2\phi_I = -\phi_I = -2\pi\rho^2 \int_0^d r^2 u(r)\, dr = K. \qquad (1.11)$$

Consider now a drop of radius R. The calculation of ϕ_I is unchanged, but the integration to obtain ϕ_S is now over an even more restricted volume because of the curvature of the surface. That is, if θ is the angle between the vector $\mathbf{r}$ and a fixed radius $\mathbf{R}$,

$$\phi_S = \pi\rho^2 \int_0^d dr\, r^2 u(r) \int_0^{\cos^{-1}(r/2R)} d\theta \sin\theta$$

$$= \pi\rho^2 \int_0^d r^2 \left(1 - \frac{r}{2R}\right) u(r)\, dr. \qquad (1.12)$$

The internal pressure in the interior of the drop is therefore

$$2\phi_S - 2\phi_1 = K + H/R = K + 2\sigma/R \qquad (1.13)$$

where H is given by (1.9). Had we taken not a spherical drop but a portion of liquid with a convex surface defined by its two principal radii of curvature R_1 and R_2, then we should have obtained an internal pressure of

$$K + \sigma(R_1^{-1} + R_2^{-1}). \qquad (1.14)$$

By a theorem of Euler,[18] the sum $(R_1^{-1} + R_2^{-1})$ is equal to the sum of the reciprocals of the radii of curvature of the surface along any two orthogonal tangents.

Since K and H are positive, and R is positive for a convex surface, it follows from (1.13) that the internal pressure in a drop is higher than that in a liquid with a plane surface. Conversely the internal pressure within a liquid bounded by a concave spherical surface is lower than that in the liquid with a plane surface since R is now negative. These results are the foundation of Laplace's theory of capillarity. The equation for the difference of pressure between p^1, that of the liquid inside a spherical drop of radius R, and p^g, that of the gas outside, is now called Laplace's equation:

$$p^1 - p^g = 2\sigma/R. \qquad (1.15)$$

It is quite general and not restricted to this molecular model. It is re-derived by purely thermodynamic arguments in Chapter 2 and used repeatedly in later chapters.

1.3 Capillary phenomena

It is interesting and useful to extend these results to a three-phase system of a solid in contact with a liquid and a vapour (again, of negligible density). The solid is of different chemical constitution from the liquid, and quite insoluble in it. We assume moreover, that it is a perfectly rigid molecularly uniform array of density ρ_2. The intermolecular potential between two molecules of the species in the liquid is denoted u_{11}, between two of those in the solid u_{22}, and that between a molecule of each u_{12}. By the argument of the last section, the force per unit area between a slab of liquid and one of solid at separation l is

$$F_{12}(l) = 2\pi\rho_1\rho_2 l \int_0^{d_{12}} r u_{12}(r)\, dr - 2\pi\rho_1\rho_2 \int_0^{d_{12}} r^2 u_{12}(r)\, dr \qquad (1.16)$$

and the work to separate the liquid and solid is

$$H_{12} = \int_0^{d_{12}} F_{12}(l)\, dl. \qquad (1.17)$$

This work is equal to the sum of the surface energies of the two new surfaces formed, liquid–gas and solid–gas, less that of the surface destroyed, liquid–solid;

$$H_{12} = \sigma^{lg} + \sigma^{sg} - \sigma^{ls}. \tag{1.18}$$

Thus the surface tension of the liquid–solid interface is

$$\sigma^{ls} = \tfrac{1}{2}H_{11} + \tfrac{1}{2}H_{22} - H_{12} \tag{1.19}$$

with $\sigma^{lg} = \tfrac{1}{2}H_{11}$ and $\sigma^{sg} = \tfrac{1}{2}H_{22}$, where, as in (1.9),

$$H_{ij} = -\tfrac{1}{4}\rho_i\rho_j \int_0^{d_{ij}} r u_{ij}(r) \, d\mathbf{r}. \tag{1.20}$$

The concept of a surface tension of an interface at which one of the phases is solid is a difficult one; it is hard even to arrive at an operationally unambiguous definition, and it is a field which we do not wish to explore further in this book.[19] However the elementary discussion above certainly emphasises the relevant physical factors that determine this surface tension or surface energy (however it be defined), namely that the magnitudes of σ^{lg}, σ^{sg}, and σ^{ls} depend on the magnitudes and ranges of the three intermolecular potentials u_{11}, d_{11}, u_{22}, d_{22}, and u_{12}, d_{12} in the way implied by (1.19) and (1.20). This was first stated by Young in a paper[20] which preceded that of Laplace by a few months, and which was on the whole a much less clear attempt to follow the same route in relating intermolecular forces to the cohesion and so surface tension of liquids. Young went further and used these results to calculate the angle of contact of a liquid and solid (Fig. 1.3). If the contact line is to move neither up nor down the solid surface then the three tensions must, he argued, be in equilibrium at that point, and so, by resolving the forces

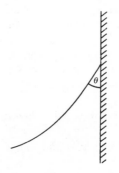

FIG. 1.3

parallel to the solid surface

$$\sigma^{sg} = \sigma^{ls} + \sigma^{lg} \cos \theta \quad \text{(Young's equation)} \tag{1.21}$$

or

$$k = (\sigma^{sg} - \sigma^{ls})/\sigma^{lg} = \cos \theta \tag{1.22}$$

where k is sometimes called the *wetting coefficient*. There is no obvious restriction on the magnitude of the positive quantities σ^{sg} and σ^{lg}, or on the magnitude and sign of σ^{ls}, since we know nothing *a priori* of the strength and ranges of the three intermolecular potentials. Hence k is apparently unrestricted in sign and size, but (1.22) has a solution for θ only if k lies between -1 and $+1$. If $k = -1$ then the liquid–solid tension is probably large and positive; the liquid does not wet the solid but, if placed on it, remains separated from it by a thin film of vapour. From (1.19) this implies weak 1–2 forces or strong 1–1 or 2–2 forces. If $-1 < k < 0$, then θ lies between π and $\pi/2$, and the solid is generally said again to be unwetted or partially wetted, since σ^{ls} is not so large as to prevent liquid–solid contact.[21] This is the case of mercury in contact with glass ($\theta \sim 140°$), a result which is attributed to particularly strong forces in the liquid. If $0 < k < +1$ then the solid is again wetted and θ lies between $\pi/2$ and 0. If $k = 1$ then the angle of contact is zero and the liquid wets completely, or spreads freely over the solid surface. This is the case with water and very clean glass, when, as Young realized, the forces between a molecule of water and of glass are stronger than those between the water molecules. (The ranges $|k| > 1$ are ruled out by an argument which we defer to § 8.3.) Gauss obtained Young's results by a variational treatment in a paper[22] which formally drew together much of the work of his predecessors.

Whether we assume, however, the existence of a fixed angle of contact between a given liquid and a given solid, or derive it from a static molecular model, as above, we can proceed at once to explain all the common capillary phenomena. The three ideas of tension at a surface, internal pressure, and angle of contact suffice; with these concepts and the key expression (1.14) for the internal pressure inside a curved surface we can solve all common equilibrium capillary problems by the methods of classical statics.[23] This is not a field we wish to explore in detail but show here only how the rise in a capillary tube is treated, since this forms the basis of the commonest method of measuring surface tension.

Consider a glass tube of small internal diameter dipped vertically into a liquid which wets the glass. A film of liquid will creep along the wall until the angle of contact satisfies Young's equation, thus causing the surface of liquid near the wall to be concave upwards. From (1.14) it follows that the internal pressure below this portion of liquid is less than

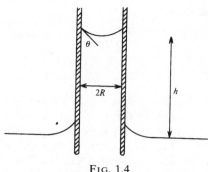

FIG. 1.4

that below the plane surface of the bulk of the liquid. This lack of equilibrium can be remedied only if the liquid rises in the tube (and to a smaller degree around the outside of the tube) until there is a difference of hydrostatic pressure sufficient to balance that of internal pressure. For simplicity we treat only the case of a tube whose bore is so fine that the internal surface of the liquid is a portion of a sphere ($\theta < \pi/2$ in Fig. 1.4), and in which we can neglect any mass of liquid above the lowest point of the meniscus. By equating the difference of pressure arising from curvature with that arising from the difference of the heights of the liquid within and without the tube we obtain at once the result that the height to which the liquid rises is

$$h = 2\sigma^{lg} \cos \theta / m\rho g R \qquad (1.23)$$

where m is the mass of a molecule, g is the acceleration due to gravity, and R is the radius of the tube. We see that the rise is proportional to the reciprocal of the radius R, and is positive only if $\theta < \pi/2$. This last point can be expressed slightly differently by introducing Young's equation (1.21), when we see that the sign of h is that of ($\sigma^{sg} - \sigma^{ls}$).

If $\theta = 0$ and if we replace ρ by the more exact expression $\Delta\rho = \rho^l - \rho^g$ then (1.23) can be expressed

$$a^2 = 2\sigma^{lg} / mg\Delta\rho = Rh \qquad (1.24)$$

where the length a, defined by the first part of this equation, is called the *capillary constant* or *capillary length* of the liquid.[23,24] For water a is 3·93 mm at 0 °C and falls steadily to zero at the critical point. This length determines the scale of many capillary phenomena, such as the shape of pendent drops and the shapes of liquid surfaces near bounding solids; it is, of course, large on a molecular scale.

At 25 °C water rises 5·87 cm in a glass tube of 0·5 mm bore ($R = 0·25$ mm). Since $\theta \sim 0$, this implies a surface tension of 72·0 mN m^{-1}

($=$dyn cm^{-1}), or in units of energy per unit area, 72·0 mJ m^{-2} ($=$erg cm^{-2}). Parallel plates at a separation of 0·5 mm produce a rise of only half that of the tube, since the surface is now part of a cylinder and so one of the two principal radii of curvature has become infinite. If the bore of the tube or the separation of the plates is much larger then the surface will no longer be spherical or cylindrical since we cannot neglect the weight of the liquid in the meniscus itself. Nevertheless (1.14) and the laws of hydrostatics are a sufficient basis for solving any problem, however complicated the geometry of the bounding solid surfaces. The problem of tubes of wide bore cannot be solved analytically but many solutions have been obtained in series and by similar methods.[25]

Most other methods[26] of measuring surface tension also depend on the production of a curved liquid surface, by the combined effect of gravity and a solid boundary, and the calculation of σ from the pressure difference given by Laplace's equation. One recent method is radically different; this is the study of the speed of capillary waves on the surface by using them to scatter light from a laser.[27] A wave of length λ on the surface of an incompressible inviscid liquid moves with a phase velocity c given by[28]

$$qc^2 = g + (\sigma/m\rho)q^2 = g[1 + \tfrac{1}{2}(aq)^2] \tag{1.25}$$

where $q = 2\pi/\lambda$ is the wave number. At long wavelengths the first term dominates and the waves are called gravity waves. At short wavelengths the second term dominates and they are called capillary waves. Again we see that the capillary length, a, is the appropriate scale of length which governs the change from one regime to the other. The optical study of capillary waves is, in principle, an attractive way of measuring σ since the liquid surface need not be touched by a solid. An accuracy of about 1 per cent in σ has been claimed for this method.[27]

An extensive summary of measurements of σ would be out of place,[29] but we show two graphs, one for argon[30] (Fig. 1.5) and the other for water (Fig. 1.6), for which we rely principally on the results of Volyak[31] for the values above 100 °C. (Below that temperature many of the best measurements are for air-saturated water in contact with a mixture of air and water vapour at a total pressure of one atmosphere.) Argon is typical of simple non-polar liquids in that both σ and $-(d\sigma/dT)$ decrease monotonically from triple to critical point. Water is unusual, firstly, in the particularly high values of σ, and secondly, in having a maximum in $-(d\sigma/dT)$ at about 200 °C. The reduction in slope at low temperatures is presumably a pale reflection of the more marked anomalies in density, compressibility, etc.[32]

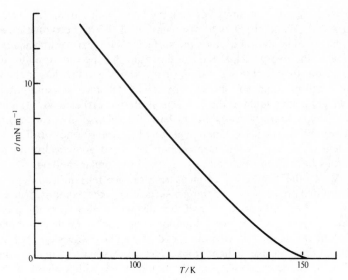

FIG. 1.5. The surface tension of argon as a function of temperature.

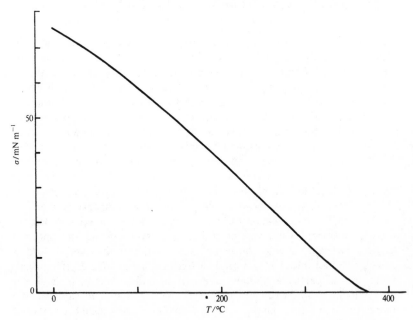

FIG. 1.6. The surface tension of water as a function of temperature.

1.4 The internal energy of a liquid

In calculating the pressure inside a curved liquid surface we introduced Rayleigh's potential 2ϕ (1.10) and remarked in passing that ϕ_1 was the cohesive energy density. This useful concept was first introduced by Dupré in 1869 as the work needed to break a piece of matter down into its constituent molecules (*le travail de désagrégation totale*). He quotes[33] a derivation of (1.11) due to his colleague F. J. D. Massieu which runs as follows. The force on a molecule near a surface, acting towards the bulk of the liquid, is the negative of that which would arise from the shaded volume in Fig. 1.7, since in the interior of the liquid the attractive force of a spherical volume of radius d is zero by symmetry. That is, the inwards force is

$$F(r) = -2\pi\rho \int_r^d ds\, s^2 f(s) \int_0^{\cos^{-1}(r/s)} d\theta \sin\theta \cos\theta$$

$$= -\pi\rho \int_r^d (s^2 - r^2) f(s)\, ds. \tag{1.26}$$

This force is positive since $f(0 < s < d) < 0$, and $F(\pm d) = 0$ since $f(s)$ is an odd function. There is no force on the molecule unless it is within distance d of the surface, on one side or the other. The work to remove one molecule from the liquid is therefore

$$\int_{-d}^d F(r)\, dr = \pi\rho \int_{-d}^d dr \int_r^d du(s)(s^2 - r^2) \tag{1.27}$$

$$= -2\pi\rho \int_{-d}^d r^2 u(r)\, dr = -\rho \int_0^d u(r)\, d\mathbf{r} \tag{1.28}$$

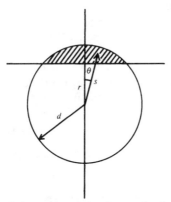

FIG. 1.7. The net inwards force on a molecule at a depth $r < d$ is the negative of the outwards force that would arise from molecules in the shaded volume if this were filled at a density ρ.

since $u(r)$ is an even function. This work is twice the negative of the energy per molecule needed to disintegrate the liquid (*twice*, to prevent the double counting of molecules, once as they are removed and once as part of the environment). That is

$$\frac{U}{N} = \tfrac{1}{2}\rho \int_0^d u(r)\,d\mathbf{r} \qquad (1.29)$$

which is a simple and indeed transparent expression for the internal energy U of a sample of liquid containing N molecules. It follows that the cohesive energy density ϕ is given by (1.10), or

$$\phi \equiv \frac{U}{V} = -K \qquad (1.30)$$

which is (1.11), after dropping the subscript I. Dupré himself[34] obtained the same result by an indirect argument. He calculated (dU/dV) by finding the work done against the intermolecular forces on expanding uniformly a cube of liquid. This gave him

$$(dU/dV) = K. \qquad (1.31)$$

Since K has the form a/V^2, (1.7) or (1.11), where the constant a (not to be confused with the capillary constant) is given by

$$a = -\tfrac{1}{2}N^2 \int_0^d u(r)\,d\mathbf{r}, \qquad (1.32)$$

then integration of (1.31) yields again (1.30).

Rayleigh[9] criticized Dupré's derivation on the grounds that it was illegitimate to consider the work done on uniform expansion from a position of balance of the cohesive and repulsive intermolecular forces by considering the cohesive forces only. This was not a step that could properly be taken without knowing more of the form of the repulsive forces. A similar criticism can, in fact, be made of many of the derivations of this chapter, but nevertheless the mechanical molecular model used here has led to a number of substantially correct equations. In particular the results that

$$K = a/V^2, \qquad U = -a/V, \qquad (dU/dV) = K, \qquad (1.33)$$

led, almost immediately after Dupré's death in 1869, to the well-known equation of state of van der Waals,[35,36]

$$\left(p + \frac{a}{V^2}\right)(V - b) = NkT. \qquad (1.34)$$

Here the term in the first bracket is the 'pressure' within the fluid, that is

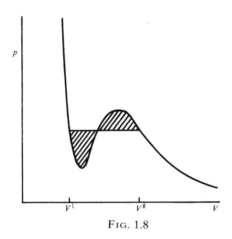

FIG. 1.8

the sum of the external pressure p and the internal pressure K. The second is the effective volume within the fluid, that is the actual volume V less the covolume,[37] b, where b is, to a first approximation, four times the actual volume of the molecules, if these are assumed to be hard spheres.

As is well known, (1.34) is a cubic in V which has three real roots if the temperature is below the critical; $NkT < NkT^c = 8a/27b$ (Fig. 1.8). Two of these roots represent the gas and liquid phases and are found by using Maxwell's equal-area rule.[38] (In introducing this, and indeed in mentioning temperature at all, we are going beyond the purely mechanical arguments of this chapter and anticipating thermodynamic considerations that properly belong to the next two chapters.) This rule determines the saturated vapour pressure, and the smallest and greatest roots, V^l and V^g. The intermediate root has no particular significance, and lies on a portion of the curve that is mechanically unstable since $(dp/dV) > 0$. Such a state cannot exist in a homogeneous fluid, but, as we shall see, the analytically continuous curve joining V^l and V^g plays an important role in some theories of capillarity. This was first recognized by James Thomson[39] in 1871, that is before the publication of van der Waals's thesis. He had used Andrews's experimental study of the critical point of carbon dioxide[40] to construct a p–V–T surface for the liquid and gaseous states, and suggested that the continuous isotherm below T^c, although clearly playing no role in describing the homogeneous fluid, might describe states that could exist in a thin intermediate and inhomogeneous layer between gas and liquid. His suggestion was taken up twenty years later by Rayleigh (§ 1.5) and by van der Waals (Chapter 3).

The fundamental equations of van der Waals's theory (1.33) survive to this day as a necessary feature[41] of what are now called mean-field approximations in the theory of liquids and their phase transitions. We return to this point in § 1.6.

1.5 The continuous surface profile

The calculation by Massieu of the work needed to remove a molecule from a liquid, (1.26)–(1.28), contains a contradiction, for if a molecule within $\pm d$ of the sharp surface experiences these forces then clearly the surface layer itself cannot be in internal equilibrium at a uniform molecular density ρ, as is supposed. In reality the density must change continuously, not as a step-function, over a distance of the order of d or larger, as one passes from the liquid state to the gas. This deficiency of the treatment of Young, Laplace, and Gauss was first pointed out by Poisson,[42] who also suggested the remedy, namely the replacement of the step-function by a continuous function $\rho(z)$, where z is the height. Unfortunately, as Maxwell[14] observed, his analysis of the revised model led him to the false conclusion that the surface tension itself would vanish if $\rho(z)$ were to approach a step-function. Maxwell's own analysis[14] avoids this fault but also leads to no useful conclusions.[44] Others tackled the problem later,[45] but the analysis we follow here is closest to that of Rayleigh,[44] who followed up James Thomson's suggestion that the density $\rho(z)$ might share some of the properties of the unstable homogeneous fluid on the van der Waals loop in Fig. 1.8.

In a planar liquid–gas surface in which the density is changing only in the z-direction, we can express the energy per molecule at height z by

$$\frac{U(z)}{N} = \pi \int_0^d dr\, r^2 u(r) \int_0^\pi d\theta\, \rho(z + r \cos \theta) \sin \theta \tag{1.35}$$

where θ is an angle measured from the z-axis, that is, from a line normal to the surface. If ρ is a continuous function we can expand it in a Taylor series about $\rho(z)$,

$$\rho(z + r \cos \theta) = \rho(z) + r \cos \theta\, \rho'(z) + \tfrac{1}{2} r^2 \cos^2 \theta\, \rho''(z) + \ldots \tag{1.36}$$

Only the even derivatives contribute to $U(z)$, as can be seen on integration over θ in (1.35).

$$U(z)/N = -l\rho(z) - \tfrac{1}{2} m\rho''(z) - \ldots \tag{1.37}$$

where the positive constants l, m, etc. are given by

$$l = -\frac{1}{2} \int u(r)\, d\mathbf{r} = a/N^2, \qquad m = -\frac{1}{6} \int r^2 u(r)\, d\mathbf{r}. \tag{1.38}$$

If $\rho(z)$ is a slowly-changing function of z on the scale of the range of $u(r)$ then we can omit the fourth and higher derivatives in (1.37).

It is the essential feature of Rayleigh's treatment (and of that of van der Waals which we examine in more detail in the next chapter) that $U(z)$ can also be expressed in terms of the equation of state of the hypothetical homogeneous fluid represented by the continuous van der Waals loops in Fig. 1.8. Thus (1.37) can be written

$$m(\mathrm{d}^2\rho/\mathrm{d}z^2) = F(\rho) \tag{1.39}$$

where $F(\rho)$ is determined by the continuous equation of state, and $F(\rho^\mathrm{l}) = F(\rho^\mathrm{g}) = 0$. Integration of (1.39) gives

$$\tfrac{1}{2}m\left(\frac{\mathrm{d}\rho}{\mathrm{d}z}\right)^2 = \int_{\rho^\mathrm{l}}^{\rho(z)} F(\rho)\,\mathrm{d}\rho \tag{1.40}$$

which can be written as a differential equation for z as a function of ρ;

$$z(\rho) = \left(\frac{m}{2}\right)^{\frac{1}{2}} \int\left[\left[\int F(\rho)\,\mathrm{d}\rho\right]^{-\frac{1}{2}}\mathrm{d}\rho \tag{1.41}$$

in which the limits of both integrations are those of (1.40). Thus we can calculate the shape of the profile $\rho(z)$ if we know the equation of state of the homogeneous fluid at all densities between ρ^g and ρ^l, if we know the zeroth and second moments of the intermolecular potential, l and m (1.38), and if we assume that ρ changes only slowly with z.

It remains to calculate the surface tension for this continuous profile. Rayleigh tackled this by calculating the pressure inside a liquid drop, bounded by a 'surface' in which ρ is a continuous function of the distance from the centre. It is easier, however, to obtain his result by using the fact that, on this molecular model, the surface tension is twice the 'excess' energy per unit area of the surface (see § 3.1). That is, it is given by

$$\sigma = \frac{2}{N} \int_{-\infty}^{\infty} \rho(z)\{U(z) - U[\rho(z)]\}\,\mathrm{d}z \tag{1.42}$$

where the 'excess' is here $\{U(z) - U[\rho(z)]\}$; $U(z)/N$ is the energy per molecule at height z, and $U[\rho(z)]/N$ is the energy per molecule in a homogeneous fluid of density $\rho(z)$; that is, it is $-l\rho(z)$. So, from (1.37),

$$\sigma = -m \int_{-\infty}^{\infty} \rho(z)\rho''(z)\,\mathrm{d}z = m \int_{-\infty}^{\infty} [\rho'(z)]^2\,\mathrm{d}z \tag{1.43}$$

in the limit of a slowly-varying profile.

1.6 The mean molecular field

The results above summarize the position reached by the molecular theory of capillarity by 1892. The attractive forces between molecules are responsible for a high, but not directly measurable, internal pressure in a liquid, K, and a state of tension in its surface, σ. K is proportional to the zeroth moment of the intermolecular potential and so to the mean internal energy per molecule; σ is proportional to the first moment, if the surface is bounded by a step-function in the density, or to the second moment if the surface density is a slowly varying function of height.

These results rest on several gross approximations which we remove as the book advances. The first, and most serious, is the neglect of molecular motion. The state of equilibrium is presumed to be one in which the molecules are at rest in positions of minimum potential energy. This assumption was a natural one at the opening of the nineteenth century but became increasingly unacceptable as time passed. After the development of classical thermodynamics and kinetic theory,[11] and the identification of temperature with the mean kinetic energy of the molecules, it became clear that a purely mechanical model was inadequate. Rayleigh's work of 1890–1892 was already out of date, as he himself admits in an aside, but then classical mechanics was his forte; one cannot read his papers without feeling that he was never fully comfortable with thermodynamic arguments. The neglect of molecular motion is, in classical thermodynamic terms, a confusion of free energy with internal energy. It can lead, in statistical mechanical terms, to integrals over the intermolecular energy rather than over the not dissimilar functions we now call the intermolecular virial function and the direct correlation function. These confusions and approximations will be removed in later chapters, but they do not destroy the usefulness of the results above in showing the essential nature of the links between the intermolecular forces and the macroscopic properties of liquids.

There is, however, one further approximation that has been used above, and that is the one we now call a *mean-field approximation*, or the assumption of the existence of a *mean molecular field*. It is an approximation which is sometimes difficult to avoid even in modern statistical mechanics, and is worth full discussion.

All the arguments above include the assumption that we can define an element of volume in a liquid which is small compared with d^3, where d is the range of the intermolecular force, but large enough for it to contain sufficient molecules for us to assume that there is within it a uniform distribution of molecules of number density ρ (see Figs 1.1 and 1.7). Laplace understood that this implies that each molecule is acted on only by a mean molecular field arising from a large number of neighbours;[46] 'Then each particle, in all positions, suffers the same attractive force and the same repulsive force of heat; it yields to the slightest pressure, and the liquid enjoys a perfect fluidity'. It is difficult to reconcile this apparently neutral equilibrium with a stable minimum energy at equilibrium which the mechanical model requires, but there are also more serious difficulties. The results to which the mean-field approximation led can themselves be used to show that the approximation is untenable.

In 1816 Young[43] observed that $2H/K$ must be a rough measure of the range of the intermolecular force since the integrals are the first and zeroth moment of $u(r)$. H can be measured directly (it is twice the surface tension) but the internal pressure K is not so easily found. Laplace realized that it was large compared with the external pressure, or with the vapour pressure of a liquid near its boiling point, but his estimate of it was quite unrealistic.[47] Young writes that 'there is reason to suppose the corpuscular forces of a section of a square inch of water to be equivalent to the weight of a column about 750 000 feet high'; that is, he puts $K = 25$ kbar but gives no reason for his choice. The way he expresses the result and his interest in the strengths and extensibilities of wires suggest that he believes K for water to be comparable with the breaking stress of a wire, but this remains a speculation. However his estimate is not an unreasonable one. There is no exact equivalent to Laplace's K in modern theory but if, with Dupré, we identify it with $(\partial U/\partial V)_T$, (1.31), then it is[32] 5·6 kbar for water at its normal boiling point and 1·7 kbar for argon. Young's guess is therefore a little high and gave him a reasonable, if somewhat low, estimate of the range of the intermolecular force, namely, in modern units $d = 0·1$ nm or 1 Å. He then went on, wrongly, to identify this distance with the mean molecular separation in steam on the point of condensation, since he thought that this was the distance at which the attractive forces must start to act if they were to bring about condensation. Since liquid water is more than a thousand times denser than steam he was able to convince himself that in the liquid the mean molecular separation is many times smaller than the range of the cohesive force, and so save the basis of the theory.

Poisson[48] was equally clear on the need for the range of forces to exceed substantially the mean molecular separation if the theory was to be tractable. Challis,[49] in reporting Poisson's views to the British Association in 1834, put it particularly clearly; 'The radii of activity of these forces are nevertheless supposed to be extremely great compared with the intervals between the molecules, and the rapid decrease [of force] to commence only at distances which are great multiples of these small intervals. Without this supposition, in bodies whose molecules are not regularly distributed, the resultant of the molecular forces, that is the total force that acts on each molecule, might be very different in magnitude and direction for two consecutive molecules, and consequently would not be subject to the law of continuity. It seems therefore necessary to make the above supposition'. He also described clearly how he believed the forces changed from attractive at large separations to repulsive at short. (Waterston,[50] nine years later, was, we believe, the first to publish an intermolecular force curve with short-ranged attractive forces and shorter-ranged repulsive forces of the kind we now draw so often.)

As the nineteenth century advanced ideas of molecular masses, sizes, and velocities became more certain, as is shown for example by Maxwell's summary[51] of the position in 1873. He, and van der Waals in the same year,[52] accepted Laplace's and Poisson's view that in a liquid there is a uniform molecular field arising from the attractive forces but neither seemed to see that this, in turn, apparently required that $u(r)$ was weak but of long-range. Van der Waals, in fact, repeated Young's calculation of the range of the force, apparently in ignorance of the original work, and obtained a range of 3–6 Å with the better results then available. He did not, at this time, point out that the best estimates of molecular separation and of the range of the forces rendered untenable Young's justification of the theory of capillarity.

Rayleigh[53] and Boltzmann[54] stated quite explicitly that the same mean-field assumption must also underlie van der Waals's justification of his equation of state. In the first chapter of the second volume of his *Vorlesungen über Gastheorie*[54] Boltzmann points out that this assumption implies that the force 'vanishes at macroscopic distances but decreases so slowly with increasing distance that it may be considered constant within distances large compared to the average separation of two neighbouring molecules'. When van der Waals saw these words before publication he objected that he had never made such an assumption and indeed, because of the smallness of the length $2H/K$, he doubted if it were true. Boltzmann's reply was: 'Nevertheless, I cannot obtain an exact foundation for his equation of state without this assumption'.

Once the problem was so clearly recognized then a solution, at least in principle, was to hand. Instead of assuming a random or uniform distribution of molecules we introduce distribution or correlation functions into our expressions for the mean interactions of molecules at two positions in the fluid. These functions are measures of the conditional probability of the occurrence of pairs (or larger groups) of molecules at specified points. Their calculation is one of the principal aims of modern theories of liquids. However there are still many problems, particularly those connected with phase transitions, which we cannot solve explicitly in terms of closed expressions for these distribution functions, and we often have recourse to mean-field approximations even today.

In this section we have set out some of the severe limitations of the simple mechanical model that descended to us from Laplace, and which reached its most complete exposition at the hands of Rayleigh. Before describing, in Chapter 3 and later, how these limitations can be lifted, it is useful to turn aside from the molecular theory and interpolate a chapter in which we consider the classical thermodynamic implications of the results so far obtained. That is, if we accept that experimental and molecular arguments justify the existence of a surface tension, or surface energy, what are the purely macroscopic implications of the existence of such an energy? To this question we now turn.

Notes and references

1. Cavendish, H. *Phil. Trans. Roy. Soc.* **88**, 469 (1798).
2. Hauksbee, F. (1709), quoted by Laplace (1806) (ref. 3, p. 3; Bowditch, ref. 3, p. 688). Hauksbee was the Demonstrator at the Royal Society and his experiments influenced the content of the much-extended 'Query 31', on the ultimate particles of matter and the forces between them, with which Newton closed the 1717 edition of his *Opticks*. See Guerlac, H. *Arch. internat. d'Hist. des Sciences* **16**, 113 (1963).
3. Laplace, [P. S.] *Traité de Mécanique Céleste; Supplément au dixième livre, Sur l'Action Capillaire*, Courcier, Paris ([1806]), and *Supplement à la Théorie de l'Action Capillaire* [1807]; reprinted in *Oeuvres complètes de Laplace*, Vol. 4, p. 349, 419, Gauthiers-Villars, Paris (1880); English translation with an extensive commentary, Bowditch, N. *Mécanique Céleste by the Marquis de la Place*, Vol. 4, p. 685, 806, Little and Brown, Boston (1839), reprinted by Chelsea Publ. Co., New York (1966).
4. Whyte, L. L. (ed.), *Roger Joseph Boscovich, S.J., F.R.S., 1711–1787, Studies of his Life and Work on the 250th Anniversary of his Birth*, Allen and Unwin, London (1961). See also Merz, J. T. *History of European Thought in the Nineteenth Century*, Vol. 1, p. 357, Blackwood, Edinburgh, (1896).
5. Williams, L. P. *Michael Faraday*. Chapman and Hall, London (1965).
6. Thompson, S. P. *Life of William Thomson, Baron Kelvin of Largs*, Vol. 2. Macmillan, London (1910).
7. A view that was the hall-mark of those who believed in the caloric theory; Fox, R. *The Caloric Theory of Gases*, Oxford University Press (1971). See also Frické, M. in *Method and Appraisal in the Physical Sciences* (ed. C. Howson), p. 277, Cambridge University Press (1976).
8. For early kinetic theories of gases, see Talbot, G. R. and Pacey, A. J. *Brit. J. Hist. Sci.* **3**, 133 (1966); and Truesdell, C. *Essays in the History of Mechanics*, Chapter 6, Springer, Berlin (1968).
9. Lord Rayleigh, *Phil. Mag.* **30**, 285, 456 (1890); reprinted in *Scientific Papers*, Vol. 3, p. 397, Cambridge University Press (1902), reprinted by Dover Publications, New York (1964). See also Fowler, R. H. and Guggenheim, E. A. *Statistical Thermodynamics*, p. 445, Cambridge University Press (1939).
10. Laplace, ref. 3, pp. 1–4. By 1819 he was discussing more specifically intermolecular repulsions which, while still ascribed to heat or caloric, have the essential property that they decrease more rapidly with distance than the attractive forces. See papers published between 1819 and 1824 in *Oeuvres complètes*, ref. 3, Vol. 13, p. 273 (1904), Vol. 14, p. 259 (1912). See also Fox, ref. 7, p. 165, and Brush, ref. 11, Chap. 11.
11. Brush, S. G. *The Kind of Motion we call Heat. A History of the Kinetic Theory of Gases in the Nineteenth Century*, 2 vol. North-Holland, Amsterdam (1976).
12. For brief accounts of the history of capillarity before Laplace, see Hardy, W. B. *Nature* **109**, 375 (1922) and Bikerman, J. J. *Arch. Hist. exact. Sci.* **18**, 103 (1978). This early period and the next thirty years were fully covered in a

review written by Challis (ref. 13) for the British Association. His review is the source on which the opening section of Maxwell's more accessible essay is based (ref. 14). Nineteenth century experiments and theories are discussed at length in the first three chapters of Bakker, G. *Kapillarität und Oberflächenspannung*, Vol. 6 of *Handbuch der Experimentalphysik* (ed. W. Wien, F. Harms, and H. Lenz), Akad. Verlags., Leipzig (1928), and by Partington, J. R. *Advanced Treatise on Physical Chemistry*, Vol. 2, pp. 134–207, Longmans, London (1951).

13. Challis, J. *Brit. Ass. Rep.* **4**, 253 (1834).

14. Maxwell, J. C. *Capillary Action, Encyclopaedia Britannica*, 9th edn. 1876; reprinted in *The Scientific Papers of James Clerk Maxwell*, Vol. 2, p. 541, Cambridge University Press (1890), reprinted by Dover Publications, New York (1952).

15. The argument, whilst still in essentials that of Laplace, is now more closely based on that of Dupré, ref. 16, p. 206 ff., and Kelvin, ref. 17.

16. Dupré, A. *Théorie mécanique de la Chaleur*. Gauthier-Villars, Paris (1869).

17. Thomson, W. *Lecture on Capillary Attraction* (1886); reprinted in *Popular Lectures and Addresses*, Vol. 1, Macmillan, London (1889).

18. See, for example, Weatherburn, C. E. *Differential Geometry of Three Dimensions*, Vol. 1, p. 72. Cambridge University Press (1927).

19. See, for example, Defay, R., Prigogine, I. and Bellemans, A. *Surface Tension and Adsorption* (trans. D. H. Everett), Chapter 17. Longmans, London (1966); Woodruff, D. P. *The Solid–Liquid Interface*, Chapter 2, Cambridge University Press (1973); and Navascués, G. *Rep. Prog. Phys.* **42**, 1131 (1979). For an account of Rayleigh's and other early work in this field, see Orowan, E. *Proc. Roy. Soc. A* **316**, 473 (1970). For attempts to measure σ^{ls} directly, see Skapski, A., Billups, R. and Rooney, A. *J. chem. Phys.* **26**, 1350 (1957), and Bailey A. I. and Kay, S. M. *Proc. Roy. Soc. A* **301**, 47 (1967).

20. Young, T. *Phil. Trans. Roy. Soc.* **95**, 65 (1805); reprinted with additions and with some unwarranted criticisms of Laplace's first paper in *Lectures on Natural Philosophy*, Vol. 2, p. 649, Johnson, London (1807), and in *Miscellaneous Works of the late Thomas Young* (ed. G. Peacock) Vol. 1, p. 418. Murrary, London (1855). Young obtained (1.15) but characteristically put it into words, not as an equation. For a defence of Young's claim to priority, see Pujado, P. R., Huh, C. and Scriven, L. E. *J. Coll. Interf. Sci.* **38**, 662 (1972), and for an attack, see Bikerman, ref. 12.

21. The terms 'non-wetting', 'wetting', and 'spreading' are not always used with the same meanings. We follow here, Padday, J. F. (ed.) *Wetting, Spreading and Adhesion*, p. 459. Academic Press, London (1978).

22. Gauss, C. F. *Comm. Soc. Reg. Sci. Gött. Rec.* Vol. 7 (1830), reprinted in *Werke*, Vol. 5, p. 29. Göttingen (1867). We are indebted to Mrs N. Rowlinson for a translation from the Latin. A German translation by R. H. Weber was published by W. Ostwald as one of the series *Klassiker der exacten Wissenschaften*, No. 135. Engelmann, Leipzig (1903).

23. We recall the IUPAC definition: 'The mechanical properties of the interfacial layer between two fluids, including the equilibrium shape of the surface, may be calculated by applying the standard mathematical techniques of mechanics to the forces associated with the surface of tension. The resulting equations—which comprise the subject of *capillarity*—form the basis of experimental methods of measuring surface tension.' IUPAC: *Manual of Symbols and Terminology for Physicochemical Quantities and Units*, App. II, Part I,

§ 1.2.1, Butterworth, London, 1971; reprinted in *Pure Appl. Chem.* **31**, 577 (1972). For a recent review of classical capillarity, see Boucher, E. A. *Rep. Prog. Phys.* **43**, 497 (1980).

24. In substance, this equation goes back to Laplace, but in this form it is due to Poisson, S. D. *Nouvelle Théorie de l'Action Capillaire*, p. 108. Bachelier, Paris (1831). Poisson was a protégé of Laplace; see Crosland, M. *The Society of Arcueil*, Heinemann, London (1967).

25. See for example Adamson, A. W. *Physical Chemistry of Surfaces*, 3rd edn., Chapter 1. Interscience, New York (1976). A thorough discussion of axially symmetric menisci is given by Padday, J. F. *Phil. Trans. Roy. Soc.* A **269**, 265 (1971).

26. See for example Alexander, A. E. and Hayter, J. B. in *Physical Methods of Chemistry*, (ed. A. Weissberger and B. W. Rossiter) Part 5, Chapter 9, Wiley, New York (1971); Pugachevich, P. P. in *Experimental Thermodynamics* (ed. B. Le Neindre and B. Vodar) Vol. 2, Chapter 20, Butterworth, London (1975).

27. Hård, S., Hamnerius, Y. and Nilsson, O. *J. appl. Phys.* **47**, 2433 (1976); Yoneda, K., Tawata, M. and Hattori, S. *Optics and Laser Tech.* **8**, 39 (1976).

28. Levich, V. G. *Physicochemical Hydrodynamics*, Chapter 11. Prentice-Hall, Englewood Cliffs, N.J. (1962).

29. Critical surveys have been made by Jasper, J. J. *J. phys. chem. Ref. Data* **1**, 841 (1972); Bonnet, J. C. and Pike, F. P. *J. chem. Engng Data* **17**, 145 (1972); and Bottomley, G. A. *Report Chem.* 23 (1973), National Physical Laboratory, Teddington. The last, although the least accessible, is useful for its more thorough discussion of surface tension at high temperatures and near critical points.

30. Stansfield, D. *Proc. phys. Soc.* **72**, 854 (1958); Sprow, F. B. and Prausnitz, J. M. *Trans. Faraday Soc.* **62**, 1097 (1966).

31. Volyak, L. D. *Dokl. Akad. Nauk. SSSR* **74**, 307 (1950). More recent results by N. B. Vargaftik, B. N. Volkov, and L. D. Volyak are in their review, *J. phys. chem. Ref. Data* **12**, 817 (1983). See also Grigull, U. and Bach, J. *Brennst.-Wärme-Kraft* **18**, 73 (1966).

32. Rowlinson, J. S. and Swinton, F. L. *Liquids and Liquid Mixtures* (3rd edn) Chapter 2. Butterworth, London (1982).

33. Dupré, ref. 16, p. 152.

34. Dupré, ref. 16, p. 145.

35. van der Waals, J. D. Thesis, *Over de Continuiteit van den Gas- en Vloeistof- toestand*, University of Leiden (1873). A German translation by F. Roth was published by Barth at Leipzig in 1881, an English translation by R. Threlfall and J. F. Adair in *Phys. Memoirs* **1**, 333 (1890), a French translation by Dommer and Pomey was published by Carré in Paris in 1894, and a second edition in German with an additional volume on mixtures was published by Barth in Leipzig in 1899 and 1900.

36. For the background and early history of van der Waals's work see Rowlinson, J. S. *Nature* **244**, 414 (1973); de Boer, J. *Physica* **73**, 1 (1974); Klein, M. J. *Physica* **73**, 28 (1974); Levelt Sengers, J. M. H. *Physica* **73**, 73 (1974), **82A**, 319 (1976); and Brush, ref. 11, Chapter 7.

37. The term is due to Dupré, ref. 16, p. 61; see also Kamerlingh Onnes, H. and Keesom, W. H. *Die Zustandsgleichung, Enc. math. Wissen.*, Vol. 5, Art. 10, p. 671. Teubner, Leipzig (1912).

38. Maxwell, J. C. *Nature* **11**, 357, 374 (1875); *J. chem. Soc.* **13**, 493 (1875), reprinted in *Scientific Papers*, Vol. 2, p. 418.
39. Thomson, J. *Proc. Roy. Soc.* **20**, 1 (1871); reprinted in *Collected Papers in Physics and Engineering*, p. 278. Cambridge University Press (1912).
40. Andrews, T. *Phil. Trans. Roy. Soc.* **159**, 575 (1869); reprinted in *The Scientific Papers of the late Thomas Andrews*, p. 296, Macmillan, London (1889). See also Rowlinson, J. S. *Nature* **224**, 541 (1969).
41. Widom, B. *J. chem. Phys.* **39**, 2808 (1963); Longuet-Higgins, H. C. and Widom, B. *Mol. Phys.* **8**, 549 (1964).
42. Poisson, ref. 24, p. 5, 8. Young (ref. 43) had remarked in 1816 that there must be a gradation of density at the surface, but had dismissed it as unimportant, the 'whole stratum being always extremely minute in comparison with any sensible radius of curvature'.
43. [Young, T.] Supplement to the Fourth, Fifth, and Sixth Editions of *Encyclopaedia Britannica*, 1816 onwards. This anonymous article is ascribed to Young, dated, and reprinted by Peacock in his edition of Young's works, ref. 20, Vol. 1, p. 454.
44. Lord Rayleigh, *Phil. Mag.* **33**, 209 (1892), or *Scientific Papers*, Vol. 3, p. 513.
45. For a summary of the work of K. Fuchs and H. Hulshof, see Chapters 1 and 15 of Bakker's book, ref. 12. The paper of Fuchs, K. *Sitzungsber. Kais. Akad. Wiss. Wien. Sekt. IIa* **98**, 1362 (1889) preceded that of Rayleigh but stops short (p. 1381) of the final integration over z; see Bakker, p. 382.
46. Laplace (1807), ref. 3, p. 68; Bowditch, ref. 3, p. 1006.
47. Laplace (1807), ref. 3, p. 72 ff.; Bowditch, ref. 3, p. 1011 ff.
48. Poisson, ref. 24, pp. 267–72.
49. Challis, ref. 13, pp. 282, 291.
50. [Waterston, J. J.] *Thoughts on the Mental Functions*, Oliver and Boyd, Edinburgh (1843); extracts from this anonymous book are reprinted in *The Collected Scientific Papers of John James Waterston* (ed. J. S. Haldane) pp. 169–72, Oliver and Boyd, Edinburgh (1928). In 1858 (*Phil Mag.* **15**, 1 or *Scientific Papers*, p. 407) he attempted to relate capillary phenomena to the latent heat of evaporation, but his attempts are less convincing than those of Dupré discussed above in § 1.4.
51. Maxwell, J. C. *Nature* **8**, 437 (1873), reprinted in *Scientific Papers*, Vol. 2, p. 361.
52. van der Waals, ref. 35, English translation, pp. 343, 361, and 435.
53. Lord Rayleigh, *Nature* **45**, 80 (1891), reprinted in *Scientific Papers*, Vol. 3, p. 469.
54. Boltzmann, L. (1896–1898), *Lectures on Gas Theory* (trans. S. G. Brush) pp. 220, 375. University of California Press, Berkeley (1964).

Addition to References

Ref. 35 For a more accessible English translation, see van der Waals, J. D. *On the continuity of the gaseous and liquid states* (ed. with an Introductory Essay by J. S. Rowlinson) *Studies in Statistical Mechanics*, Vol. 14, North-Holland, Amsterdam (1988).

THERMODYNAMICS

2.1 Thermodynamics and kinetic theory

The cohesion of liquids and the concomitant tension in the gas–liquid surface are almost self-evident properties of liquids. We have seen in the last chapter how they can be explained in terms of forces between stationary molecules, and how this explanation can tell us something of the strength and range of these forces. By the middle of the nineteenth century it was clear that the static molecular model was becoming untenable and not only because of the internal contradictions discussed in § 1.6. The properties of a liquid change drastically with temperature and cannot therefore be understood quantitatively in terms of a model in which temperature plays no role. The theoretical basis of the understanding of the physical world had to be expanded beyond classical mechanics before further progress could be made. The two great advances were the development of classical thermodynamics between 1824 and 1860, and the kinetic theory of matter from the 1840s onwards. The two subjects were often loosely conflated by the physicists of the time under the title of the Mechanical Theory of Heat.

It was the essence of the kinetic theory that molecules move, and that the temperature of a piece of matter is a measure of their mean kinetic energy. The equivalence of temperature and kinetic energy was accepted first for gases; the realization that it applied also to liquids came more slowly and obscurely. It lies behind Rankine's distinction[1] between *real* and *observed* specific heats in his vortex theory of atoms (1851), and behind Clausius's distinction[2] between *heat in the body* and *disgregation* (1862). In 1863 Tyndall[4] was still uncertain as to 'the exact nature of the motion of heat', but the equivalence with kinetic energy appears clearly in the final chapter, on the Molecular Theory of the Constitution of Bodies, of Maxwell's *Theory of Heat*[5] (1871). The kinetic theory required the replacement of the molecular statics of the last chapter by molecular dynamics, but before this science could be applied to anything but a dilute gas some awkward statistical problems had to be overcome, for the solution of Newton's equations of motion for a vast number of molecules was not practicable. J. J. Thomson[6] wrote in 1888 'Considering our almost complete ignorance of the structure of the bodies which form most of the dynamical systems with which we have to deal in physics, it might seem a somewhat unpromising undertaking to attempt to apply dynamics to such a system.' He did nevertheless make the attempt[6] but produced no results of lasting value. A direct attack on the properties of liquids by using classical molecular dynamics became possible only in the 1950s, with the development of computers; it is discussed in Chapter 6. The second half of the nineteenth century was not the right time for an advance in a purely molecular or kinetic theory of liquids.

The interrelations of such phenomena and properties as heat, work, internal energy, and temperature had, however, been found to follow strictly some very general rules which were not, for the moment, interpretable in molecular terms. The establishment of these rules, that is of the laws of classical thermodynamics, is one of the finest examples of abstract reasoning from a few simple facts that science has seen. Throughout the second half of the nineteenth century they proved extremely fruitful in analysing and interrelating the physical and chemical properties of matter in equilibrium; their application to the properties of gases, liquids, and their interfaces was one of the most important.

The application to surfaces was the work of one man, J. Willard Gibbs.[7] Not only did he carry through the original treatment but he did it with such care, thoroughness, and elegance that later workers have been able to do little but extend the range of applications or devise alternative routes to the same equations.

Since classical thermodynamics eschews all attempts at a molecular interpretation, so also is it silent on the absolute magnitudes of energies, pressures, surface tensions, etc. It does however lead to a deep set of relations between these different macroscopic properties, some of which we explore in this chapter.

2.2 The thermodynamics of the surface

The mechanical properties of a system comprising a liquid in equilibrium with its vapour are, as we have seen, consistent with the interface being treated as if it were a surface in which there resides a tension, that is, as a taut membrane of zero mass and zero thickness. In the hands of Young and Laplace the theory of capillarity combined this simple picture with attempts at its molecular justification, in the way described in Chapter 1, but this conflation of macroscopic and microscopic views is foreign to the methods of classical thermodynamics, which are based solely on the two hypotheses above, namely, the existence of a surface in which there is a tension, and the position of that surface in laboratory space. Our assumption is, therefore, that the mechanical behaviour of a real interfacial system is the same as that of a model system comprising a mathematical surface S subject to a tension σ. Let us therefore first summarize the mechanical consequences of this assumption, so obtaining again some of the results of Chapter 1 free (as a thermodynamic purist would see it) from the dubiety of their molecular justifications.

Consider a spherical gas bubble of radius R within the body of a liquid. The tension in the surface will make the bubble collapse unless the pressure inside exceeds that outside by, say, Δp. The work of a virtual increase in R vanishes at equilibrium, so $\Delta p\,dV$ equals $\sigma\,dA$, where dV and dA are the increases in volume and area of the sphere. In a d-dimensional space $dV/dA = (d-1)^{-1}R$, and so, for $d=3$

$$\Delta p = 2\sigma/R. \tag{2.1}$$

Had we taken an element of surface of principal radii of curvature R_1 and R_2 then, in place of (2.1), we should have obtained Laplace's result, (1.14) and (1.15)

$$\Delta p = \sigma(R_1^{-1} + R_2^{-1}). \tag{2.2}$$

The radius of the bubble was introduced above as if it were a well-defined quantity, but if we accept (as we must) the Poisson–Rayleigh view that the real gas–liquid surface is not sharp at a molecular level, then we must ask precisely how we define the radius. The answer, at a macroscopic level of argument, is that it is the distance that makes (2.1) the correct relation between Δp and σ (§ 2.4). Such a surface between two phases is called the *surface of tension*; it is the second macroscopic property, the first being the tension itself, which enters into the mechanical and thermodynamic discussion. In liquids below their normal boiling points the surface is found to be optically sharp, that is, sharp on a scale of length of 100 nm or less, and the surface of tension coincides, within this limit, with the sharp surface that is seen.

Let us now replace the experiment shown in Fig. 1.2 by a more carefully designed thought-experiment. Figure 2.1 represents a horizontal vessel of rectangular cross-section containing two (possibly multicomponent) fluid phases, α and β, e.g. a liquid and its vapour. It is closed at the right-hand end by a thin but rigid piston which can be moved horizontally and have its inclination changed from the vertical by adjusting the forces in the two springs. There are two smaller auxiliary pistons at the left-hand end. The internal volume is V and the pressure, as measured on either of

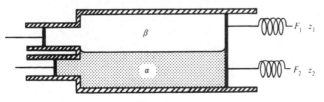

FIG. 2.1

the auxiliary pistons is p. From (2.2) this pressure is the same in both bulk phases since the surface is plane. The external pressure is zero. The force on each of the auxiliary pistons is the product of p and each of their areas, but that on the main piston is less than this product by an amount which is proportional to the length of line of contact of the α, β-surface with the face of the piston. The constant of proportionality is, by definition, the tension of the α, β-surface, or $\sigma^{\alpha\beta}$.

The main piston is maintained at a fixed position and a fixed (vertical) inclination by adjusting the forces, F_1 and F_2, exerted by the springs at heights z_1 and z_2. If we ignore any effects of the solid walls, a measurement of the sum $(F_1 + F_2)$ determines σ, since the pressure is known, and a measurement of F_1 and F_2, and the heights z_1 and z_2 determines also z_s, the height of the surface of tension. We learn no more about the system by attaching three springs to the rigid piston. If the piston were flexible then attaching more than two springs might appear to be a way to learn more of the distribution of pressure and tension over its face, but this is not so unless the flexibility is total down to a scale of length comparable with the thickness of the gas–liquid interface, and such pistons do not exist. Moreover the uncertainty with which we can measure the heights z_1 and z_2 is greater than the thickness of the interface which, as we shall see later, is comparable with the range of the inter-molecular forces. This lack of precision is not merely the obvious mechanical one of making springs with a wire of a diameter of, say, 0·5 nm, but the theoretical one that such a device would not measure a steady tension because of uncontrolled molecular fluctuations or 'noise'. To measure z_s to 0·1 nm with a piston of 1 cm square and a liquid of $\sigma = 50$ mN m^{-1}, requires measuring a torque of $5 \cdot 10^{-6}$ N m with a precision of $5 \cdot 10^{-14}$ N m. One of the smallest torques measured is that by Spero et al.[8] of $4 \cdot 10^{-13}$ N m. Thus only the zeroth and an approximate value of the first moment of the pressure $p(z)$ acting on the piston can be determined; the higher moments are inaccessible to experiment. We return in § 4.4 to the theoretical side of the problem of determining z_s precisely, and consider then the influence of the walls.

Let us now compress the system reversibly by an infinitesimal movement of the main piston to the left with a small change in its inclination, and with simultaneous adjustments of the auxiliary pistons to keep constant the areas of contact of both phases with the solid material of the assembly. Let the increase in volume be $dV < 0$, and the change of area of α, β-surface be $dA < 0$. The work done on the system, through all three pistons, is

$$dW = -p \, dV + \sigma^{\alpha\beta} \, dA. \tag{2.3}$$

This result is obtained however the change in volume is distributed between the two phases (by changing the inclination of the main piston) and is independent of the height of the surface of tension. Thus we see that this height, although a mechanical property of the system, measurable to a certain precision, is irrelevant to the thermodynamic description of a planar surface. It is, however, necessary for a thermodynamic description of a curved surface (§ 2.4).

If the system is insulated so that no heat passes to or from the fluids during the compression then, by the first law, the work dW is equal to the increase of energy dU. A reversible adiabatic compression[†] is one carried out at a fixed value of the entropy, S; hence (dropping the superscript on σ)

$$dU = -p(dV)_S + \sigma(dA)_S \qquad (2.4)$$

or

$$p = -(\partial U/\partial V)_{S,A,\mathbf{n}} \quad \text{and} \quad \sigma = (\partial U/\partial A)_{S,V,\mathbf{n}}. \qquad (2.5)$$

(The subscript $\mathbf{n}$ denotes that the amounts of all c components, $n_1, \ldots, n_c$, in the system remain constant.)

If the compression is isothermal and reversible then the work is equal to the increase of the Helmholtz free energy, dF,

$$dF = -p(dV)_T + \sigma(dA)_T \qquad (2.6)$$

or

$$p = -(\partial F/\partial V)_{T,A,\mathbf{n}} \quad \text{and} \quad \sigma = (\partial F/\partial A)_{T,V,\mathbf{n}}. \qquad (2.7)$$

Thus the naive equating of surface tension with surface energy that followed from contemplation of Fig. 1.2, is shown to be correct, but which surface energy we need depends on the external constraints. In the mechanical models of Chapter 1 there was no distinction between adiabatic and isothermal processes, and so no distinction between U and F. In practice, most changes of surface area are carried out at fixed temperatures, so that σ is the rate of increase of F with A, (2.7).

If the reversible compression is neither adiabatic nor isothermal but accompanied by a heat input dq then we have

$$dU - dW = dU + p\, dV - \sigma\, dA \qquad (2.8)$$

$$= dq = T\, dS \qquad (2.9)$$

or

$$dU = T\, dS - p\, dV + \sigma\, dA. \qquad (2.10)$$

We have assumed so far that the whole system is of fixed composition, but now extend the discussion to cover open multicomponent

[†] There is here the minor problem that an adiabatic compression results in a change of temperature and so of a change in the state of the fluids adsorbed on the solid walls. We can either suppose that the walls are made from a material for which this effect is zero, or proceed directly to the case of isothermal compression, which is free from this complication.

systems in which both phases can exchange molecules with external reservoirs. For the exchange to be reversible, a component must be present in the reservoir at the same chemical potential, μ, that it has in the system. The extension of (2.10) to cover changes in the amounts in both phases of each of c components $n_1, \ldots, n_c$, is,

$$dU = T\,dS - p\,dV + \sigma\,dA + \mu . dn \qquad (2.11)$$

where the scalar-product notation is a convenient way of writing the sum

$$\sum_{i=1}^{c} \mu_i\,dn_i \equiv \mu . dn. \qquad (2.12)$$

Each term on the right-hand side of the fundamental equation (2.11) has the form $x\,dY$ where Y, like U, is an extensive thermodynamic function and x is a field.[9] At equilibrium the fields T and μ are the same in each bulk phase, as also is the pressure if the surface boundaries are plane $(T^\alpha = T^\beta, \mu^\alpha = \mu^\beta, p^\alpha = p^\beta)$ and $\sigma^{\alpha\beta}$ is uniform over the whole of the $\alpha\beta$-surface. We need not ask what are the values of p, T, and μ at the interface; indeed the question has no meaning within the description of the system now being used. We return to this question, however, when we discuss the point-thermodynamic description in § 2.5.

By a well-known rearrangement we can write the fundamental equation (2.11) in terms of the Helmholtz free energy F,

$$dF = -S\,dT - p\,dV + \sigma\,dA + \mu . dn. \qquad (2.13)$$

A third form of the fundamental equation will be useful when we come to discuss the statistical mechanics in Chapter 4. The free energy F is the proper thermodynamic function to associate with the (petit) canonical ensemble, but it is often more convenient to work with the grand canonical ensemble, for which the appropriate potential is Ω, where

$$\Omega = F - \mu . n. \qquad (2.14)$$

Differentiation and the elimination of dF between the resulting equation and (2.13) gives

$$d\Omega = -S\,dT - p\,dV + \sigma\,dA - n . d\mu. \qquad (2.15)$$

So a third thermodynamic definition of σ to be set alongside (2.5) and (2.7) is

$$p = -(\partial\Omega/\partial V)_{T,A,\mu} \quad \text{and} \quad \sigma = (\partial\Omega/\partial A)_{T,V,\mu}. \qquad (2.16)$$

Each of the equations (2.3)–(2.16) is the same as the familiar thermodynamics of systems without an interface, except for the addition of the term $\sigma\,dA$, and for the presence of the new partial derivatives in (2.5), (2.7), and (2.16). The formal thermodynamics of the system is therefore only a trivial change from what is familiar, but this new term has interesting consequences in multicomponent systems.

2.3 Surface functions

The differential thermodynamic equations of the last section can now be integrated so that we can identify the specifically surface terms in the total functions U, F, etc.

Let us consider a two-phase system (α and β) of volume V, temperature T, with a planar interface of area A, and with amount n_i of component i. We do not define the volumes of the two separate phases, V^α and V^β, by the height of the surface of tension since that height is thermodynamically irrelevant for a planar surface, and since we do not want to be restricted to that choice of surface. There is, however, never any doubt about the directions of the normals to the surface whether it be planar or curved; at any point $\mathbf{r}$ where there is a gradient of n_i, the normal is that direction along which $\nabla n_i(\mathbf{r})$ has its maximum value, for any component i. We can therefore always choose a parallel set of surfaces which are everywhere perpendicular to these normals. Any one of these mathematical surfaces can be chosen to be what Gibbs calls the *dividing surface*, and it is convenient to choose one that is close to the position of the interface revealed by casual observation. Once the dividing surface is chosen, V^α and V^β are fixed, and satisfy

$$V^\alpha + V^\beta = V. \tag{2.17}$$

Away from the dividing surface each phase has well-defined densities,[9] namely for phase α the ratios $\rho_i^\alpha = \delta n_i/\delta V^\alpha$, $\phi^\alpha = \delta U/\delta V^\alpha$, etc. where δn_i, δU, etc. are the amounts of substance, energy, etc. in a small (but molecularly large) volume, δV^α, of phase α, chosen at some point remote from the interface. We now define the extensive properties of phase α, that is, $\mathbf{n}^\alpha$, U^α, etc., by the equations

$$\mathbf{n}^\alpha = \rho^\alpha V^\alpha \qquad U^\alpha = \phi^\alpha V^\alpha, \quad \text{etc.} \tag{2.18}$$

In general we find that the sums $(\mathbf{n}^\alpha + \mathbf{n}^\beta)$, $(U^\alpha + U^\beta)$, etc. are not equal to the total values of the whole system $\mathbf{n}$, U, etc. The differences are called the surface contributions $\mathbf{n}^s$, U^s, etc.

$$\mathbf{n}^\alpha + \mathbf{n}^\beta + \mathbf{n}^s = \mathbf{n} \qquad U^\alpha + U^\beta + U^s = U, \quad \text{etc.} \tag{2.19}$$

Since the position of the dividing surface is arbitrary, so also are $\mathbf{n}^s$, U^s, etc.; they may be positive or negative. In a single-component system we can choose the surface to make $n^s = 0$, a choice which is called the *equimolar surface*. Once this choice is made then all other extensive surface functions, U^s, F^s, S^s, etc., are determined, and are proportional to the area A. In a mixture no choice will make all n_i^s vanish, but convenience often dictates that we make either one of them vanish, e.g. $n_1^s = 0$, or that we make the sum vanish, $\sum_i \mu_i n_i^s = \boldsymbol{\mu} \cdot \mathbf{n}^s = 0$. Once the choice is made $(c-1)$ surface amounts, and all other extensive surface functions are fixed.

The Helmholtz free energy of the whole system, F, is an extensive function; more formally it is a homogeneous function of first-order in V, A, and $\mathbf{n}$. With a suitably chosen geometry we integrate (2.13) at fixed wall area to give

$$F = -pV + \sigma A + \boldsymbol{\mu} \cdot \mathbf{n}. \qquad (2.20)$$

Similarly for each of the two homogeneous phases, defined as above,

$$F^\alpha = -pV^\alpha + \boldsymbol{\mu} \cdot \mathbf{n}^\alpha, \qquad F^\beta = -pV^\beta + \boldsymbol{\mu} \cdot \mathbf{n}^\beta. \qquad (2.21)$$

By subtraction from (2.20)

$$F^s = \sigma A + \boldsymbol{\mu} \cdot \mathbf{n}^s. \qquad (2.22)$$

Thus if, and only if, we choose the dividing surface so that the sum $\boldsymbol{\mu} \cdot \mathbf{n}^s = 0$, do we have the simple result that the surface tension is the surface free energy per unit area,

$$\sigma = F^s / A. \qquad (2.23)$$

The equivalent equations in terms of the grand potential are even simpler. Integration of (2.15) gives

$$\Omega = -pV + \sigma A \qquad (2.24)$$

where

$$\Omega^\alpha = -pV^\alpha, \qquad \Omega^\beta = -pV^\beta, \qquad (2.25)$$

and so

$$\sigma = \Omega^s / A \qquad (2.26)$$

for any position of the dividing surface.

Irrespective of the choice of the dividing surface, we define *excess* or *surface densities* by dividing $\mathbf{n}^s$, U^s, etc. by the surface area. For these we use the following symbols

$$\boldsymbol{\Gamma} = \mathbf{n}^s / A, \qquad \phi^s = U^s / A, \qquad \psi^s = F^s / A,$$
$$\eta^s = S^s / A, \qquad \sigma = \Omega^s / A. \qquad (2.27)$$

The surface density of component i, Γ_i, is called the *adsorption* of i (by tradition it carries no superscript s), since a large positive value of Γ_i is evidence that component i is found in the surface layer in higher concentrations than is to be expected from its concentrations in the bulk phases. This justification of the name is deliberately phrased loosely since the magnitude of Γ_i depends also on the choice of dividing surface. We follow Gibbs in using the symbol $\Gamma_{i(j)}$ for the adsorption of component i when the dividing surface is chosen so that $\Gamma_j = 0$.

Before going further with the thermodynamic discussion it is worth emphasizing several related points. First, the surface or excess functions are merely differences between properties of the whole system and those that we should expect if each bulk phase continued unchanged up to the dividing surface, and so are defined in terms of operationally accessible functions. There is no implication that the actual interface is sharp.

It might seem to be more realistic to start by defining three phases rather than two. Two of these, α and β, would be strictly homogeneous in reality, and not merely as above, by definition. The third or surface phase would lie between α and β and would contain all regions in which there were non-zero gradients of n, ϕ, etc. This mode of treatment was followed by the Dutch school, notably Bakker,[10] and later by Guggenheim.[11] Its advantages are illusory for we cannot measure or define unambiguously and independently the thermodynamic properties of the surface phase; it is, in Defay's words,[12] *a non-autonomous phase.* This important point of principle is discussed further in §§ 2.5 and 4.10. We can evade the difficulty only by defining the properties of the surface phase as differences between those of the whole system and those of phases α and β. But then the treatment differs only trivially from that of Gibbs, and indeed leads to the same experimental consequences.

A further point worth emphasizing is related to the last one, and has already been touched on in the last section, namely that the values of the potentials p, T, and μ which enter into the arguments above are those of the bulk phases. We do not have to ask if they have the same value at the surface as they have in the bulk. The question does not arise in Gibbs's treatment; it does arise and must be answered, if a third phase is introduced. It is true that Gibbs[13] specifically states that μ has the same value at the surface as in the bulk, but he neither uses this result nor provides an operational definition of μ at the surface. This and similar questions about 'local' thermodynamic functions are best answered by using the methods of statistical thermodynamics (§ 4.10).

A final point to be discussed before passing on to the consequences of these equations is the role of gravity. It is clear from (2.13) that, at fixed V, T, and n, the minimum of F is at a minimum of A. The surface of minimum area, for fixed volume of both phases, is a sphere, and so the planar surface specified at the beginning of this section can be realized only if there is a gravitational field. It is, therefore, an implicit assumption of the thermodynamic treatment that there exists a gravitational field which is strong enough to permit us to discuss the equilibrium properties of a plane surface, but weak enough not to distort the gradients of **n**, which we therefore assume to have the same values as in the absence of the field. We can temporarily evade this issue by supposing that we go to the thermodynamic limit of infinite **n**, V, and A, in such a way that the radius of the spherical drop of (say) phase α becomes infinite, and its surface becomes a plane. However the more subtle question of the behaviour of a finite surface in the limit of a gravitational field going to zero is again one which can be handled only by statistical thermodynamic arguments (§ 4.9).

By subtraction from (2.13) and (2.15) of the equivalent equations for the two bulk phases,

$$
\begin{aligned}
dF^\alpha &= -S^\alpha \, dT - p \, dV^\alpha + \boldsymbol{\mu} \cdot \mathbf{dn}^\alpha, \\
d\Omega^\alpha &= -S^\alpha \, dT - p \, dV^\alpha - \mathbf{n}^\alpha \cdot \mathbf{d\mu}, \\
dF^\beta &= -S^\beta \, dT - p \, dV^\beta + \boldsymbol{\mu} \cdot \mathbf{dn}^\beta, \\
d\Omega^\beta &= -S^\beta \, dT - p \, dV^\beta - \mathbf{n}^\beta \cdot \mathbf{d\mu},
\end{aligned}
\tag{2.28}
$$

we obtain the fundamental equations for the surface functions

$$
\begin{aligned}
dF^s &= -S^s \, dT + \sigma \, dA + \boldsymbol{\mu} \cdot \mathbf{dn}^s, \\
d\Omega^s &= -S^s \, dT + \sigma \, dA - \mathbf{n}^s \cdot \mathbf{d\mu}.
\end{aligned}
\tag{2.29}
$$

By subtraction of these equations from those obtained on general differentiation of (2.20) and (2.24) we obtain the Gibbs–Duhem equation for the surface

$$
S^s \, dT + A \, d\sigma + \mathbf{n}^s \cdot \mathbf{d\mu} = 0
\tag{2.30}
$$

or, dividing by A,

$$
d\sigma + \eta^s \, dT + \boldsymbol{\Gamma} \cdot \mathbf{d\mu} = 0
\tag{2.31}
$$

where η^s and $\boldsymbol{\Gamma}$ are defined by (2.27). This important equation, the *Gibbs adsorption equation*, relates a change of σ to changes in the potentials T and $\boldsymbol{\mu}$ of the bulk phases. The sum $\boldsymbol{\Gamma} \cdot \mathbf{d\mu}$ is invariant with change of position of the dividing surface. This can be seen by considering a displacement of the surface by a distance δz towards phase β. From the definition of $\boldsymbol{\Gamma}$ it follows that

$$
\delta \boldsymbol{\Gamma} + \Delta \boldsymbol{\rho} \, \delta z = 0
\tag{2.32}
$$

where $\Delta \boldsymbol{\rho} = \boldsymbol{\rho}^\alpha - \boldsymbol{\rho}^\beta$ (we use below the same contraction for differences of other densities). Equilibrium between the phases requires that at fixed p and T

$$
\Delta \boldsymbol{\rho} \cdot \mathbf{d\mu} = 0
\tag{2.33}
$$

(from the Gibbs–Duhem equation) so that

$$\delta(\mathbf{\Gamma} . \mathbf{d\mu}) = 0. \tag{2.34}$$

It is sometimes convenient to rewrite the adsorption equation in other more symmetrical forms. In a system of c components we can specify the state by the values of $(c+1)$ independent densities, for example, the c molar densities and the entropy density, η. Let *all* these densities be denoted by the symbol $\mathbf{\rho}$, a vector of $(c+1)$ components, and their surface excesses by $\mathbf{\Gamma}$. Similarly let $\mathbf{\mu}$ be the conjugate set of fields.[†] In the example given the first c of these would be the chemical potentials and the last would be the temperature. The adsorption equation (2.31) can then be written in more symmetric form

$$d\sigma + \mathbf{\Gamma} . \mathbf{d\mu} = 0 \tag{2.35}$$

and the condition of equilibrium (2.33), whilst formally unchanged, acquires an extra term, $\Delta\rho_{c+1} \, d\mu_{c+1}$ where $\rho_{c+1} = \eta$ and $\mu_{c+1} = T$. In this extended sense it is a form of the Clapeyron equation.

Another choice of independently variable densities might be the c molar densities and the energy density ϕ. The conjugate fields $\mathbf{\nu}$ are now the ratios μ_i/T; and $-1/T$. In these variables the Clapeyron equation is

$$\Delta\mathbf{\rho} . \mathbf{d\nu} = 0 \tag{2.36}$$

and the adsorption equation

$$d(\sigma/T) + \mathbf{\Gamma} . \mathbf{d\nu} = 0. \tag{2.37}$$

The experimental consequences of the adsorption equation are best brought out by eliminating one of the $(c+1)$ terms in (2.35) or (2.37) by means of (2.33) or (2.36). If we eliminate $d\mu_i$, and multiply through by $\Delta\rho_i = (\rho_i^\alpha - \rho_i^\beta)$, then from (2.35)

$$\Delta\rho_i \, d\sigma = \sum_i (\Delta\rho_j \Gamma_i - \Delta\rho_i \Gamma_j) \, d\mu_j. \tag{2.38}$$

In a one-component system ρ_i can be taken to be the number density and the remaining density to be the entropy density η; (2.38) is then

$$\Delta\rho \, d\sigma = (\Delta\eta\Gamma - \Delta\rho\eta^s) \, dT. \tag{2.39}$$

Similarly, from the alternative form of the adsorption equation (2.37),

$$\Delta\rho \, d(\sigma/T) = -(\Delta\phi\Gamma - \Delta\rho\phi^s) \, d(1/T). \tag{2.40}$$

If we choose an equimolar dividing surface $(\Gamma = n^s/A = 0)$ then these

---

[†] Appendix 1 gives a short account of the equations of thermodynamics in terms of densities and their conjugate fields.

equations yield

$$\eta^s = -(d\sigma/dT), \qquad \phi^s = d(\sigma/T)/d(1/T) = \sigma - T(d\sigma/dT),$$
$$\psi^s = \phi^s - T\eta^s = \sigma. \tag{2.41}$$

These equations have, however, little physical content, they merely define the surface free energy, entropy, and energy of a one-component system consistent with the choice of an equimolar dividing surface. Since surface tensions usually fall with rising temperature (and always do so near the gas–liquid critical point) it follows that η^s is positive and ϕ^s is larger than ψ^s. Near the critical temperature σ is proportional to $(T^c - T)^\mu$, and ϕ^s to $(T^c - T)^{\mu-1}$, where the index μ (not the chemical potential) is about $1\cdot26$ (§ 9.3). Hence $(d\sigma/dT)$ vanishes at T^c, as is shown in Figs 1.5 and 1.6, but $(d\phi^s/dT)$ is negative infinite. If there is a temperature at which $(d^2\sigma/dT^2)$ is zero (as for water, Fig. 1.6) then ϕ^s has there its maximum value.

If we choose a dividing surface for which η^s or ϕ^s vanishes, then we obtain the (number) adsorption with respect to these surfaces,

$$\Gamma_{\rho(\eta)} = \frac{\Delta\rho}{\Delta\eta}\left(\frac{d\sigma}{dT}\right), \qquad \Gamma_{\rho(\phi)} = -\frac{\Delta\rho}{\Delta\phi}\left(\frac{d(\sigma/T)}{d(1/T)}\right), \tag{2.42}$$

and

$$\Delta\eta\Gamma_{\rho(\eta)} + \Delta\rho\Gamma_{\eta(\rho)} = \Delta\phi\Gamma_{\rho(\phi)} + \Delta\rho\Gamma_{\phi(\rho)} = 0. \tag{2.43}$$

Since the number density ρ is uniquely defined it follows from (2.41) that there is no arbitrariness about the adsorptions with respect to the equimolar surface, that is $\Gamma_{\eta(\rho)} = \eta^s$ and $\Gamma_{\phi(\rho)} = \phi^s$. However the values of $\Delta\eta$ and $\Delta\phi$ depend on the zeros of entropy and energy, so it follows from (2.42) and (2.43) that $\Gamma_{\rho(\eta)}$ and $\Gamma_{\rho(\phi)}$ are not uniquely defined. It is convenient to consider only the configurational part of entropy and energy.

In a mixture it is more useful to restrict (2.38) to isothermal changes of composition of one phase. Thus if there is a change in n_i^α at fixed temperature and all other amounts, $(\mathbf{n}^\alpha)'$, then (2.38) becomes

$$\Delta\rho_i\left(\frac{\partial\sigma}{\partial n_i^\alpha}\right)_{T,(\mathbf{n}^\alpha)'} = \sum_j \{\Delta\rho_i\Gamma_i - \Delta\rho_i\Gamma_j\}\left(\frac{\partial\mu_j}{\partial n_i^\alpha}\right)_{T,(\mathbf{n}^\alpha)'}. \tag{2.44}$$

All derivatives are independent and can be measured. The terms in braces are invariant with change of dividing surface, and are the only combination of adsorptions that can be obtained by purely thermodynamic measurements. If, as is often the case, phase α is a liquid and phase β a vapour of negligible density, then this equation can be simplified to give

$$\rho_i^l\left(\frac{\partial\sigma}{\partial n_i^l}\right)_{T,(\mathbf{n}^l)'} = \sum_j (\rho_j^l\Gamma_i - \rho_i^l\Gamma_j)\left(\frac{\partial\mu_j}{\partial n_i^l}\right)_{T,(\mathbf{n}^l)'}. \tag{2.45}$$

The derivative on the left is the change of surface tension with liquid composition, and those on the right can be measured from the changes of vapour pressure and composition with liquid composition.[14] We can also, by careful analysis of the whole system and of each bulk phase separately, measure directly the functions $(\rho_j^l \Gamma_i - \rho_i^l \Gamma_j)$, but this is difficult unless we are dealing with a binary system in which one of the components is strongly adsorbed at the gas–liquid surface, for example, higher alcohols and fatty acids which form adsorbed monolayers at the water–vapour interface. In these extreme cases the Gibbs adsorption equation, in the form of (2.45), has been verified, although the errors of measuring the adsorptions do not make the tests very searching.[15] We return to this point in Chapter 6 when we describe attempts to verify the equation for much simpler model liquids by the method of computer simulation.

An isothermal extremum in σ, that is a state of the liquid system at which $(\partial\sigma/\partial n_i^l)_{T,(\mathbf{n}^l)'}$ is zero for all components i, is the surface analogue of an azeotrope,[16] at which the vapour pressure is at an extremum. (We use here the conventional sense of the word azeotrope, see Appendix 1.) It follows from (2.45) that surface azeotropy implies that all relative adsorptions, $(\rho_j^l \Gamma_i - \rho_i^l \Gamma_j)$, are zero.

A further property that is accessible to purely macroscopic measurement is the separation of the dividing surfaces at which the adsorptions of two different generalized components are zero. It follows from (2.38), or from the more specialized (2.39)–(2.45), that $\Gamma_{i(j)}$ is a thermodynamically accessible quantity. In terms of these adsorptions we can rewrite (2.35) as

$$d\sigma + \sum_{i \neq j} \Gamma_{i(j)}\, d\mu_i = 0 \qquad (2.46)$$

where the sum is over c of the $(c+1)$ independent densities and hence where the variations $d\mu_i$ are now all independent. Thus

$$\Gamma_{i(j)} = -(\partial\sigma/\partial\mu_i)_{\mu_k} \qquad (2.47)$$

where the index k runs over $(c-1)$ of the $(c+1)$ generalized components, excluding the two cases $k=i$ and j. From (2.32) we have

$$\Gamma_i = \Delta\rho_i(z_i - z) \qquad (2.48)$$

when the dividing surface is at z, and where z_i is the dividing surface at which $\Gamma_i = 0$. (The value of Γ_i may, however, be subject to a conventional choice of zero, as explained above.) Hence

$$\Gamma_{i(j)} = \Delta\rho_i(z_i - z_j) \qquad (2.49)$$

and so $(z_i - z_j)$ is also an experimentally accessible quantity.[17] Thus for argon at its triple point $\Gamma_{\phi(\rho)}$ or ϕ^s is $35 \cdot 0\,\text{mJ m}^{-2}$, $(\Delta\phi/\Delta\rho)$ is

$-5940 \, \text{J mol}^{-1}$ and so, from (2.41) and (2.42), $\Gamma_{\rho(\phi)}$ is $5 \cdot 89 \times 10^{-6} \, \text{mol m}^{-2}$. ($\phi$ is here the configurational energy density.) Since $\Delta\rho$ is $35 \cdot 5 \times 10^{3} \, \text{mol m}^{-3}$ (if α is the liquid phase) it follows from (2.49) that $(z_{\rho} - z_{\phi})$ is $0 \cdot 167 \, \text{nm}$, which is somewhat smaller than the size of an argon atom. The surface z_{ϕ} is on the liquid side of z_{ρ}. The separation increases as the temperature rises and becomes infinite at the critical point (§ 9.3).

The discussion above is based principally on the choice of the Helmholtz free energy and the grand potential as the fundamental thermodynamic functions. It is not useful to introduce the Gibbs free energy since pressure is not as convenient an independent variable as density. Indeed there is no agreed definition of G; both $(F + pV)$ and $(F + pV - \sigma A)$ are used.[18]

2.4 The spherical surface

The thermodynamics of curved surfaces is more subtle than that of planar, and we discuss only the most important case, the spherical surface, for which the two principal radii of curvature are equal. The extension of the argument to other curved surfaces is beset with difficulties into which we do not enter.[19] The spherical surface, the bubble or the drop, is the only one that is stable in the absence of an external field. The original analysis of Gibbs[7] was clarified and its consequences worked out by Tolman,[20] whose work Koenig[21] extended to multicomponent systems. Buff,[22] Hill[23], and Kondo[24] describe explicitly how the surface tension depends on the position of the dividing surface to which it is referred, or at which it is calculated.

From a spherical drop of phase α surrounded by phase β we choose a system formed of a conical section of solid angle ω (Fig. 2.2). The side

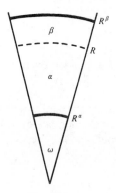

FIG. 2.2

walls of the cone serve only to define the system and do not affect the fluid. The two phases are deemed to be separated by a dividing surface of radius R. Until this surface is chosen we cannot define V^α and V^β (as was the case also for the planar surface), and we cannot even define the area of the surface. It is this last point on which curved surfaces differ radically from planar, and which here makes the use of a dividing surface a necessity, and not merely a convenience. From the geometry of the cone

$$A = \omega R^2, \qquad V^\alpha = \tfrac{1}{3}\omega[R^3 - (R^\alpha)^3], \qquad V^\beta = \tfrac{1}{3}\omega[(R^\beta)^3 - R^3],$$
(2.50)

so that four geometrical variables, ω, R, R^α, and R^β, are needed to define the sizes of the two phases and the surface area. It is convenient to use in place of these four the set R, A, V^α, and V^β, so that the generalization of (2.11) is

$$dU = T\, dS - p^\alpha\, dV^\alpha - p^\beta\, dV^\beta + \sigma\, dA + C\, dR + \boldsymbol{\mu}\cdot\mathbf{dn} \qquad (2.51)$$

where σ and C, the coefficients of the area and curvature terms, are defined by this equation. The parallel equation for the free energy, cf. (2.13), is

$$dF = -S\, dT - p^\alpha\, dV^\alpha - p^\beta\, dV^\beta + \sigma\, dA + C\, dR + \boldsymbol{\mu}\cdot\mathbf{dn}. \qquad (2.52)$$

This can be integrated at fixed T and R by changing ω;

$$F = -p^\alpha V^\alpha - p^\beta V^\beta + \sigma A + \boldsymbol{\mu}\cdot\mathbf{n}. \qquad (2.53)$$

The grand potential is

$$\Omega = F - \boldsymbol{\mu}\cdot\mathbf{n} = -p^\alpha V^\alpha - p^\beta V^\beta + \sigma A. \qquad (2.54)$$

For a planar surface $p^\alpha = p^\beta$ and (2.53) and (2.54) reduce to (2.20) and (2.24).

The free energies F and Ω are independent of the choice of the position of the dividing surface since this choice is quite arbitrary. Let differentials in square brackets denote changes in functions that follow from a notional change in the position of this surface. Such a change is called notional since it affects only the description of the system, and does not correspond to any physical change. We continue to use round brackets (or no brackets) for differentials that describe some real physical change in the state of the system, e.g. a drop becoming larger or smaller. From (2.52) and (2.53) or (2.54),

$$[dF] = -(p^\alpha - p^\beta)\omega R^2[dR] + \sigma 2\omega R[dR] + C[dR] = 0, \qquad (2.55)$$

$$[dF] = [d\Omega] = -(p^\alpha - p^\beta)\omega R^2[dR] + \sigma 2\omega R[dR] + \omega R^2[d\sigma] = 0. \qquad (2.56)$$

The solutions of these equations are

$$p^\alpha - p^\beta \equiv \Delta p = \frac{2\sigma}{R} + \left[\frac{d\sigma}{dR}\right],\tag{2.57}$$

$$C = A\left[\frac{d\sigma}{dR}\right].\tag{2.58}$$

The first is a generalization of Laplace's equation for a surface tension σ measured at, or referred to, an arbitrary dividing surface. Since Δp is invariant with respect to the choice of R, it follows that, in general, σ must formally be a function of R, which we write $\sigma[R]$. Moreover if we define the surface of tension as the radius at which the tension acts, in conformity with the model described in the opening paragraph of § 2.2, then a comparison of (2.1) with (2.57) shows that the second term vanishes at this surface. That is,

$$[d\sigma/dR]_{R=R_s} = 0.\tag{2.59}$$

It follows that $\sigma[R_s]$ is an extremum, which, we shall see, is a minimum. The curvature term vanishes at this surface.

The formal derivative $[d\sigma/dR]$ can be expressed in terms of the adsorptions. Consider the change in F that follows from real isothermal changes in the variables of (2.52),

$$(dF)_T = -p^\alpha(dV^\alpha)_T - p^\beta(dV^\beta)_T + \sigma(dA)_T + A[d\sigma/dR](dR)_T + \boldsymbol{\mu} \cdot (\mathbf{dn})_T.\tag{2.60}$$

For the separate phases

$$(dF^\alpha)_T = -p^\alpha(dV^\alpha)_T + \boldsymbol{\mu} \cdot (\mathbf{dn}^\alpha)_T, \qquad (dF^\beta)_T = -p^\beta(dV^\beta)_T + \boldsymbol{\mu} \cdot (\mathbf{dn}^\beta)_T.\tag{2.61}$$

Subtraction gives

$$(dF^s)_T = \sigma(dA)_T + A[d\sigma/dR](dR)_T + \boldsymbol{\mu} \cdot (\mathbf{dn}^s)_T.\tag{2.62}$$

From (2.53)

$$F^s = \sigma A + \boldsymbol{\mu} \cdot \mathbf{n}^s\tag{2.63}$$

so that we have also

$$(dF^s)_T = \sigma(dA)_T + A(d\sigma)_T + \boldsymbol{\mu} \cdot (\mathbf{dn}^s)_T + \mathbf{n}^s \cdot (\mathbf{d\mu})_T\tag{2.64}$$

or

$$[d\sigma/dR] = (\partial\sigma/\partial R)_T + \boldsymbol{\Gamma} \cdot (\partial \boldsymbol{\mu}/\partial R)_T.\tag{2.65}$$

We now specialize the discussion to a one-component system. If σ_s is the value of σ at the surface of tension, where $[d\sigma/dR]$ is zero, then the adsorption at that surface Γ_s is

$$\Gamma_s = -(\partial\sigma_s/\partial\mu)_T.\tag{2.66}$$

Conversely if σ_e is the value of σ at the equimolar dividing surface, then

$$[d\sigma/dR]_{R=R_e} = (\partial\sigma_e/\partial R_e)_T. \tag{2.67}$$

From (2.66) we obtain the (real) dependence of σ_s on R_s. For an isothermal change (we drop the suffix T),

$$d\sigma_s = -\Gamma_s \, d\mu = -\Gamma_s \, dp^\alpha/\rho^\alpha = -\Gamma_s \, dp^\beta/\rho^\beta. \tag{2.68}$$

From the last two of these equations

$$d(\Delta p) = \Delta\rho \, d\mu \tag{2.69}$$

and so from (2.57) and (2.59)

$$d\sigma_s = -\Gamma_s(\Delta\rho)^{-1} \, d(2\sigma_s/R_s). \tag{2.70}$$

For a planar surface $\Gamma_s/\Delta\rho$ is $(z_e - z_s)$, the separation of the equimolar surface and the surface of tension (2.48). For a curved surface[20]

$$\frac{\Gamma_s}{\Delta\rho} = \delta\left(1 + \frac{\delta}{R_s} + \frac{1}{3}\frac{\delta^2}{R_s^2}\right), \qquad \delta = R_e - R_s, \tag{2.71}$$

but the terms in (δ/R_s) and $(\delta/R_s)^2$ can be omitted to the order of accuracy we need below. So from (2.70)

$$d\sigma_s = -\frac{2\delta}{R_s}\left(d\sigma_s - \frac{\sigma_s}{R_s}dR_s\right) \tag{2.72}$$

or

$$\frac{d\ln\sigma_s}{dR_s} = \frac{1}{R_s} - \frac{1}{R_s + 2\delta} \tag{2.73}$$

and by integration from the planar limit $R_s \to \infty$,

$$\frac{\sigma_s}{\sigma_\infty} = \frac{R_s}{R_s + 2\delta} = \left(1 - \frac{2\delta}{R_s} + \dots\right). \tag{2.74}$$

This equation describes the variation of σ at R_s with the radius of the drop, as measured by R_s. Clearly the change is small since δ is of molecular size. The sign of the change depends on the relative disposition of z_e and z_s at a planar surface; we return to this point below and in §§ 4.4, 5.7, 6.5, and 7.6.

The formal change of σ with the position R to which it is referred can be found from (2.57) which can be written

$$\left[\frac{d}{dR}\right]R^2\sigma[R] = R^2(p^\alpha - p^\beta). \tag{2.75}$$

On integrating from R_s to an arbitrary surface of radius R, this gives

$$\frac{\sigma[R]}{\sigma[R_s]} = 1 + \left(\frac{R - R_s}{R}\right)^2 \frac{(R_s + 2R)}{3R_s} \tag{2.76}$$

so that $\sigma[R_s]$ is the minimum tension. It is therefore only to terms of order (δ/R) that σ is independent of the formal choice of dividing surface, and so in principle a useful physical property. That is, (2.74) cannot be extended to higher powers by including the higher terms in (2.71). The integration of the formally more complete function has been carried out by Tolman[20] and others,[25] but leads to no new physical information. For drops or bubbles that are so small that $(\delta/R) \sim 1$ the surface tension is not definable, nor indeed can the equations of bulk thermodynamics be used for the internal phase α.

The definition and dependence on curvature of the surface tension are closely related, through the Laplace equation, to the pressure difference $p^\alpha - p^\beta$, but we have not yet determined the two pressures themselves in terms of p_∞, the vapour pressure above the planar surface.[26] This can be done by invoking the condition that the chemical potential is the same in each phase both for the planar system ($\mu_\infty^\alpha = \mu_\infty^\beta$) and for the system of arbitrary curvature ($\mu_R^\alpha = \mu_R^\beta$). From the condition

$$\mu_R^\alpha - \mu_\infty^\alpha = \mu_R^\beta - \mu_\infty^\beta \qquad (2.77)$$

we have

$$\int_{p_\infty}^{p^\alpha} v^\alpha \, dp = \int_{p_\infty}^{p^\beta} v^\beta \, dp \qquad (2.78)$$

where v is the molar volume. This equation, together with that of Laplace, determines p^α and p^β. If the liquid phase α is incompressible ($v^\alpha = v^l = $ constant), and if the vapour phase β is a perfect gas ($v^\beta = N_A kT/p^\beta$), then the vapour pressure p^β is given by

$$N_A kT \ln(p^\beta/p_\infty) = v^l(p^\beta - p_\infty + 2\sigma/R) \simeq 2\sigma v^l/R. \qquad (2.79)$$

This is Kelvin's equation for the vapour pressure of a drop of radius R; the second (approximate) form of which holds only if $N_A kT \gg v^l p_\infty$. It is inapplicable to drops containing only a few molecules since they have no uniform interior for which the calculation of p is meaningful. For a drop of water of radius 1 mm, $\ln(p^\beta/p_\infty)$ is 10^{-6}, and for a drop of radius 1 μm, it is 10^{-3}. Hence in a mist the very small drops evaporate and the larger drops grow. All are unstable with respect to the pool of liquid that ultimately forms.

The equation has been confirmed by Fisher and Israelachvili[27] for a meniscus of cyclohexane held between mica spheres that are almost in contact. Here R is negative and p^β is less than p_∞. Their values of $(-R)$ ranged from 19 nm down to 4·2 nm and agreement with theory is perhaps marginally improved for $(-R)$ less than about 10 nm if it is assumed that σ itself is a decreasing linear function of R^{-1}, but the evidence for this change of σ is not strong, and is not confirmed by measurements of the pressure of adhesion between two solid bodies with a liquid film between them.[27]

In a mixture the approximate form of Kelvin's equation is

$$N_A kT \ln[p_i^\beta/(p_i)_\infty] \simeq 2\sigma v_i/R \qquad (2.80)$$

where p_i is the partial pressure of component i in the vapour and v_i its partial molar volume in the liquid. In this form the equation has been confirmed experimentally by La Mer and Gruen.[28]

2.5   Quasi-thermodynamics—a first look

Gibbs's treatment of the thermodynamics of surfaces leads, as do all unadulterated classical thermodynamic arguments, to an equation be-tween experimentally observable properties. It is natural to want to push behind such a formalism and to try to interpret what is happening at the surface in terms of local thermodynamic functions, without having to use the apparently more difficult, but truly molecular methods of statistical mechanics. There have been many such attempts of varying degrees of sophistication, some of which are discussed later; they go generally under the name of *quasi-* or *local thermodynamics*. Here we obtain two general results common to all quasi-thermodynamic methods, and then consider briefly only the simplest, and ultimately untenable treatment of this kind to which we give the name of *point-thermodynamics*.

The assumption on which all these methods rest is that it is possible to define unambiguously (or, at least, consistently) local values of the thermodynamic fields, p, T, and μ, and of the densities ρ, ϕ, ψ, and η, even in an inhomogeneous system. At a planar fluid–fluid interface these fields and densities are functions only of the height, z. We shall see later that this assumption is generally wrong, §§ 4.3 and 4.8.

Consider a one-component system of two phases confined to a cube of side l, and hence of cross-section $A = l^2$. The two phases are separated by a horizontal surface, and the height z is measured from the mid-height of the vessel. The amount of substance n and the Helmholtz free energy F are given by

$$n = A \int_{-l/2}^{l/2} \rho(z)\, \mathrm{d}z, \qquad F = A \int_{-l/2}^{l/2} \psi(z)\, \mathrm{d}z. \qquad (2.81)$$

The pressure in either bulk phase is the scalar quantity p. Near the interface it is a tensor (the negative of the stress tensor) since its tangential components (parallel with a dividing surface) include the tension of the surface, and so differ from the normal component. Mechanical stability requires that the gradient of this tensor is zero everywhere in the fluid,[29]

$$\nabla \cdot \mathbf{p} = 0. \tag{2.82}$$

The symmetry of the surface requires that $\mathbf{p}$ is a diagonal tensor

$$\mathbf{p}(\mathbf{r}) = \mathbf{e}_x \mathbf{e}_x p_{xx}(\mathbf{r}) + \mathbf{e}_y \mathbf{e}_y p_{yy}(\mathbf{r}) + \mathbf{e}_z \mathbf{e}_z p_{zz}(\mathbf{r}) \tag{2.83}$$

with

$$p_{xx}(\mathbf{r}) = p_{yy}(\mathbf{r}),$$

where $\mathbf{e}_x$ is a unit vector in the x-direction. Substitution of (2.83) into (2.82) shows that p_{xx} and p_{yy} are functions only of z, and that p_{zz} is a constant:

$$p_{xx}(z) = p_{yy}(z) = p_T(z) \qquad p_{zz}(z) = p_N(z) = p \tag{2.84}$$

where p_N and p_T are the normal and tranverse components of the pressure.

If a side-wall of the cube is displaced isothermally and reversibly so as to increase the area by δA, then the (tangential) work done on the system is

$$\delta W_T = -\delta A \int_{-l/2}^{l/2} p_T(z) \, dz. \tag{2.85}$$

A similar displacement of the top or bottom wall by a distance $\delta A/l$ serves to maintain the volume constant, and requires further (normal) work

$$\delta W_N = A(\delta A/l)p = l \, \delta A \, p. \tag{2.86}$$

Thus the total work done, and hence the increase in free energy is

$$\delta F = \delta W_T + \delta W_T = \delta A \int_{-\infty}^{\infty} [p - p_T(z)] \, dz. \tag{2.87}$$

The limits can be taken as $\pm\infty$ since $p_T(z)$ differs from p only near the interface. This increase in free energy per unit area is, by (2.13), equal to the surface tension

$$\sigma = \int_{-\infty}^{\infty} [p - p_T(z)] \, dz. \tag{2.88}$$

This equation is usually called the mechanical definition of surface tension. Its form is virtually self-evident, for if we can define satisfactorily a local function $p_T(z)$ then its integral through the surface must, by purely mechanical arguments, be related to the tension by (2.88). It should be emphasized, however, that this equation should not be regarded as a definition of σ, in the way that (2.11), (2.13), or (2.15) can be so regarded, until we have a satisfactory way of determining $p_T(z)$. An elementary calculation shows just how bizarre a function this is.[30] The surface of a typical liquid at its normal boiling point is about 1 nm or 10 Å thick (Chapters 6 and 7); that is, the change of density from ρ^l to ρ^g is essentially complete over such a distance. A typical value of the surface tension is $20 \, \text{mN m}^{-1}$, hence the average value of $[p - p_T(z)]$ over the thickness of the surface is about $2 \times 10^7 \, \text{N m}^{-2}$ or 200 bar. Since $p \simeq 1 \cdot \text{bar}$ this means that the average value of $p_T(z)$ is about -200 bar.

By taking moments about a point z we see that the height of the surface of tension z_s, that is the height at which the surface tension appears to act, is given by[31]

$$\sigma z_s = \int_{-\infty}^{\infty} z[p - p_T(z)] \, dz. \tag{2.89}$$

We shall see later (§ 4.4) that such mechanical arguments are not to be trusted at the microscopic level and that, although (2.88) is a satisfactory equation, (2.89) is not, because of the difficulty of defining $p_T(z)$ uniquely.

The simplest way of defining $p_T(z)$, etc., which we may call the *point-thermodynamic* approximation, is to assume that local functions depend on local variables in just the same way as they do in homogeneous systems. Thus if the free energy density ψ is a known function of density and temperature in a homogeneous system, $\psi(\rho, T)$, then it is assumed that

$$\psi(z) = \psi[\rho(z), T]. \tag{2.90}$$

This assumption was made by Tolman[31] and by Ono and Kondo[32] in spite of a better local thermodynamic prescription made much earlier by van der Waals, which we describe in the next chapter. (Many examples can be cited[33] of first-class experimental and theoretical work on the physics of liquids done between 1870 and 1914 being forgotten by physicists and unknown to chemists and engineers until the 1950s.)

We can use this assumption to relate $p_T(z)$ to other thermodynamic functions. Consider a small closed system formed of a slice of the cubic box above, and of thickness dz. Let the side-wall be displaced isothermally to increase the area by δA but now without compensating adjustment of the top or bottom boundary of the slice. The work done on the system is

$$\delta W = -\delta A \, dz \, p_T(z) \tag{2.91}$$

and since the system is closed

$$\delta(A\rho(z)) = 0 \quad \text{or} \quad \delta A \, \rho(z) + A \, \delta\rho(z) = 0. \tag{2.92}$$

The free energy of the system is

$$F = \psi(z)A\,dz. \tag{2.93}$$

Hence the change of free energy is

$$\delta F = \delta\psi(z)A\,dz + \psi(z)\delta A\,dz = -\delta A\,dz p_T(z). \tag{2.94}$$

On substitution from (2.92), and letting the increment δ become infinitesimal,

$$\rho(z)\left(\frac{\partial\psi(z)}{\partial\rho(z)}\right)_T = \psi(z) + p_T(z) = \rho(z)\mu(z) \tag{2.95}$$

where we have used the familiar (bulk) thermodynamic results

$$(\partial\psi/\partial\rho)_T = \mu = \rho^{-1}(\psi + p) \tag{2.96}$$

and have invoked the hypothesis of point-thermodynamics. We see that it is $p_T(z)$, and not $p_N(z)$, that is related to the free energy by the usual differentiation.[34]

Let us now return to the complete system, that is, the contents of the cube, and consider a third small change, namely an arbitrary redistribution of the local density $\rho(z)$. Since the system is closed we have, by (2.81),

$$\delta n = A\int_{-l/2}^{l/2}\delta\rho(z)\,dz = 0, \tag{2.97}$$

$$\delta F = A\int_{-l/2}^{l/2}\delta\psi(z)\,dz = A\int_{-l/2}^{l/2}\left(\frac{\partial\psi(z)}{\partial\rho(z)}\right)_T\delta\rho(z)\,dz$$

$$= A\int_{-l/2}^{l/2}\mu(z)\delta\rho(z)\,dz. \tag{2.98}$$

If the system is at equilibrium then F is at a minimum with respect to such arbitrary redistribution. A comparison of (2.97) and (2.98) shows that $\delta F = 0$ is consistent with $\mu(z) = \mu$, a chemical potential which is constant throughout the system. Minimizing F and imposing constancy on μ are equivalent operations in quasi-thermodynamic treatments.

So far the analysis is satisfactory and self-consistent, but it breaks down if we ask *what* distribution of matter $\rho(z)$ leads to a minimum value of F. If we have a system with fixed n, V, and A, but where V is so large that any redistribution of matter near the interface cannot affect p or μ in the bulk phases, then we see from (2.20) that a minimum value of F requires a minimum value of σ. If there are to be two phases and an interface then σ cannot be negative. But from (2.88) it follows that σ can be zero if $p_T(z)$ is everywhere equal to p; that is, if the interface is a step-function in density so that $\rho(z)$ is everywhere equal to ρ^α or to ρ^β. Thus minimization of the free energy under the point-thermodynamic approximation leads to a density profile $\rho(z)$ that is a step-function at all temperatures, and to the vanishing of the surface tension.

The necessity of taking a less restrictive definition of local functions than that of this point-thermodynamic treatment (that is, of (2.90) and its analogues) was known·to van der Waals (see Chapter 3), but was brought forward independently in more recent times by Hill.[35] The treatment of van der Waals was discussed by Bakker[36] in 1928, but his book, although much quoted, is apparently less often read.

It is interesting to note that it is only within the confines of point-thermodynamics that a step-function in $\rho(z)$ leads to the vanishing of surface tension. As we saw in the last chapter, Poisson thought that this vanishing was always implied by the assumption of a step-function, but he was wrong, and the treatment of Laplace quite properly leads to a non-vanishing value of σ because his integral H is not zero. Point-thermodynamics would be correct only if the range of the intermolecular forces were zero, but then H would also be zero, and so it is not surprising that minimizing the free energy under this approximation leads to zero surface tension.

Notes and references

1. Rankine, W. J. M. *Phil. Mag.* **2,** 509 (1851), reprinted in *Miscellaneous Scientific Papers of W. J. Macquorn Rankine* (ed. W. J. Millar) p. 16, Griffin, London (1881).
2. Clausius, R. *Pogg. Ann. Phys.* **116,** 73 (1862); English translation in *Phil. Mag.* **24,** 81, 201 (1862) and in *Mechanical Theory of Heat with its Applications to the Steam-Engine* (ed. T. A. Hirst) p. 215, Van Voorst, London (1867). Clausius compares his views with those of Rankine in *The Mechanical Theory of Heat* (trans. W. R. Browne) p. 345, Macmillan, London (1879). Gibbs discusses the disgregation of Clausius in his memoir of him, *Proc. Amer. Acad. Arts. Sci.* **16,** 458 (1889), reprinted in his *Collected Works*, ref. 3, Vol. 2, p. 261. See also the more recent discussion of Gibbs's views by Klein, M. J. *Historical Studies in the Physical Sciences*, Vol. 1, p. 127, University of Pennsylvania Press, Phildelphia (1969).
3. [Gibbs, J. W.], *The Collected Works of J. Willard Gibbs*, 2 vol. Longmans, Green, New York (1928).
4. Tyndall, J. *Heat considered as a Mode of Motion*, Lecture 3. Longman, London (1863).
5. Maxwell, J. C. *Theory of Heat*. Longmans, Green, London (1871).
6. Thomson, J. J. *Applications of Dynamics to Physics and Chemistry*, p. 5, Macmillan, London (1888); reprinted by Dawsons, London, 1968. See also Topper, D. R. *Arch. Hist. exact Sci.* **1,** 393 (1971). The idea that the end of the nineteenth century was marked by a rapid advance in the applications of classical thermodynamics and a relative stagnation in kinetic theory has been put forward by P. Clark in *Method and Appraisal in the Physical Sciences* (ed. C. Howson) p. 41, Cambridge University Press (1976), and disputed by Smith, C. *Hist. Sci.* **16,** 231 (1978).

7. Gibbs, J. W. *On the Equilibrium of Heterogeneous Substances, Trans. Connecticut Acad.* **3**, 108, 343 (1875–1878), republished in *The Scientific Papers of J. Willard Gibbs*, Vol. 1, p. 55, Longmans, Green, New York (1906), reprinted by Dover Publications, New York, 1961. The paper was published again, with the same pagination as in the *Scientific Papers*, in *The Collected Works*, ref. 3, which is the source to which references are given here. The section on surfaces, called *Theory of Capillarity* in the Synopsis (p. 350), is on pp. 219–331. See also Rice, J. in *A Commentary on the Scientific Writings of J. Willard Gibbs* (ed. F. G. Donnan and A. Haas) Vol. 1, p. 505, Yale University Press, New Haven (1936). A clear modern resumé of Gibbs's work (and a criticism of Rice's) is to be found in Sections 1–8 of Tolman, R. C. *J. chem. Phys.* **16**, 758 (1948).

8. Spero, R., Hoskins, J. K., Newman, R., Pellam, J. and Schultz, J. *Phys. Rev. Lett.* **44**, 1645 (1980).

9. We adopt the nomenclature of Griffiths, R. B. and Wheeler, J. C. *Phys. Rev.* **A2**, 1047 (1970); see Appendix 1.

10. Bakker, G. *Kapillarität und Oberflächenspannung*, Vol. 6 of *Handbuch der Experimentalphysik* (ed. W. Wien, F. Harms, and H. Lenz) Chapter 10, Akad. Verlags., Leipzig (1928); see also Verschaffelt, J. E. *Acad. roy. Belg., Bull Class. sci.* **22**, 373, 390, 402 (1936).

11. Guggenheim, E. A. *Trans. Faraday Soc.* **36**, 397 (1940); *Thermodynamics*, 5th edn p. 45. North-Holland, Amsterdam (1967); see also R. S. Hansen, *J. phys. Chem.* **66**, 410 (1962).

12. Defay, R. (1934), see Defay, R., Prigogine, I. and Bellemans, A. *Surface Tension and Adsorption* (trans. D. H. Everett) pp. 56, 365. Longmans, Green, London (1966).

13. Gibbs, Ref. 3, Vol. 1, p. 223.

14. For example, Rowlinson, J. S. and Swinton, F. L. *Liquids and Liquid Mixtures*, 3rd edn, Chapter 4. Butterworth, London (1982).

15. Adamson, A. W. *Physical Chemistry of Surfaces*, 3rd edn, Chapter 2, p. 76, Wiley, New York (1976); see also the bibliography compiled by Defay *et al.*, ref. 12, pp. 29, 92.

16. Rowlinson and Swinton, ref. 14, Chapter 4; Malesiński, W. *Azeotropy and other Theoretical Problems of Vapour-Liquid Equilibrium*, Interscience, London (1965). McLure, I. A., Edmonds, B., and Lal, M. *Nature, Phys. Sci.* **241**, 71 (1973) have proposed the word *aneotrope* for the surface analogue of an azeotrope.

17. Lovett, R. A. *Thesis*, Part I, Univ. Rochester, 1966; Buff, F. P. and Lovett, R. A. in *Simple Dense Fluids* (ed. H. L. Frisch and Z. W. Salsburg) Chapter 2, Academic Press, New York (1968); Good, R. J. *Pure appl. Chem.* **48**, 427 (1976).

18. Defay *et al.*, ref. 12, pp. 53–4.

19. Boruvka, L. and Neumann, A. W. *J. chem. Phys.* **66**, 5464 (1977).

20. Tolman, R. C. *J. chem. Phys.* **17**, 118, 333 (1949).

21. Koenig, F. O. *J. chem. Phys.* **18**, 449 (1950).

22. Buff, F. P. *J. chem. Phys.* **19**, 1591 (1951).

23. Hill, T. L. *J. phys. Chem.* **56**, 526 (1952).

24. Kondo, S. *J. chem. Phys.* **25**, 662 (1956).

25. See, for example, Nonnenmacher, T. F. *Chem. Phys. Lett.* **47**, 507 (1977).

26. Thomson, W. *Phil. Mag.* **42**, 448 (1871). See also Goodrich, F. C. in *Surface and Colloid Science* (ed. E. Matijević), Vol. 1, p. 1. Wiley, New York (1969); Defay, *et al.*, ref. 12, Chapter 15; Everett, D. H. and Haynes, J. M. *Colloid*

Science (ed. D. H. Everett) Vol. 1, Chapter 4. Spec. Per. Rep., Chemical Society, London (1973); McElroy, P. J. *J. Coll. Interf. Sci.* **72,** 147 (1979).

27. Fisher, L. R. and Israelachvili, J. N. *Nature* **277,** 548 (1979); *Chem. Phys. Lett.* **76,** 325 (1980).
28. La Mer, V. K. and Gruen, R. *Trans. Faraday Soc.* **48,** 410 (1952); see also Defay *et al.*, ref. 12, Chapter 16.
29. See, for example, Landau, L. D. and Lifshitz, E. M. *Theory of Elasticity* (trans. J. B. Sykes and W. H. Reid) 2nd edn, p. 7. Pergamon Press, Oxford (1970).
30. See also Carey, B. S., Scriven, L. E. and Davis, H. T. *J. chem. Phys.* **69,** 5040 (1978).
31. Tolman, ref. 7 and 20.
32. Ono, S. and Kondo, S. *Encyclopedia of Physics* (ed. S. Flügge) Vol. 10, p. 134, Springer, Berlin (1960). They use the term *quasi-thermodynamic* for what we here call *point-thermodynamic*.
33. Rowlinson, J. S. *Nature* **244,** 414 (1973); *Speaking of Science, 1977*, p. 269, Royal Institution, London (1978); Levelt Sengers, J. M. H. *Physica* **82A,** 319 (1976).
34. Motomura, K. *Adv. Coll. Interf. Sci.* **12,** 1 (1980).
35. Hill, T. L. *J. chem. Phys.* **19,** 261 (1951).
36. Bakker, ref. 10, Chapter 15.

THE THEORY OF VAN DER WAALS

3.1 Introduction

'I am more than ever an admirer of van der Waals'; so wrote Lord Rayleigh[1] in a private letter in 1891. It was a view shared by many of the best nineteenth-century physicists and chemists. Twenty years earlier Maxwell had taught himself Dutch in order to be able to read van der Waals's thesis, while Ostwald translated his work for republication in German. What commanded the admiration of such men was his development of powerful and far-sighted methods of approximation in the newly developing subject of statistical thermodynamics. We have already touched in Chapter 1 on his contribution to the development of the mean-field approximation; here we describe his application of these ideas to the state of matter within the gas–liquid surface.

In the quasi-thermodynamics of the last chapter we introduced a local free-energy density $\psi(z)$, although its identification in (2.90) proved unsatisfactory; it led to a vanishing interfacial tension σ and to a density profile $\rho(z)$ that was a step-function. What is missing in (2.90) is a term that imposes a characteristic length on the fluid, and thus leads, after minimization of the integrated excess free-energy density, to a density profile consistent with an interface of non-zero thickness. The theory was given such a form by van der Waals;[2] Landau and Lifshitz's theory of magnetic domain structure[3] is equivalent, and the theory has been reformulated and extended by Cahn and Hilliard.[4] In this chapter we present van der Waals's theory and some of its elaborations.

Let $\psi(z)$ continue to represent the local free-energy density of (initially) a one-component system, though now not necessarily given by (2.90). Whatever the correct $\psi(z)$, let $\Psi(z)$ be the free-energy-density excess that is due to the inhomogeneity in the region of the interface; that is, let

$$\Psi(z) = \psi(z) - \psi(\rho^{\alpha,\beta}, T) = \psi(z) - \psi^{\alpha,\beta} \qquad (3.1)$$

where on the α-side of the dividing surface $\rho^{\alpha,\beta}$ and $\psi^{\alpha,\beta}$ mean the density ρ^{α} and the Helmholtz free-energy density ψ^{α} of the bulk phase α, while on the β-side they mean ρ^{β} and ψ^{β}. Choose the dividing surface to be that of vanishing adsorption (vanishing superficial density of matter, $\Gamma_{\rho} = 0$);

$$\int_{-\infty}^{\infty} [\rho(z) - \rho^{\alpha,\beta}]\, dz = 0. \qquad (3.2)$$

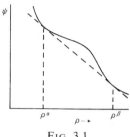

FIG. 3.1

With this dividing surface, the surface or excess Helmholtz free energy is
the surface tension, as in (2.23). Thus, from the definition of $\Psi(z)$ in (3.1)

$$\sigma = \int_{-\infty}^{\infty} \Psi(z)\, dz, \tag{3.3}$$

provided (3.2) holds.

Figure 3.1 shows the thermodynamic free-energy density $\psi(\rho, T)$ as a
function of ρ for a fixed temperature T in the region of two-phase
coexistence. The common-tangent construction, which yields the densities
ρ^α and ρ^β of the coexistent phases, is equivalent to the equal-areas
construction in Fig. 1.8. The curve is the analytic $\psi(\rho, T)$ given by a
mean-field approximation to the equation of state. In that sense it is like
the analytic, unreconstructed $p(v, T)$ of the van der Waals equation
shown in Fig. 1.8. (The meaning and origin of such functions will be
discussed in § 3.5.) The distance of the curve $\psi(\rho, T)$ above its double
tangent is a function we shall call $-W(\rho)$ (suppressing, for brevity, explicit
reference to T). Its negative, $W(\rho)$, is shown in Fig. 3.2; it has two equal
maxima at $\rho = \rho^\alpha$ and ρ^β, where $W = 0$; elsewhere, $W < 0$. On account of
(3.2)

$$\int_{-\infty}^{\infty} \{\psi[\rho(z), T] - \psi^{\alpha,\beta}\}\, dz = \int_{-\infty}^{\infty} -W[\rho(z)]\, dz \tag{3.4}$$

where the integral on the right-hand side is manifestly independent of the
choice of dividing surface. Identifying the free-energy density $\psi(z)$ with

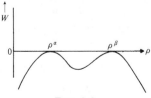

FIG. 3.2

$\psi[\rho(z), T]$, as in (2.90), is seen from (3.4) to be equivalent to identifying the excess free-energy density $\Psi(z)$ with $-W[\rho(z)]$, in the sense that they have the same integral.

Van der Waals augmented the excess free-energy density by terms that are large when the density gradient is large. In the simplest version of the theory (and retaining the same symbol Ψ for this augmented function) we have

$$\Psi(z) = -W[\rho(z)] + \tfrac{1}{2}m\rho'(z)^2, \qquad (3.5)$$

where $\rho'(z)$ is the gradient $d\rho/dz$, and where m is a coefficient independent of $\rho'(z)$ and of any higher derivatives. His reasons for the introduction of this term were essentially the same as (but apparently arrived at independently of) those that led Rayleigh to introduce the same term into the local energy density; his expression for m in terms of the intermolecular potential is the same as that of Rayleigh, (1.38). Where he went beyond Rayleigh is that he added this term not merely to the energy but to the free energy. This involved him in a conscious decision to assume, without any good grounds, that the local entropy density did not contain such a term; he held it to be a function of $\rho(z)$ alone. We return to this point later in the chapter. A general formula for m is obtained by the methods of statistical mechanics in the next chapter, and we shall see concrete realizations of this parameter in idealized model systems in Chapter 5. For now we shall merely take (3.5), and several of its generalizations to be discussed below, as the basis of a phenomenological theory, and study its implications.

Once Ψ is identified as in (3.5) (or even, incorrectly, by the first term alone), rather than as in (3.1), the integral on the right-hand side of (3.3) is independent of the choice of dividing surface, and it is no longer necessary to specify that (3.2) holds; we may choose it to hold, or not, as we please. The difference between (3.5) and (3.1) in that regard is that the Ψ in (3.1) is an excess Helmholtz free-energy density, the integral of which, per unit area, is the surface tension σ only when (3.2) holds, cf. (2.22) and (2.23); whereas $-W$, which also has the dimensions of a free-energy density, is the negative of an excess pressure (as we shall see in a moment); that is, it is the excess density of the grand potential Ω; and, by (2.26), the integral of that excess, per unit area, is the surface tension σ without qualification. To see that $-W$ is the negative of an excess pressure, recall its definition as the height of the $\psi(\rho, T)$ isotherm above the double tangent (Fig. 3.1). Then

$$-W(\rho) = p - \mu\rho + \psi(\rho, T), \qquad (3.6)$$

where μ, the equilibrium chemical potential, is the slope of the double tangent, and p is the equilibrium pressure—that in the interior of the bulk phases when the interface is planar. Thus, $-W(\rho)$ is the negative of the excess of the formal pressure $\mu\rho - \psi(\rho, T)$, an analytic function of ρ, over the equilibrium pressure p. When augmented by the term $\frac{1}{2}m\rho'(z)^2$, according to (3.5), it becomes the negative excess pressure $\Psi(z)$ which, when integrated as in (3.3), yields the surface tension σ.

Equation (3.5) may be thought of as the beginning of an expansion in powers of the gradient $\rho'(z)$ or higher derivatives. If Ψ is supposed analytic in the gradient, then its necessary invariance to the choice of the direction of positive z requires that there be only even powers,[5] and if the gradient is supposed small (this is not generally a good approximation, as we shall see, but suffices for the purpose at hand), then the expansion may be truncated at the $\rho'(z)^2$ term. A term in Ψ proportional to the second derivative, $\rho''(z)$, would lead to the same integrated excess free energy as a term proportional to $\rho'(z)^2$ with an appropriate coefficient $m(\rho)$, as may be seen upon integrating by parts; so, if m is suitably ρ-dependent, (3.5) would not be essentially more general if it included terms of the former kind.

We have seen that if $\Psi(z)$ were $-W[\rho(z)]$ alone, the resulting $\rho(z)$ would be a step-function, and the surface tension σ would vanish. By contrast if $\Psi(z)$ were $\frac{1}{2}m\rho'(z)^2$ alone, the excess free energy per unit area, as given by (3.3), would be minimized by a density profile that was infinitely diffuse; that is, by a $\rho(z)$ that required an infinite distance over which to make the transition from ρ^α to ρ^β; and the minimum value of σ would again be 0. But with both terms present in $\Psi(z)$, as in (3.5), the $\rho(z)$ that minimizes σ in (3.3) is a compromise between those two extremes, (a) and (b) in Fig. 3.3, and is as in (c), corresponding to an interface of non-zero but finite thickness.

That is seen most readily if, for purposes of illustration, we take the coefficient m of the square-gradient term in (3.5) to be independent of ρ. Then the profile $\rho(z)$ we seek, being that which minimizes σ in (3.3), may be obtained by analogy with the dynamics of a particle moving on a line: ρ is the 'coordinate', z the 'time', m the 'mass', $W(\rho)$ the 'potential energy', Ψ the 'Lagrangian', and σ the 'action'. By Hamilton's principle,[6] the $\rho(z)$ that minimizes σ is the coordinate of the particle, as a function of the time, as the particle moves between ρ^α and ρ^β, subject to the potential $W(\rho)$ of Fig. 3.2. The particle's total energy is equal to the common value of $W(\rho^\alpha)$ and $W(\rho^\beta)$, so the particle moves infinitely slowly at ρ^α and ρ^β and takes infinite time to reach either point, corresponding to the limits $\pm\infty$ for the time in the action integral in (3.3). The particle's velocity $\rho'(z)$ is maximal at the intermediate ρ at which W has its intermediate minimum (Fig. 3.2). The profile $\rho(z)$ is thus just as in (c) in Fig. 3.3.

With the m in (3.5) still imagined to be independent of ρ, the

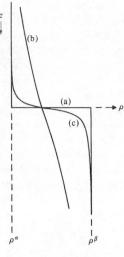

FIG. 3.3

Euler–Lagrange equation[7] for the $\rho(z)$ that minimizes σ is

$$m\, d^2\rho/dz^2 = -dW(\rho)/d\rho, \tag{3.7}$$

which is Newton's law in the dynamical analogy. Since $-W(\rho)$ is defined to be the distance of the analytic $\psi(\rho, T)$ isotherm above its double-tangent line (Fig. 3.1), the 'force' $-dW(\rho)/d\rho$ is $M(\rho) - \mu$, where $M(\rho) = [\partial\psi(\rho, T)/\partial\rho]_T$ is the analytic chemical potential as a function of density at the temperature T, and μ is the uniform, equilibrium chemical potential in the two-phase system.

The analytic $M(\rho)$ for a fixed $T < T^c$ is shown schematically in Fig. 3.4, together with the equal-areas construction that yields the equilibrium

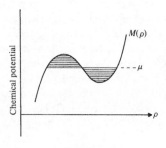

FIG. 3.4

μ. This is the equivalent, in the μ, ρ-plane, of the equal-areas construction in the p, v-plane shown in Fig. 1.8, and is the once differentiated version of the double-tangent construction in Fig. 3.1.

Equation (3.7) is to be solved subject to the boundary conditions $\rho = \rho^{\alpha,\beta}$ at $z = \pm\infty$. From that equation and its boundary conditions, together with the form of $M(\rho) - \mu$ given in Fig. 3.4, it is again clear that the resulting $\rho(z)$ is as in Fig. 3.3(c). The origin of z remains arbitrary. It may (although it need not) be chosen to be the location of the dividing surface $\Gamma_\rho = 0$, for which (3.2) holds. Equation (3.7) has as a first integral

$$W(\rho) + \tfrac{1}{2}m\rho'(z)^2 = 0, \tag{3.8}$$

corresponding to a fixed total energy of 0 [$= W(\rho^\alpha) = W(\rho^\beta)$; see Fig. 3.2] for the moving particle in the dynamical analogy. It follows from (3.5) and (3.8) that the equilibrium value of $\Psi(z)$ is $m\rho'(z)^2$, or $-2W(\rho)$; that is, $\Psi(z)$ is twice the distance of $\psi[\rho(z)]$ above the common tangent of Fig. 3.1. (This is the factor of two that was introduced arbitrarily into (1.42) in our account of Rayleigh's similar theory.)

The profile $\rho(z)$, in the form of the inverse function $z(\rho)$, follows by quadrature:

$$z = \pm\left(\frac{m}{2}\right)^{\frac{1}{2}}\int[-W(\rho)]^{-\frac{1}{2}}\,d\rho. \tag{3.9}$$

The sign is arbitrary; its choice determines the direction of increasing z. There is also an arbitrary constant of integration, the choice of which determines the origin of the scale of z. In effect, (3.7) and (3.8) were anticipated in (1.39) and (1.40), with $F(\rho) = -dW(\rho)/d\rho$.

Either (3.3) and (3.5), or (3.7), may be taken to be the primitive form of the van der Waals theory, but it is often necessary to consider more general versions. First, as already mentioned, derivatives of higher order, and higher powers of derivatives, may appear as additional terms in (3.5) and (3.7). Secondly, an important generalization of (3.7) which we shall encounter in Chapters 4 and 5 is to an integral equation, in which $m\,d^2\rho/dz^2$ is seen to be merely an approximation, holding for slowly varying $\rho(z)$, to an integral that is itself a functional of the whole of $\rho(z)$, over the full range $-\infty < z < \infty$, and so is non-local. Thirdly, even in the primitive form of the theory, (3.5), the coefficient m, as already mentioned, may depend on ρ, in which case (3.7) is more generally

$$m\,d^2\rho/dz^2 + \tfrac{1}{2}(dm/d\rho)(d\rho/dz)^2 = -dW(\rho)/d\rho, \tag{3.10}$$

of which (3.8) is still a first integral. We shall include such a generalization in § 3.2 below, where we consider the formulae for the equilibrium surface tension that follow from (3.3) in this form of the theory. Finally, we have so far, for purposes of illustration, considered only a one-component system, with only one density, $\rho(z)$, varying through the interface. It is often necessary to extend the theory to c-component systems, and also to allow more than one density to vary independently through the interface.[8] The generalization to more than one independently varying density may be necessary even in a one-component system, because in an interface the energy density $\phi(z)$ and matter density $\rho(z)$ are not necessarily related to each other in the same way as in a bulk phase. In a c-component system there are $c+1$ such densities whose independent variations through the interface must be considered. In § 3.3, the theory based on (3.5) or (3.7) is generalized to treat two or more variable densities.

Even in the primitive versions of the van der Waals theory with m independent of ρ, that coefficient may still depend on the temperature T (or, equivalently, on the chemical potential μ) at which the phases are in equilibrium. While in Chapter 5 we shall see some examples, or limiting idealized cases, in which m is a fixed constant, independent of T, and is determined by the intermolecular forces alone, as in (1.38), it will, more generally, depend on T; and in that event, as we shall see in § 3.4, the connection between this theory and the Gibbs adsorption equation (2.31) is not entirely straightforward and requires discussion.

Finally, in § 3.5, we discuss briefly the meaning and origin of the analytic functions p, ψ, and M shown as the smooth curves in Figs 1.8, 3.1, and 3.4, and some related matters, all lying at the foundations of the theory.

3.2 The surface tension

From (3.3), (3.5), and (3.8), it follows that the surface tension in this primitive form of the van der Waals theory, with a general ρ-dependent m, may be calculated from any of the alternative formulae

$$\sigma = \int_{-\infty}^{\infty} m[\rho(z)]\rho'(z)^2 \, dz \tag{3.11}$$

$$= -2 \int_{-\infty}^{\infty} W[\rho(z)] \, dz \tag{3.12}$$

$$= \int_{\rho^\alpha}^{\rho^\beta} [-2m(\rho)W(\rho)]^{\frac{1}{2}} \, d\rho. \tag{3.13}$$

Of these, (3.13) is often the most useful, for it allows σ to be calculated from $m(\rho)$ and $W(\rho)$ alone, without prior knowledge of $\rho(z)$. In the dynamical analogy it is the expression of the action as the integral of the momentum over the coordinate.

Equation (3.11) with constant m is Rayleigh's formula, (1.43). Allowing m to be ρ-dependent can be important for obtaining an accurate value of the surface tension.[9]

3.3 Independently variable densities

The equation of state of a one-component system may be expressed as a relation among the energy, entropy, and matter densities ϕ, η, and ρ, in the form $\phi = \phi(\rho, \eta)$, for example. Alternatively, it may be expressed as a relation among three fields instead of among three densities (we again use the nomenclature of Griffiths and Wheeler[10]);† for example, $p = p(\mu, T)$, giving the pressure as a function of chemical potential and temperature. The energy density ϕ is the thermodynamic potential for a one-component fluid of prescribed ρ and η, and likewise p is the potential for prescribed μ and T; all intensive thermodynamic properties are derivable from $\phi(\rho, \eta)$ or from $p(\mu, T)$ by differentiation (see Appendix 1).

The equilibrium ρ and η in a fluid of prescribed μ and T are those which minimize

$$\phi(\rho, \eta) - \mu\rho - T\eta$$

at the given μ and T. This is a consequence of the thermodynamic identities $(\partial\phi/\partial\eta)_\rho = T$ and $(\partial\phi/\partial\rho)_\eta = \mu$ (Appendix 1). Then at the equilibrium ρ and η, the function

$$-W(\rho, \eta) = p(\mu, T) + \phi(\rho, \eta) - \mu\rho - T\eta \qquad (3.14)$$

is minimal for the prescribed μ and T, and its minimal value is 0, on account of the identity $p = \mu\rho + T\eta - \phi$ (Appendix 1).

Now let $\phi(\rho, \eta)$ be the analytic ϕ as it might be given by a mean-field theory—not the reconstructed ϕ that is obtained from it by the convex-envelope construction,[11,12] which is the generalization, to two or more densities, of the equal-areas construction of Figs 1.8 and 3.4 and the double-tangent construction of Fig. 3.1. Then the $-W(\rho, \eta)$ defined by (3.14) is a two-density generalization of the earlier $-W(\rho)$. That $-W(\rho)$ is the minimum, over η, of the present $-W(\rho, \eta)$; it is thus given by

$$-W(\rho) = p(\mu, T) + \psi(\rho, T) - \mu\rho, \qquad (3.15)$$

where $\psi(\rho, T)$ is $\phi(\rho, \eta) - T\eta$ with η determined as a function of ρ and T by $\partial\phi(\rho, \eta)/\partial\eta = T$. Equation (3.15) agrees with the original definition of

---

† A field has the same value in two coexistent phases, while a density does not—or, when it does, it is recognized as an azeotropy and hence exceptional (see Appendix 1).

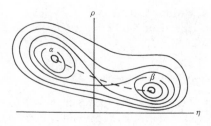

FIG. 3.5

$-W$ as the distance of the $\psi(\rho, T)$ isotherm over its double-tangent line [Fig. 3.1 and eqn (3.6)], because of the thermodynamic identities $p = \mu\rho - \psi$ and $(\partial\psi/\partial\rho)_T = \mu$ which hold in the equilibrium phases.

When two phases are in equilibrium, μ and T cannot be prescribed independently; they must be related in such a way as to make $W(\rho^\alpha, \eta^\alpha) = W(\rho^\beta, \eta^\beta)$, or, in the earlier theory, $W(\rho^\alpha) = W(\rho^\beta)$.

Figure 3.5 shows schematic contours of constant W in the ρ, η-plane; it is the analogue, for $W(\rho, \eta)$, of Fig. 3.2 for $W(\rho)$. The equilibrium bulk phases α and β occur at the densities ρ^α, η^α and ρ^β, η^β where W has two equal maxima with the common value 0; elsewhere, $W < 0$.

In the simplest form of this extended van der Waals theory, the excess free-energy density $\Psi(z)$ is given by a generalization of (3.5)

$$\Psi(z) = -W(\rho, \eta) + \tfrac{1}{2}m_{11}\rho'(z)^2 + m_{12}\rho'(z)\eta'(z) + \tfrac{1}{2}m_{22}\eta'(z)^2 \quad (3.16)$$

or still more generally by

$$\Psi(z) = -W(\rho, \eta) + K, \quad (3.17)$$

where K, the kinetic energy in the dynamical analogy, is a functional of $\rho(z)$ and $\eta(z)$. If K is as in (3.16), and if $m_{11}m_{22} > m_{12}^2$, then by simple transformation the problem of determining the $\rho(z)$ and $\eta(z)$ which minimize

$$\sigma = \int_{-\infty}^{\infty} (-W + K)\,dz \quad (3.18)$$

can be made equivalent to that of finding the trajectory of a single particle on the ρ, η-plane when the particle is subject to the potential energy $W(\rho, \eta)$, has total energy $0[= W(\rho^\alpha, \eta^\alpha) = W(\rho^\beta, \eta^\beta)]$, and moves between ρ^α, η^α and ρ^β, η^β. The resulting trajectory might be like that shown as the dashed curve in Fig. 3.5. If $m_{11}m_{22} < m_{12}^2$, or if K is a functional of a form other than that in (3.16), the analogy to the dynamics of a single particle may be lost; but σ in (3.18) may still be taken to be the action of a dynamical system of Lagrangian $-W + K$, with ρ and η then generalized coordinates. In Chapter 5 we shall see an example in which K is of the form $mx'(z)y'(z)$, so that the momentum conjugate to the coordinate x is $my'(z)$ and that conjugate to y is $mx'(z)$.

If η varied with ρ through the interface in the same way as in the bulk, then η would always be that function of ρ which, at the given T, satisfied $\partial\phi(\rho, \eta)/\partial\eta = T$, or from (3.14),

$$\partial W(\rho, \eta)/\partial\eta = 0. \qquad (3.19)$$

But (3.19) is not, in general, satisfied by the trajectory, shown schematically in Fig. 3.5, which is obtained by solving the Euler–Lagrange equations (the equations of motion, in the dynamical analogy) associated with the variational integral in (3.18); say, for example,

$$m_{11}\rho''(z) + m_{12}\eta''(z) = -\partial W(\rho, \eta)/\partial\rho$$
$$m_{12}\rho''(z) + m_{22}\eta''(z) = -\partial W(\rho, \eta)/\partial\eta, \qquad (3.20)$$

which are the Euler–Lagrange equations when Ψ is given by (3.16). Equation (3.19) is merely an approximation to the true trajectory; in an interface η need not vary with ρ as it would in a bulk phase. The surface tension calculated from (3.18) with the approximation of (3.19) is necessarily greater than the true surface tension, by the variational principle.

The ϕ, η, and μ in (3.14) may be limited to their configurational parts: ϕ may be taken to be the potential-energy density, η the density of configurational entropy, and μ the configurational chemical potential; for the parts of the total ϕ, η, and μ that arise from the molecular momenta, such as the kinetic-energy-density component of ϕ, cancel in the combination $\phi - \mu\rho - T\eta$.

It is a peculiarity of the lowest order of the mean-field approximation that the configurational η and ϕ are functions of ρ alone, $\eta = \eta(\rho)$ and $\phi = \phi(\rho)$, so that the points representing single-phase states in the ρ, η, ϕ space lie on a curve instead of on a surface $\phi = \phi(\rho, \eta)$. This was the original assumption of van der Waals which was later justified by Ornstein[13] for a system composed of molecules with long-ranged but weak attractive forces. The assumption is not exact for more realistic potentials. With this assumption, $\eta = \eta(\rho)$ and $\phi = \phi(\rho)$ are then also, in particular, the entropy and energy densities in a phase that is of mass density ρ and is one of two coexisting phases; so the same relations $\eta = \eta(\rho)$ and $\phi = \phi(\rho)$ are then also the equations of the two-phase coexistence curve in the ρ, η, ϕ space, $\eta = \eta(\rho)$ is the equation of the coexistence curve in the η, ρ plane, etc. As coexistence curves these are in no way peculiar; the only peculiarity is that the single-phase states—in this version of this approximation—have collapsed onto the coexistence curve instead of being represented by the points of an extended two-dimensional region of which the coexistence curve is merely a boundary. There is then no proper two-density description of the one-phase states of a one-component system in the lowest order of the mean-field approximation. There is such a description of the two-phase states, where not even in mean-field approximation is there any discernible peculiarity; but in practice the potential W for the two-density form of the van der Waals theory is then not constructed by the prescription in (3.14) but by other means. We shall see an example in § 9.1, where a $W(x, y)$ arising from mean-field theory is given, in which the densities x and y are linear combinations (one almost arbitrary, the other almost unique) of the deviations of η and ρ, say, from their values at the critical point. Also, it will be seen there that as the critical point is approached the one- and two-density theories become equivalent—not as another peculiarity of the mean-field approximation, but as a general property of a critical point.

In a c-component system, $c + 1$ densities—the densities $\rho_1, \ldots, \rho_c$ of the c components and the entropy density η, say—may vary independently through the interface. (As before we use a vector notation $\boldsymbol{\rho}$ for the set $\rho_1, \ldots, \rho_c$.) We have the analytic, unreconstructed energy density $\phi = \phi(\boldsymbol{\rho}, \eta)$ and the related $-W$,

$$-W(\boldsymbol{\rho}, \eta) = p(\boldsymbol{\mu}, T) + \phi(\boldsymbol{\rho}, \eta) - T\eta - \boldsymbol{\mu} \cdot \boldsymbol{\rho} \tag{3.21}$$

where the $\boldsymbol{\mu}$ and p are the equilibrium chemical potentials and the pressure. The function W has two equal maxima, where it takes the value 0, at the densities $\boldsymbol{\rho}^\alpha$, η^α and $\boldsymbol{\rho}^\beta$, η^β of the two equilibrium phases. In the simplest form of the $(c+1)$-density van der Waals theory,[14,15] the excess free-energy density $\Psi(z)$ is again $-W + K$, as in (3.17), with K some functional of the $\boldsymbol{\rho}(z)$ and of $\eta(z)$, often a quadratic form in the derivatives $\boldsymbol{\rho}'(z)$ and $\eta'(z)$, as in (3.16). The equilibrium interfacial tension σ is then obtained as the minimum over all $\boldsymbol{\rho}(z), \eta(z)$ of the variational integral in (3.18). The approximation in which one imagines any ρ_i merely to follow the other densities in the interface as it would in a bulk phase, is $\partial W/\partial \rho_i = 0$; the tension calculated from this or any other approximation to the true trajectory in the $(c + 1)$-dimensional density space is greater than the true tension.

Aside from the question of accuracy, there are qualitative features of some interfaces, especially in systems of more than one component, that require two or more independently varying densities for their description. An example is strong positive or negative adsorption of a component i, associated with a non-monotonic profile $\rho_i(z)$ such as that in Fig. 3.6(a) or (b). In the one-density theory based on the approximations $\partial W/\partial \eta = 0$ and $\partial W/\partial \rho_j = 0$ for all $j \neq i$, the resulting one-density $W(\rho_i)$ is like that in Fig. 3.2, and so, as we may conclude from the dynamical analogy, leads inevitably to a monotonic $\rho_i(z)$. In a theory based on two or more densities, by contrast, we may have a trajectory with which is associated

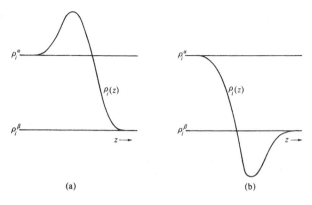

(a)   (b)

FIG. 3.6

non-monotonic behaviour of one or more of the components, like that of $\rho_i(z)$ in Fig. 3.7 if we suppose $\rho_j(z)$ to be monotonic. We must also allow the independent variation of at least two densities if we are to account, within the van der Waals theory, for non-zero contact angles in three-phase equilibrium, as we shall see in Chapter 8.

The whole theory may be reformulated[14,15] with the thermodynamic potentials $(p/T)(\mathbf{\mu}/T, -1/T)$ and $\eta(\mathbf{\rho}, \phi)$ in place of $p(\mathbf{\mu}, T)$ and $\phi(\mathbf{\rho}, \eta)$. Then σ/T is obtained as the minimum, over $\mathbf{\rho}(z)$, $\phi(z)$, of

$$\frac{\sigma}{T} = \int_{-\infty}^{\infty} (-\tilde{W} + \tilde{K}) \, dz \qquad (3.22)$$

with

$$-\tilde{W}(\mathbf{\rho}, \phi) = (p/T)(\mathbf{\mu}/T, -1/T) - \eta(\mathbf{\rho}, \phi) + \phi/T - (\mathbf{\mu}/T) \cdot \mathbf{\rho} \qquad (3.23)$$

and with $\tilde{K}$ some functional of $\mathbf{\rho}(z)$, $\phi(z)$, typically and most simply a quadratic form in the derivatives $\mathbf{\rho}'(z)$, $\phi'(z)$. The advantage of this over

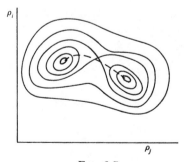

FIG. 3.7

the earlier formulation is that the energy density ϕ is often more readily identifiable microscopically than the entropy density η, for ϕ, like the matter densities $\boldsymbol{\rho}$, is a purely mechanical quantity. Then $\eta(\boldsymbol{\rho}, \phi)$ is often a more convenient, and is in a sense a more fundamental, thermodynamic potential than the inverse function $\phi(\boldsymbol{\rho}, \eta)$.

3.4   Gibbs adsorption equation in the van der Waals theory

In the form of the theory given by (3.18) and (3.21), or by (3.22) and (3.23), with K or $\tilde{K}$ a functional of $\rho(z)$ and of $\eta(z)$ or $\phi(z)$, respectively, we ask what the status is of the Gibbs adsorption equation,[15,16] (2.35) or (2.37).

If K does not depend explicitly on any of the fields $\boldsymbol{\mu}$, T—for example, if K is a quadratic form in the $\rho_i'(z)$ and $\eta'(z)$, as in (3.16), with constant, field-independent coefficients m_{ij}—then (2.35), it will be seen, holds with a straightforward identification of the surface excesses Γ_i.

With K not explicitly dependent on the fields, there are only two potential sources of dependence on $\boldsymbol{\mu}$ or T of the tension in (3.18): first, $-W$, as given in (3.21), depends explicitly on the fields; second, $-W$ and K depend on the densities, and the $\rho(z)$ and $\eta(z)$ that minimize σ in (3.18) depend, in turn, on the fields. But by the variational principle, σ is stationary with respect to the variation of the densities about their equilibrium forms; so, in the end, the field-dependence of σ comes entirely from the explicit field-dependence of $-W$. By the Gibbs–Duhem equation $p(\boldsymbol{\mu}, T)$ is such that

$$\mathrm{d}p = \boldsymbol{\rho}^\alpha \cdot \mathrm{d}\boldsymbol{\mu} = \boldsymbol{\rho}^\beta \cdot \mathrm{d}\boldsymbol{\mu} \qquad (3.24)$$

where as in (2.35) the component ρ_{c+1} of the vector $\boldsymbol{\rho}$ is the entropy density η while its conjugate μ_{c+1} in $\boldsymbol{\mu}$ is the temperature T. Then from (3.18), (3.21), and (3.24),

$$\mathrm{d}\sigma = \left\{ \int_{-\infty}^{\infty} [\boldsymbol{\rho}^{\alpha,\beta} - \boldsymbol{\rho}(z)] \, \mathrm{d}z \right\} \cdot \mathrm{d}\boldsymbol{\mu} \qquad (3.25)$$

where $\boldsymbol{\rho}^{\alpha,\beta}$ means $\boldsymbol{\rho}^\alpha$ on the α-side of an arbitrarily chosen dividing surface and $\boldsymbol{\rho}^\beta$ on the β-side. This is the adsorption equation, (2.35), with the adsorption Γ identified as

$$\Gamma = \int_{-\infty}^{\infty} [\boldsymbol{\rho}(z) - \boldsymbol{\rho}^{\alpha,\beta}] \, \mathrm{d}z. \qquad (3.26)$$

The right-hand sides of (3.26) are indeed the surface excesses of the quantities whose densities are $\boldsymbol{\rho}(z)$, so we have found the van der Waals theory, in the form of (3.18) and (3.21) with a functional K that does not depend explicitly on the fields, to be consistent with the Gibbs adsorption equation.

Similarly, in the form of the theory based on (3.22) and (3.23), if the functional $\tilde{K}$ does not depend explicitly on the fields, then the alternative form of the adsorption equation, that in (2.37), holds, with Γ again given by (3.26), but with ρ_{c+1} now the energy density ϕ, and with $\nu_{c+1} = -1/T$ its conjugate.

But K and $\tilde{K}$ cannot both be independent of the fields. For example, when both are quadratic forms in the density gradients, as in (3.16), if $\frac{1}{2}m$, say, is the coefficient of some $\rho'(z)^2$ in K, then $\frac{1}{2}m/T$ is the coefficient of that same $\rho'(z)^2$ where it occurs in $\tilde{K}$. We must therefore see how the foregoing analysis is altered when K or $\tilde{K}$ depends explicitly on the fields.

If the functional K is explicitly field-dependent, it provides one more source of field-dependence in the σ of (3.18). Then an additional term $\partial K/\partial \mu_i$ appears in the integrand of the ith term in (3.25); and so, for consistency with the adsorption equation, (2.35), the adsorption Γ must be identified as

$$\Gamma = \int_{-\infty}^{\infty} [\boldsymbol{\rho}(z) - \partial K/\partial \boldsymbol{\mu} - \boldsymbol{\rho}^{\alpha,\beta}]\,dz \qquad (3.27)$$

rather than by (3.26). Similarly, if $\tilde{K}$ is explicitly field-dependent, then for consistency with the adsorption equation in the form of (2.37), the adsorption Γ must be identified as

$$\Gamma = \int_{-\infty}^{\infty} [\boldsymbol{\rho}(z) - \partial \tilde{K}/\partial \boldsymbol{\nu} - \boldsymbol{\rho}^{\alpha,\beta}]\,dz. \qquad (3.28)$$

To see the implications of these observations, let us take for illustration a one-component system, with

$$
\begin{aligned}
-W(\rho, \eta) &= p(\mu, T) + \phi(\rho, \eta) - \mu\rho - T\eta \\
-\tilde{W}(\rho, \phi) &= (p/T)(\mu/T, -1/T) - \eta(\rho, \phi) - (\mu/T)\rho + \phi/T
\end{aligned}
\qquad (3.29)
$$

$$K = \tfrac{1}{2}m\rho'(z)^2 \qquad \tilde{K} = \tfrac{1}{2}(m/T)\rho'(z)^2. \qquad (3.30)$$

We take $-W$ and K in one form of the theory, or $-\tilde{W}$ and $\tilde{K}$ in the other. If m is a constant, then the surface excesses of matter and entropy are given correctly by (3.26), and the local densities of matter and entropy may be identified simply as the functions $\rho(z)$ and $\eta(z)$ that minimize the variational integral in (3.18). But the local energy density is not then correctly identified as the $\phi(z)$ that minimizes the variational integral in (3.22); it is rather $\phi(z) + \frac{1}{2}m\rho'(z)^2$, because of the term $-\partial \tilde{K}/\partial(-1/T) = \frac{1}{2}m\rho'(z)^2$ in the integrand in (3.28). This circumstance, in which there is no density-gradient term in the local entropy density while there is one in the local energy density, is typical; though if m were proportional to T the density-gradient term would appear in the local entropy density rather than in the energy density; and with a more generally temperature-dependent m, such a term would appear in both.

Once there is a square-density-gradient term in the local free-energy density, as there is in even the simplest form of the theory in (3.5), that term must be apportioned in some way among the constituent parts of the local free-energy density. It may be found in the local energy density, in the local entropy density, or partly in both. It could even, by a formal device, be made to appear in neither the local entropy nor energy densities, but rather in the $\mu\rho$-component of the local free-energy density. Thus, we may at will choose to regard m or m/T not as a function of T or of $-1/T$, but rather as a function of μ or of μ/T, since for two-phase equilibrium in this one-component system μ and T are not independent. Then the thermodynamically consistent local densities of entropy and matter or of energy and matter would be $\eta(z)$ and $\rho(z)-\frac{1}{2}(dm/d\mu)\rho'(z)^2$, or $\phi(z)$ and $\rho(z)-\frac{1}{2}[d(m/T)/d(\mu/T)]\rho'(z)^2$, respectively. There are therefore no generally correct identifications of local densities, but once the otherwise arbitrary choice is made of the terms in which to formulate the theory, that then dictates what are the thermodynamically consistent identifications of the local densities.

In any local density $\rho_i(z)$ there is always the additional arbitrariness of an added term, a function of z, whose integral from $-\infty$ to $+\infty$ vanishes; for by (3.26) or (3.27), an added term of that kind in $\rho_i(z)$ contributes nothing to the adsorption Γ_i and so cannot affect the consistency of the theory.

3.5 Constraints on the range of fluctuations

We have seen that the analytic functions $p(v)$, $\psi(\rho)$, and $M(\rho)$, not reconstructed by the equal-areas and double-tangent rules—i.e., the functions shown as the smooth isothermal curves in Figs 1.8, 3.1, and 3.4—are central to the van der Waals one-density theory; and that the analytic functions $\phi(\mathbf{\rho}, \eta)$ and $\eta(\mathbf{\rho}, \phi)$, not reconstructed by the convex-envelope rule, are likewise central to the $(c+1)$-density generalization of that theory. We now ask how those functions arise in the theory and what physical quantities they represent.

They are the pressure, Helmholtz free-energy density, etc., in a fluid that is constrained to have uniform values of the densities that are the arguments of those functions. For example, $p(v)$ is what, hypothetically, the pressure in the fluid would be if at the given temperature the fluid were in equilibrium under the constraint that its specific or molar volume be uniform with the prescribed value v. How and why such functions arise in the van der Waals theory will be seen presently.

Uniformity can only be defined with respect to some length, call it L. The number density ρ, for example, would be said to be uniform on the scale of L if volumes L^3 were constrained always to contain ρL^3 molecules. Within the volume L^3 the fluctuations in the spatial locations of the ρL^3 molecules may be the same as in the unconstrained fluid; the constraint only suppresses fluctuations over distances greater than L.

Such constraints on the range of fluctuations—hence, the imposition of uniformity over distances greater than that range—lead to van der Waals loops in $p(v)$ and $M(\rho)$.[17] If the fluid were unconstrained, and if at the temperature in question its average molar volume were between v^l and v^g (see Fig. 1.8), it would not be stable as a homogeneous phase of that intermediate v, but would separate into two phases, one of molar volume v^l and one of molar volume v^g. Only when subject to the constraint of uniformity can the fluid be stable as a single, homogeneous phase. Phase separation implies non-uniformity over macroscopic distances, and is therefore suppressed by a constraint of uniformity on the scale of any microscopic length L.

But what the pressure $p(v)$, chemical potential $M(\rho)$, etc., in the constrained system are, depend on the distance L that defines the constraint. If L is very large, the fluctuations within L^3 can almost amount to phase separation; the van der Waals loops in $p(v)$ and $M(\rho)$ would then enclose only small areas, and the analytic functions $p(v)$, $\psi(\rho)$, etc., would be close to the non-analytic functions obtained from them by the equal-areas, double-tangent, or convex-envelope constructions. The effect of the constraint with such large L is minimal; and in the limit in which L is macroscopic the thermodynamic properties become those of the unconstrained fluid. But when L is small, the deviation of $p(v)$ from the equilibrium pressure in the unconstrained system at that temperature is considerable, and similarly for the other thermodynamic functions.

To what length must L be compared to determine if it is large or small? Unconstrained fluctuations in the density or composition of an otherwise homogeneous phase have a coherence length, ξ, which is such that the fluctuations at two points distant less than ξ are strongly correlated while the fluctuations at two points separated by a few multiples of ξ are almost independent. Far from a critical point, ξ may be little greater than the range of the intermolecular forces themselves, perhaps a few Å; but ξ diverges on approach to a critical point,[18] and in practice, even in fluids of small molecules interacting with short-range forces, may become as great as a few hundreds or even thousands of Å at easily attainable proximities to the critical point (the region of strong critical opalescence). In a dense fluid, all thermodynamically relevant fluctuations are found represented in any volume of the fluid that is much greater than

ξ^3; such a volume is virtually macroscopic, and the properties of the fluid contained in it are accurately describable by conventional thermodynamics. Volumes much less than ξ^3 are clearly microscopic. The borderline between large and small L in a dense fluid is $L = \xi$. In a dilute gas, where $\xi \ll \rho^{-\frac{1}{3}}$, the mean distance between neighbouring molecules, the borderline L is instead the latter.

In mean-field theories of the thermodynamic properties of homogeneous fluids, the approximation is made that within the range of the attractive forces of every molecule there is found the same number of neighbouring molecules, a number which in the lowest order of that approximation is determined simply by the average density ρ of the fluid: say ρL^3, with L the range of the forces. (Had the attractive forces been of macroscopically long range, that constraint of uniformity would then have been on the scale of a macroscopic L, which, as we have seen, amounts to no constraint at all; that is the reason mean-field theories become exact in the limit in which the attractive forces are long ranged.) The analytic functions $p(v)$, etc., yielded by mean-field theories are then the thermodynamic functions of a hypothetical fluid constrained to be uniform over distances greater than the range of the attractive forces.

In the van der Waals theory of the interface, the free-energy density $\psi[\rho(z), T]$ of a hypothetical fluid that is constrained on some scale to have a uniform density $\rho(z)$ equal to that of the inhomogeneous fluid at the height z, arises in a natural way in the description of that inhomogeneous fluid. In the mean-field approximation of van der Waals's original theory,[2] and in that with which we treat the model systems in Chapter 5, the scale on which the analytic $\psi(\rho)$ is defined is again the range of the attractive forces. But unlike in the corresponding mean-field approximation to the thermodynamics of homogeneous fluids, the theory does take account of, and indeed yields a description of, non-uniformities that may extend beyond that range. Those non-uniformities are in the profiles $\rho(z)$, and also manifest themselves in the local thermodynamic properties, through the square-gradient terms, for example.

Near the critical point, where the coherence length ξ is very large, the mean-field approximation to the thermodynamics of a homogeneous fluid—which, as we saw, amounts to a constraint of uniformity on a fixed, microscopic scale—can be greatly in error, and is indeed known to give an incorrect account of the thermodynamic singularities at such a point.[18] If the constraint were instead one of uniformity on a scale of ξ, which diverges on approach to the critical point, the theory would be much improved; and the corresponding van der Waalsian theory of the inhomogeneous fluid would be essentially exact near the critical point. That was remarked by Langer,[19] who also pointed out that it is the basis of the renormalization-group theory[20] of critical phenomena.

More precisely, Langer remarks that if the density at the position $\mathbf{r}$ in the fluid, coarse-grained over regions of volume ξ^3, is prescribed to be $\rho(\mathbf{r})$—so that if the fluid is imagined divided into cells of volume ξ^3 the number of molecules in the cell centred at $\mathbf{r}$ is prescribed to be $\xi^3\rho(\mathbf{r})$—then the free-energy density of the fluid so constrained, is

$$\psi(z) = \psi[\rho(\mathbf{r}), T] + \tfrac{1}{2}m\,|\nabla\rho|^2, \tag{3.31}$$

where $\psi(\rho, T)$ is the free-energy density of the fluid constrained (on the scale of ξ) to be of uniform density ρ, and is qualitatively like that in Fig. 3.1. Both the function $\psi(\rho, T)$ and the coefficient m are in principle calculable from the renormalization-group theory. Alternatively, their most important features can be inferred[21] from what is known in other ways about thermodynamic singularities and the correlation of fluctuations in a fluid near its critical point. That is how we shall proceed in Chapter 9, when we treat the interface between near-critical phases.

With $\rho(\mathbf{r})$ a function of z alone, (3.31) is in effect (3.5), the formal beginning of the van der Waals theory as discussed in this chapter. A theory with the same structure as that just outlined, but in which the cells with respect to which the density $\rho(\mathbf{r})$ is coarse-grained are of fixed size instead of increasing with proximity to the critical point, is the basis of van Kampen's derivation of the van der Waals equation of state (a derivation that he ascribes to Ornstein, 1908), and of his closely related mean-field theory of the inhomogeneous fluid.[17,22]

Notes and references

1. Strutt, R. J. (4th Lord Rayleigh), *John William Strutt, Third Baron Rayleigh*, p. 183. Arnold, London (1924).
2. van der Waals, J. D. *Verhandel. Konink. Akad. Weten., Amsterdam (Sect. 1)* **1**, No. 8, 56 pp. (1893). German and French translations with new material in five appendices were published in *Zeit. phys. Chem.* **13**, 657 (1894) and *Arch. Néerl.* **28**, 121 (1895), and an English translation with a brief historical introduction in *J. stat. Phys.* **20**, 197 (1979). See also van der Waals, J. D. and Kohnstamm, Ph. *Lehrbuch der Thermodynamik*, Vol. 1, § 4, Maas and van Suchtelen, Amsterdam (1908).
3. Landau, L. and Lifshitz, E. *Phys. Zeit. Sowjetunion* **8**, 153 (1935), reprinted in English in *Collected Papers of L. D. Landau* (ed. D. ter Haar) p. 101, Pergamon Press, Oxford (1965). See also Landau, L. D. and Lifshitz, E. M. *Electrodynamics of Continuous Media*, § 39, p. 158, Pergamon Press, Oxford (1960).
4. Cahn, J. W. and Hilliard, J. E. *J. chem. Phys.* **28**, 258 (1958).
5. Hopper, R. W. and Uhlmann, D. R. *J. chem. Phys.* **56**, 4043 (1972).
6. See, for example, Goldstein, H. *Classical Mechanics*, p. 30 ff. Addison-Wesley, Reading, Mass. (1950).
7. Goldstein, ref. 6, p. 38; Margenau, H. and Murphy, G. M. *Mathematics of Physics and Chemistry*, 2nd edn, pp. 200, 270, Van Nostrand, Princeton, New Jersey (1956).

8. Such a generalization was first suggested by Cahn and Hilliard, ref. 4, footnote ‖ on p. 260.
9. Harrington, J. M. and Rowlinson, J. S. *Proc. Roy. Soc.* A **367**, 15 (1979); Telo da Gama, M. M. and Evans, R. *Mol. Phys.* **38**, 367 (1979); see §§ 5.6 and 7.4.
10. Griffiths, R. B. and Wheeler, J. C. *Phys. Rev.* **A2**, 1047 (1970).
11. Buckingham, M. J. in *Phase Transitions and Critical Phenomena* (ed. C. Domb and M. S. Green) Vol. 2, Chapter 1, Academic Press, London (1972). The convex-envelope construction was used first by Gibbs, J. W. *Trans. Conn. Acad.* **2**, 382 (1873), reprinted in *Collected Works*, Vol. 1, p. 33.
12. Lebowitz, J. L. and Penrose, O. *J. math. Phys.* **7**, 98 (1966).
13. Ornstein, L. S. *Proc. Roy. Acad. Sci., Amsterdam* **11**, 526 (1909).
14. Widom, B. in *Statistical Mechanics and Statistical Methods in Theory and Application* (ed. U. Landman) pp. 33–71. Plenum, New York (1977).
15. Widom, B. *Physica* **95A**, 1 (1979).
16. Cahn, J. W. *J. chem. Phys.* **66**, 3667 (1977).
17. van Kampen, N. G. *Phys. Rev.* **135A**, 362 (1964); Hansen, J. P. and Verlet, L. *Phys. Rev.* **184**, 151 (1969); Penrose, O. and Lebowitz, J. L. *J. stat. Phys.* **3**, 211 (1971).
18. For reviews of this and other aspects of critical points, see, for example, Fisher, M. E. *Rep. Prog. Phys.* **30**, 615 (1967), and Stanley, H. E. *Introduction to Phase Transitions and Critical Phenomena.* 2nd edn, Oxford University Press (in preparation).
19. Langer, J. S. *Physica* **73**, 61 (1974). See also Kawasaki, K., Imaeda, T. and Gunton, J. D. in *Perspectives in Statistical Physics* (ed. H. J. Raveché) Chapter 12, which is Vol. 9 of *Studies in Statistical Mechanics* (ed. E. W. Montroll and J. L. Lebowitz), North-Holland, Amsterdam (1981).
20. Wilson, K. G. *Phys. Rev.* B **4**, 3174, 3184 (1971). For an elementary account of the theory see Widom, B. in *Fundamental Problems in Statistical Mechanics III* (ed. E. G. D. Cohen) pp. 1–45, North-Holland, Amsterdam (1975).
21. Fisk, S. and Widom, B. *J. chem. Phys.* **50**, 3219 (1969); Luban, M. in *Phase Transitions and Critical Phenomena* (ed. C. Domb and M. S. Green) Vol. 5a, Chapter 2, Academic Press, London (1976).
22. ter Haar, D. *Lectures on Selected Topics in Statistical Mechanics* pp. 71–80. Pergamon Press, Oxford (1977).

STATISTICAL MECHANICS OF THE
LIQUID-GAS SURFACE

4.1 Introduction

The mechanical, thermodynamic, and quasi-thermodynamic arguments of
the last three chapters take us a long way towards our goal of understanding the behaviour of matter near a surface, but they stop short of a full
interpretation at a molecular level. For that we need to extend the usual
methods of statistical mechanics to cover the properties of equilibrium
systems in which there are strong inhomogeneities in the local densities.
We follow two earlier expositions of this subject.[1,2] Our account owes
much to the second of them, that by Evans,[2] but we aim at a less
condensed exposition, and take account of work done since his admirable
review was written.

4.2 Distribution and correlation functions

In this section we define the molecular distribution and correlation
functions[3] we need for the discussion of the gas–liquid surface, and set
out the more important equations that relate these functions to each
other and to the thermodynamic properties. We restrict ourselves to
states of equilibrium but not necessarily to homogeneous ones; we
therefore introduce at once an arbitrary external potential $\mathcal{V}$ which acts
independently on each molecule and which is a function of position $\mathbf{r}$
within a vessel of fixed shape and volume V. The molecules are assumed
to be without internal structure, of spherical symmetry, and all of one
species. None of these restrictions is important at this stage, but without
them the notation becomes intolerably clumsy; each can usually be lifted
when the occasion arises.

Consider then an open system specified by fixed values of μ, V, and
T, and by the external potential $\mathcal{V}$. The momentum† and position of
molecule i are $\mathbf{p}_i$ and $\mathbf{r}_i$, and sets of n of these vectors are denoted $\mathbf{p}^n$ and
$\mathbf{r}^n$. We define the distribution function $f^{(n, N)}(\mathbf{p}^n, \mathbf{r}^n)$, a dimensionless
quantity, by the statement that

$$h^{-3n}f^{(n, N)}(\mathbf{p}^n, \mathbf{r}^n)\, d\mathbf{p}^n\, d\mathbf{r}^n$$

† The momentum, a vector $\mathbf{p}$, is not to be confused with the pressure which is, in general, a
tensor $\mathbf{p}$, or in a uniform fluid, a scalar p.

is the probability that in a system at equilibrium there is a molecule in each of the n elements of phase space $(d\mathbf{p}_i\, d\mathbf{r}_i/h^3)$ where h is Planck's constant, when the whole system contains N molecules. By integration over all coordinates of one molecule we obtain the distribution function of next lower order,

$$h^{-3} \int\int f^{(n,N)}\, d\mathbf{p}_n\, d\mathbf{r}_n = (N-n+1)f^{(n-1,N)}. \qquad (4.1)$$

Hence, by repeated integration, we have the normalization

$$h^{-3n} \int\int f^{(n,N)}\, d\mathbf{p}^n\, d\mathbf{r}^n = N!/(N-n)! \qquad (f^{(0,N)} \equiv 1). \qquad (4.2)$$

The equilibrium distribution in the canonical ensemble, a closed system containing just N molecules, is implied by the proportionality of $f^{(N,N)}$ to the Boltzmann factor of the Hamiltonian $\mathscr{H}_N$,

$$\mathscr{H}(\mathbf{p}^N, \mathbf{r}^N) \equiv \mathscr{H}_N = \sum_{i=1}^{N} \mathbf{p}_i^2/2m + \mathscr{U}(\mathbf{r}^N) + \mathscr{V}. \qquad (4.3)$$

The configurational energy $\mathscr{U}$ depends on the position of the N molecules and satisfies the usual conditions[4] of short-range, etc. needed for the system to have a thermodynamics. The external potential $\mathscr{V}$ is a field whose value is specified at all points within the volume V, but since we are not interested in its value at places where there are no molecules we can write

$$\mathscr{V} = \mathscr{V}_N = \sum_{i=1}^{N} v(\mathbf{r}_i). \qquad (4.4)$$

The proportionality of the distribution function to the Boltzmann factor of the Hamiltonian can be expressed

$$f^{(N,N)} = Z_N^{-1} \exp(-\mathscr{H}_N/kT) \qquad (4.5)$$

where the constant of proportionality, Z_N^{-1}, is the reciprocal of the canonical partition function. By integration of (4.5) over all momenta and positions of the molecules, and use of (4.2),

$$Z_N = \frac{1}{h^{3N}N!} \int\int \exp(-\mathscr{H}_N/kT)\, d\mathbf{p}^N\, d\mathbf{r}^N. \qquad (4.6)$$

Since the momenta are independent, (4.3), we can integrate over them to give

$$Z_N = \frac{1}{\Lambda^{3N}N!} \int \exp(-(\mathscr{U}_N + \mathscr{V}_N)/kT)\, d\mathbf{r}^N \qquad (4.7)$$

where $\Lambda = (h^2/2\pi mkT)^{\frac{1}{2}}$ is the characteristic length of the quantal translational partition function. The average value of a function $X(\mathbf{r}^N)$ in the canonical ensemble is

$$\langle X \rangle = \frac{1}{\Lambda^{3N} N! \, Z_N} \int X(\mathbf{r}^N) \exp(-(\mathcal{U}_N + \mathcal{V}_N)/kT) \, d\mathbf{r}^N. \tag{4.8}$$

By repeated integration of (4.5)

$$f^{(n, N)} = \frac{1}{h^{3(N-n)}(N-n)! \, Z_N} \int\!\!\int \exp(-\mathcal{H}_N/kT) \, d\mathbf{p}_{n+1} \dots d\mathbf{p}_N \; d\mathbf{r}_{n+1} \dots d\mathbf{r}_N. \tag{4.9}$$

In the grand ensemble[5] the probability that a system contains just N molecules is proportional to $\lambda^N Z_N$, where λ is the activity

$$\mu = kT \ln \lambda. \tag{4.10}$$

(The zero of μ and so of $\ln \lambda$ is discussed later.) Hence

$$f^{(n)} = \sum_{N \geqslant n} f^{(n, N)}(\lambda^N Z_N/\Xi) \tag{4.11}$$

where the term in parentheses is the probability that a system contains just N molecules; the constant of proportionality here is the reciprocal of the grand partition function Ξ. By normalizing the sum of probabilities, or putting $n = 0$ in (4.11)

$$\Xi = \sum_{N \geqslant 0} \lambda^N Z_N \tag{4.12}$$

and so

$$f^{(n)} = \Xi^{-1} \sum_{N \geqslant n} \frac{\lambda^N}{h^{3(N-n)}} \frac{1}{(N-n)!}$$
$$\times \int\!\!\int \exp(-\mathcal{H}_N/kT) \, d\mathbf{p}_{n+1} \dots d\mathbf{p}_N \; d\mathbf{r}_{n+1} \dots d\mathbf{r}_N. \tag{4.13}$$

(The first term of this sum is $\lambda^n \exp(-\mathcal{H}_n/kT)$; there is no integration.) The Helmholtz free energy F and the grand potential Ω are related to their respective partition functions by the parallel equations

$$F(N, V, T) = -kT \ln Z_N, \qquad \Omega(\mu, V, T) = -kT \ln \Xi. \tag{4.14}$$

By integrating (4.11) over $d\mathbf{p}^n$ and $d\mathbf{r}^n$ we have

$$h^{-3n} \int\!\!\int f^{(n)} \, d\mathbf{p}^n \, d\mathbf{r}^n = \sum_{N \geqslant n} (N!/(N-n)!)(\lambda^N Z_N/\Xi)$$
$$= \overline{N!/(N-n)!} \tag{4.15}$$

where the bar denotes an average in the grand ensemble. Hence

$$h^{-6} \iint [f^{(2)}(\mathbf{p}_1, \mathbf{p}_2, \mathbf{r}_1, \mathbf{r}_2) - f^{(1)}(\mathbf{p}_1, \mathbf{r}_1)f^{(1)}(\mathbf{p}_2, \mathbf{r}_2)]\, d\mathbf{p}_1\, d\mathbf{p}_2\, d\mathbf{r}_1\, d\mathbf{r}_2$$

$$= \int [\rho^{(2)}(\mathbf{r}_1, \mathbf{r}_2) - \rho(\mathbf{r}_1)\rho(\mathbf{r}_2)]\, d\mathbf{r}_1\, d\mathbf{r}_2$$

$$= \overline{N^2} - \bar{N} - \bar{N}^2 \tag{4.16}$$

where $\rho^{(n)}(\mathbf{r}^n)$ is the configurational part of $f^{(n)}$,

$$\rho^{(n)}(\mathbf{r}^n) = h^{-3n} \int f^{(n)}(\mathbf{p}^n, \mathbf{r}^n)\, d\mathbf{p}^n. \tag{4.17}$$

The $\rho^{(n)}$ are generalized densities, with dimensions of V^{-n}; conventionally we omit the superscript from the singlet function $\rho^{(1)}$. So from (4.13)

$$\rho^{(n)} = \Xi^{-1} \sum_{N \geq n} \frac{(\lambda/\Lambda^3)^N}{(N-n)!} \int \exp(-(\mathcal{U}_N + \mathcal{V}_N)/kT)\, d\mathbf{r}_{n+1} \ldots d\mathbf{r}_N. \tag{4.18}$$

From (4.12)

$$\left(\frac{\partial \ln \Xi}{\partial \ln \lambda}\right)_{V,T} = \bar{N}, \tag{4.19}$$

$$kT\left(\frac{\partial \bar{N}}{\partial \mu}\right)_{V,T} = \left(\frac{\partial^2 \ln \Xi}{\partial (\ln \lambda)^2}\right)_{V,T} = \overline{N^2} - \bar{N}^2, \tag{4.20}$$

and, by a standard manipulation,

$$\left(\frac{\partial \bar{N}}{\partial \mu}\right)_{V,T} = -\left(\frac{\bar{N}}{V}\right)^2 \left(\frac{\partial V}{\partial p}\right)_{\bar{N},T}. \tag{4.21}$$

Thus both the integral in (4.16) and the mean-square fluctuation of the number of molecules in a system (4.20) can be expressed[6] in terms of the isothermal compressibility $\kappa \equiv -V^{-1}(\partial V/\partial p)_T$.

For a uniform fluid ($\mathcal{V} = 0$), the singlet distribution function ρ is independent of $\mathbf{r}$ and equal to $\bar{N}/V$; we denote it ρ_u. The integrand of (4.16) is then a function only of the scalar separation r_{12}, and an integration over one of the pair $d\mathbf{r}_1\, d\mathbf{r}_2$ can be made at once; $\int d\mathbf{r}_1 = V$. From (4.16), (4.20), and (4.21),

$$1 + \rho_u \int h(\mathbf{r}_{12})\, d\mathbf{r}_{12} = kT\rho_u\kappa \tag{4.22}$$

where the total correlation function $h(\mathbf{r}_1, \mathbf{r}_2)$ is defined by

$$\rho^{(2)}(\mathbf{r}_1, \mathbf{r}_2) - \rho(\mathbf{r}_1)\rho(\mathbf{r}_2) = \rho(\mathbf{r}_1)\rho(\mathbf{r}_2)h(\mathbf{r}_1, \mathbf{r}_2). \tag{4.23}$$

As r_{12} becomes infinite the pair function $\rho_u^{(2)}(r_{12})$ becomes equal to ρ_u^2 and so $h(r_{12} \to \infty) = 0$.

The equivalent pair function in the canonical ensemble differs from zero by terms of $O(1/N)$, and so is generally less useful,[7] particularly in a system of two phases. If $\mathcal{V} = 0$ in this ensemble then the two phases, α and β, are on average uniformly distributed throughout the total space of volume V. The ensemble average $\rho(\mathbf{r})$ is independent of $\mathbf{r}$ and is everywhere (N/V), a mean density which lies between ρ^α and ρ^β. The pair function $h(r_{12})$ has the meaning that $\rho[h(r_{12}) + 1]$ is the mean local density at $\mathbf{r}_2$, given that there is a molecule at $\mathbf{r}_1$; this density is again independent of $\mathbf{r}_1$. It can be shown[8] that $h(r_{12} \to \infty)$ is

$$-\rho^{-2}(\rho - \rho^\alpha)(\rho - \rho^\beta) > 0.$$

In the grand ensemble with no external field the system is in a state specified by the variables V, T, and μ. If $\mu^{l,g}$ is the common value of the chemical potential of the coexisting liquid and gaseous phases, then the system will be one gaseous phase if $\mu < \mu^{l,g}$, and one liquid phase if $\mu > \mu^{l,g}$. In each the fluctuations of the number of molecules N is of the order of $\sqrt{N}$. If $\mu = \mu^{l,g}$ exactly, then the system may be liquid or gas or any arbitrary combination of them; that is, the fluctuations are of order N. Clearly the liquid–gas surface is quite undefined and so cannot be studied in such a system. We therefore add an external field whose local value is $v(\mathbf{r})$ in order to separate the phases. The total chemical potential μ, which is the parameter held constant in the specification of the grand ensemble, can be expressed[2] by means of an important equation whose derivation is given later [see (4.132)],

$$\mu = \mu_{\text{int}}(\mathbf{r}) + v(\mathbf{r}) \tag{4.24}$$

where μ_{int} is the 'intrinsic' chemical potential. In a uniform fluid this is just the potential specified by the local values of the other fields, i.e. p and T for a one-component system. If $v(\mathbf{r})$ is a slowly varying function of $\mathbf{r}$ and if these fields are not close to their values at a phase transition, then μ_{int} is extremely close to its local value. This is the case, for example, in the earth's atmosphere in a gravitational field, $v = mgz$, where m is the mass of a molecule, g the acceleration due to gravity, and z the height. In a highly inhomogeneous system $\mu_{\text{int}}(\mathbf{r})$ is not a purely local function but depends also on the state of the system near but not at $\mathbf{r}$; a more complete definition is given in § 4.5.

If μ is close to $\mu^{l,g}$ then there may be parts of the system where, from (4.24), $\mu_{\text{int}}(\mathbf{r})$ is greater than $\mu^{l,g}$ and parts where it is less. The former will contain liquid, and the latter gas. Thus if $\mu = \mu^{l,g}$ and if the external potential is gravitational, $v = mg(z - z^*)$, then that part of the volume for which $z < z^*$ contains liquid, and that part for which $z > z^*$

contains gas. The singlet density $\rho(\mathbf{r})$ defined by (4.18) with $n = 1$, is now a function of $\mathbf{r}$ and equal to ρ^{l}, if $\mathbf{r}$ lies in the liquid, or ρ^{g} if $\mathbf{r}$ lies in the gas. If $\mathbf{r}$ lies in the interface then ρ is a function of z which it is our aim to calculate. The pair function $\rho^{(2)}(\mathbf{r}_1, \mathbf{r}_2)$ approaches $\rho(\mathbf{r}_1)\rho(\mathbf{r}_2)$ as r_{12} becomes infinite, and so $h(r_{12} \to \infty) = 0$. The values of all $\rho^{(n)}$ almost certainly change only slowly with g over wide ranges of this field, and it is these plateau values, or perhaps their linear extrapolation from the plateau to $g = 0$, if that extrapolation exists, which are used in this chapter. We ignore the changes in $\rho^{(n)}$ which set in at extremely low values of g and which are characteristic of the true limit $g = 0$ or $v(\mathbf{r}) = 0$ discussed above, and to which we return in § 4.9.

It is convenient now to introduce the direct correlation function of a uniform fluid $c(r_{12})$ which is defined by its relation to the total function $h(r_{12})$ by the integral equation of Ornstein and Zernike,[6]

$$h(r_{12}) = c(r_{12}) + \rho_u \int c(r_{13})h(r_{32}) \, d\mathbf{r}_3. \tag{4.25}$$

Integration of both sides over $d\mathbf{r}_2$ leads to an immediate factorization of the convolution integral,

$$\int h(r_{12}) \, d\mathbf{r}_2 = \left[1 + \rho_u \int h(r_{12}) \, d\mathbf{r}_2 \right] \int c(r_{12}) \, d\mathbf{r}_2 \tag{4.26}$$

and so to the complement of (4.22)

$$1 - \rho_u \int c(r_{12}) \, d\mathbf{r}_{12} = (kT\rho_u \kappa)^{-1}. \tag{4.27}$$

In an inhomogeneous fluid the direct correlation function is defined by

$$h(\mathbf{r}_1, \mathbf{r}_2) = c(\mathbf{r}_1, \mathbf{r}_2) + \int c(\mathbf{r}_1, \mathbf{r}_3)\rho(\mathbf{r}_3)h(\mathbf{r}_3, \mathbf{r}_2) \, d\mathbf{r}_3. \tag{4.28}$$

If $\mathcal{U}_N = 0$ then the equations above can be greatly simplified since the integrations over $d\mathbf{r}_i$ become independent. So for a perfect gas,

$$Z_N = \frac{1}{\Lambda^{3N}N!} \left[\int e^{-v(\mathbf{r})/kT} \, d\mathbf{r} \right]^N$$

$$\Xi = \exp\left[\frac{\lambda}{\Lambda^3} \int e^{-v(\mathbf{r})/kT} \, d\mathbf{r} \right] \tag{4.29}$$

$$\rho(\mathbf{r}) = (\lambda/\Lambda^3) e^{-v(\mathbf{r})/kT}.$$

The last equation is the *barometric law*; it expresses the proportionality of the local density $\rho(\mathbf{r})$ to the Boltzmann factor of the local external field

$v(\mathbf{r})$. The constant of proportionality (λ/Λ^3) is the uniform density of the perfect gas in a field-free environment $(\mathcal{V} = 0)$, a result which serves to establish the zero for $\ln \lambda$. That is, we define the activity in any fluid so that it is proportional to the gas density as this density goes to zero. The constant of proportionality is here the microscopic volume Λ^3; later we shall find it convenient to use other microscopic volumes, but to retain the proportionality to the limiting gas density.

If $\mathcal{U}_N$ is not zero then $\rho(\mathbf{r})$ is not solely a function of $v(\mathbf{r})$ but we can introduce a new function $c^{(1)}(\mathbf{r})$ defined for any fluid, gas or liquid, by

$$\rho(\mathbf{r}) = (\lambda/\Lambda^3) \exp[-v(\mathbf{r})/kT + c^{(1)}(\mathbf{r})]. \tag{4.30}$$

Clearly the one-body potential $-kTc^{(1)}(\mathbf{r})$ is a functional of $\mathcal{U}(\mathbf{r}^N)$ and we see below that it is the singlet function analogous to the pair function $c(\mathbf{r}_1, \mathbf{r}_2)$ already introduced.

The next most simple assumption after $\mathcal{U}_N = 0$, is to assume that $\mathcal{U}$ is composed of pair potentials

$$\mathcal{U}(\mathbf{r}^N) = \sum_{i<j} u(r_{ij}). \qquad \text{(pp)} \quad (4.31)$$

(We denote equations that are valid only under this form of $\mathcal{U}_N$ by the mark (pp) before the equation number. The vector $\mathbf{r}_{ij}$ is defined to be $\mathbf{r}_j - \mathbf{r}_i$, and so $r_{ij} = |\mathbf{r}_j - \mathbf{r}_i|$, and similarly for the Cartesian components; $z_{ij} = z_j - z_i$, etc.) We now differentiate (4.18) with respect to the position of one molecule of the set n, say with respect to $\mathbf{r}_1$. To do this we divide (4.31) into three classes of terms

$$\mathcal{U}(\mathbf{r}^N) = \sum_{1 < i \leqslant n} u(r_{1i}) + \sum_{n < i} u(r_{1i}) + \sum_{1 < i < j} u(r_{ij}). \qquad \text{(pp)} \quad (4.32)$$

On differentiating (4.18) we have three terms on the right-hand side, one from $v(\mathbf{r}_1)$ and two from the first two sums in (4.32),

$$-kT \boldsymbol{\nabla}_1 \rho^{(n)}(\mathbf{r}^n) = \rho^{(n)}(\mathbf{r}_n)\boldsymbol{\nabla}_1 v(\mathbf{r}_1) + \rho^{(n)}(\mathbf{r}^n) \sum_{i>1}^{n} \boldsymbol{\nabla}_1 u(r_{1i})$$

$$+ \int \rho^{(n+1)}(\mathbf{r}^{n+1})\boldsymbol{\nabla}_1 u(r_{1,n+1}) \, d\mathbf{r}_{n+1}. \quad \text{(pp)} \quad (4.33)$$

This set of equations $(n \geqslant 1)$ is usually called the Yvon–Born–Green (YBG) hierarchy.[9] The first is

$$-kT \boldsymbol{\nabla}_1 \rho(\mathbf{r}_1) = \rho(\mathbf{r}_1)\boldsymbol{\nabla}_1 v(\mathbf{r}_1) + \int \rho^{(2)}(\mathbf{r}_1, \mathbf{r}_2)\boldsymbol{\nabla}_1 u(r_{12}) \, d\mathbf{r}_2. \quad \text{(pp)} \quad (4.34)$$

If $u_{12} = 0$, we recover the barometric law (4.29).

If the molecules are not spherical then (4.34) has a trivial generalization in which molecule 1 has a fixed orientation and in which the integration over the position of the molecule 2 becomes an integration over both position and orientation. There is, however, an additional equation that follows on differentiating the generalization of (4.18) with respect to orientation[10]

$$-kT \, \boldsymbol{\nabla}_{\omega_1} \rho(\mathbf{r}_1, \omega_1) = \rho(\mathbf{r}_1, \omega_1) \boldsymbol{\nabla}_{\omega_1} v(\mathbf{r}_1, \omega_1)$$
$$+ \int \rho^{(2)}(\mathbf{r}_1, \omega_1, \mathbf{r}_2, \omega_2) \boldsymbol{\nabla}_{\omega_1} u(\mathbf{r}_1, \omega_1) \, d\mathbf{r}_2 \, d\omega_2 \quad \text{(pp)} \quad (4.35)$$

where, for linear molecules

$$\boldsymbol{\nabla}_\omega = \mathbf{r} \times \boldsymbol{\nabla} = -\mathbf{e}_\theta \frac{1}{\sin\theta} \frac{\partial}{\partial\phi} + \mathbf{e}_\phi \frac{\partial}{\partial\theta} \quad (4.36)$$

and $\mathbf{e}_\theta$ and $\mathbf{e}_\phi$ are unit vectors orthogonal to $\mathbf{r}$.

Differentiation of the equation for spherical molecules (4.30) gives

$$kT\rho(\mathbf{r}_1)\boldsymbol{\nabla}_1 c^{(1)}(\mathbf{r}_1) = kT \, \boldsymbol{\nabla}_1 \rho(\mathbf{r}_1) + \rho(\mathbf{r}_1)\boldsymbol{\nabla}_1 v(\mathbf{r}_1)$$
$$= -\int \rho^{(2)}(\mathbf{r}_1, \mathbf{r}_2)\boldsymbol{\nabla}_1 u(r_{12}) \, d\mathbf{r}_2. \quad \text{(pp)} \quad (4.37)$$

If $\mathcal{V} = 0$ and if we have a homogeneous fluid then all parts of (4.34) and (4.37) vanish (the integral by symmetry), but if $\mathcal{V}$ approaches zero in such a way that $\boldsymbol{\nabla}\rho(\mathbf{r})$ is not everywhere zero then we have useful equations which relate the gradients of $\rho(\mathbf{r})$ and $c^{(1)}(\mathbf{r})$ to the gradient of the intermolecular potential $u(r_{12})$. Such is the case at the gas–liquid surface.

The second member of the hierarchy (4.33), $n = 2$) is non-trivial even for a uniform fluid with $\mathcal{V} = 0$. With a suitable approximation for $\rho^{(3)}$ it is the basis of the YBG theory of homogeneous liquids. We do not pursue this line further.[11]

Other useful results follow from consideration of the response of the equilibrium density to small changes in the external potential;

$$\delta\mathcal{V} = \sum_{i=1}^N \delta v(\mathbf{r}_i). \quad (4.38)$$

This causes a change in $\rho(\mathbf{r})$ of (4.18, $n = 1$) through changes both in the integrals and in Ξ. That in the integral for N molecules is

$$-\frac{1}{kT} \int \exp(-(\mathcal{U}_N + \mathcal{V}_N)/kT)[\delta v(\mathbf{r}_1) + (N-1)\delta v(\mathbf{r}_2)] \, d\mathbf{r}_2 \dots d\mathbf{r}_N$$

and that in the corresponding term in Ξ is

$$-\frac{N}{kT} \int \exp(-(\mathcal{U}_N + \mathcal{V}_N)/kT)\delta v(\mathbf{r}_2) \, d\mathbf{r}^N.$$

Hence

$$-kT\delta\rho(\mathbf{r}_1) = \rho(\mathbf{r}_1)\delta v(\mathbf{r}_1) + \int [\rho^{(2)}(\mathbf{r}_1, \mathbf{r}_2) - \rho(\mathbf{r}_1)\rho(\mathbf{r}_2)]\delta v(\mathbf{r}_2) \, d\mathbf{r}_2$$

$$= \int \delta v(r_2)[\rho(\mathbf{r}_2)\delta(\mathbf{r}_1 - \mathbf{r}_2) + \rho^{(2)}(\mathbf{r}_1, \mathbf{r}_2) - \rho(\mathbf{r}_1)\rho(\mathbf{r}_2)] \, d\mathbf{r}_2 \qquad (4.39)$$

where $\delta(\mathbf{r}_1 - \mathbf{r}_2)$ is the Dirac delta-function which is used here as a device to bring the first term within the integral.† The term in square brackets in the integrand of (4.39) is a measure of that part of the change $\delta\rho(\mathbf{r}_1)$ that arises from the change of $\mathcal{V}$ at $\mathbf{r}_2$; that is, it is the functional derivative[12] of $-kT\rho(\mathbf{r}_1)$ with respect to $v(\mathbf{r}_2)$,

$$-kT\frac{\delta\rho(\mathbf{r}_1)}{\delta v(\mathbf{r}_2)} = \rho(\mathbf{r}_2)\delta(\mathbf{r}_1 - \mathbf{r}_2) + \rho^{(2)}(\mathbf{r}_1, \mathbf{r}_2) - \rho(\mathbf{r}_1)\rho(\mathbf{r}_2). \qquad (4.40)$$

The right-hand side of this equation is called the density–density correlation function, since it is a measure of the correlation of fluctuations in numbers of molecules at $\mathbf{r}_1$ and $\mathbf{r}_2$; cf. (4.16). The last two terms can be expressed in terms of the total correlation function of (4.23),

$$h(\mathbf{r}_1, \mathbf{r}_2) = -\frac{kT}{\rho(\mathbf{r}_1)\rho(\mathbf{r}_2)} \frac{\delta\rho(\mathbf{r}_1)}{\delta v(\mathbf{r}_2)} - \frac{\delta(\mathbf{r}_1 - \mathbf{r}_2)}{\rho(\mathbf{r}_1)}. \qquad (4.41)$$

We call this the first Yvon equation.[13] The second is its inverse

$$c(\mathbf{r}_1, \mathbf{r}_2) = \frac{1}{kT} \frac{\delta v(\mathbf{r}_1)}{\delta\rho(\mathbf{r}_2)} + \frac{\delta(\mathbf{r}_1 - \mathbf{r}_2)}{\rho(\mathbf{r}_1)}, \qquad (4.42)$$

a result which can be verified by seeing that (4.41) and (4.42) satisfy the general form of the Ornstein–Zernike equation (4.28). This check requires the use of the identity

$$\int \left(\frac{\delta v(\mathbf{r}_1)}{\delta\rho(\mathbf{r}_3)}\right)\left(\frac{\delta\rho(\mathbf{r}_3)}{\delta v(\mathbf{r}_2)}\right) d\mathbf{r}_3 = \delta(\mathbf{r}_1 - \mathbf{r}_2) \qquad (4.43)$$

which is the functional analogue of the chain-rule of ordinary differentiation.

Similarly, by considering the change in the grand potential at constant λ caused by a small change in the pair potentials, we have

$$\delta\Omega = -kT(\delta\Xi/\Xi) = \frac{1}{2} \int \rho^{(2)}(\mathbf{r}_1, \mathbf{r}_2)\delta u(r_{12}) \, d\mathbf{r}_1 \, d\mathbf{r}_2 \qquad \text{(pp)} \quad (4.44)$$

or

$$\delta\Omega/\delta u(r_{12}) = \tfrac{1}{2}\rho^{(2)}(\mathbf{r}_1, \mathbf{r}_2). \qquad \text{(pp)} \quad (4.45)$$

† The properties of this 'function' are set out in Appendix 2.

(The factor of $\frac{1}{2}$ arises from the fact that there are $\frac{1}{2}N(N-1)$ pair terms $u(r_{ij})$ in $\mathcal{U}_N$ (4.31), but there is a factor of $N(N-1)$ between $(N-2)!$ in $\rho^{(2)}$ (4.18) and $N!$ in Ξ, (4.6) and (4.12).) In a uniform fluid (4.44) can be written

$$\delta\Omega = \frac{1}{2}V \int \rho^{(2)}(r_{12})\delta u(r_{12}) \, d\mathbf{r}_{12}. \qquad \text{(pp)} \qquad (4.46)$$

If we re-write (4.30) as

$$c^{(1)}(\mathbf{r}_1) = v(\mathbf{r}_1)/kT + \ln(\Lambda^3 \rho(\mathbf{r}_1)/\lambda) \qquad (4.47)$$

and differentiate with respect to $\rho(\mathbf{r}_2)$ we obtain, after using the second Yvon equation,

$$\frac{\delta c^{(1)}(\mathbf{r}_1)}{\delta\rho(\mathbf{r}_2)} = c(\mathbf{r}_1, \mathbf{r}_2). \qquad (4.48)$$

Thus the pair direct correlation function is the functional derivative of the singlet function. The sequence can be continued;[14] a triplet function $c(\mathbf{r}_1, \mathbf{r}_2, \mathbf{r}_3)$ is the functional derivative of $c(\mathbf{r}_1, \mathbf{r}_2)$ with respect to $\rho(\mathbf{r}_3)$, etc., but we shall not need these higher results.

The equilibrium density is a functional of the field $v(\mathbf{r})$; that is, if we specify $v(\mathbf{r})$ at every point then $\rho(\mathbf{r})$ is determined at every point.[2] So if we displace the field by a vector $\mathbf{s}$ then the density is similarly displaced. Formally, if $\rho(\mathbf{r}_1, [v(\mathbf{r})]) = \rho(\mathbf{r}_1)$, then $\rho(\mathbf{r}_1, [v(\mathbf{r}+\mathbf{s})]) = \rho(\mathbf{r}_1 + \mathbf{s})$. Therefore the first term of a functional Taylor expansion gives

$$\rho(\mathbf{r}_1+\mathbf{s}) = \rho(\mathbf{r}_1) + \int [v(\mathbf{r}_2+\mathbf{s}) - v(\mathbf{r}_2)] \left[\frac{\delta\rho(\mathbf{r}_1+\mathbf{s})}{\delta v(\mathbf{r}_2)}\right]_{s=0} d\mathbf{r}_2 \qquad (4.49)$$

or, as s goes to zero

$$\nabla_1 \rho(\mathbf{r}_1) = \int \frac{\delta\rho(\mathbf{r}_1)}{\delta v(\mathbf{r}_2)} \nabla_2 v(\mathbf{r}_2) \, d\mathbf{r}_2. \qquad (4.50)$$

Substitution from the first Yvon equation (4.41) gives

$$\nabla_1 v(\mathbf{r}_1)/kT + \nabla_1 \ln \rho(\mathbf{r}_1) = -\int h(\mathbf{r}_1, \mathbf{r}_2)\rho(\mathbf{r}_2)\nabla_2(v(\mathbf{r}_2)/kT) \, d\mathbf{r}_2 \qquad (4.51)$$

which is a particular form of (4.39) when the arbitrary change δv is specialized to the gradient ∇v.

Conversely,[2] since $v(r)$ is also a unique functional of $\rho(\mathbf{r})$, a functional Taylor expansion of $v(\mathbf{r}_1+\mathbf{s})$ and the use of the second Yvon equation (4.42) gives

$$\nabla_1 v(r_1)/kT + \nabla_1 \ln \rho(\mathbf{r}_1) = \int c(\mathbf{r}_1, \mathbf{r}_2)\nabla_2\rho(\mathbf{r}_2) \, d\mathbf{r}_2 \qquad (4.52)$$

whilst the first part of (4.37), which is not restricted to pair potentials, leads to

$$\boldsymbol{\nabla}_1 c^{(1)}(\mathbf{r}_1) = \int c(\mathbf{r}_1, \mathbf{r}_2)\boldsymbol{\nabla}_2\rho(\mathbf{r}_2)\,d\mathbf{r}_2. \tag{4.53}$$

If $v(\mathbf{r})$ is zero then (4.51) appears to give $\boldsymbol{\nabla}\rho(\mathbf{r}) = 0$; that is, it appears that the fluid is uniform; we return to the implications of this equation in § 4.9. Equation (4.52) does not, however, necessarily lead to this trivial result although $\boldsymbol{\nabla}\rho(\mathbf{r}) = 0$ is always a solution. Like the YBG equation (4.34), it becomes an integral equation for $\boldsymbol{\nabla}\rho(\mathbf{r})$ which can help determine the density profile at a gas–liquid surface. The pair of equations, (4.51) and the more useful (4.52), were obtained by Triezenberg,[15] by Lovett, Mou, and Buff,[16] and by Wertheim.[17]

Equations (4.51) and (4.52) can be generalized to cover non-spherical molecules by integrating on the right-hand sides over all orientations of the second molecule. As for the YBG equation, a more interesting consequence of the dependence of the energy on molecular orientations is the appearance of new integral equations.[10] The new equation that is the analogue of (4.51) is not of interest unless the external field v is a function of molecular orientation as well as position. The analogue of (4.52) is (omitting any term in $\boldsymbol{\nabla}_\omega v$)

$$(\mathbf{r}_1 \times \boldsymbol{\nabla}_{\mathbf{r}_1} + \boldsymbol{\nabla}_{\omega_1})\ln \rho(\mathbf{r}_1, \omega_1) = \int c(\mathbf{r}_1, \omega_1, \mathbf{r}_2, \omega_2)(\mathbf{r}_2 \times \boldsymbol{\nabla}_{\mathbf{r}_2} + \boldsymbol{\nabla}_{\omega_2})\rho(\mathbf{r}_2, \omega_2)\,d\mathbf{r}_2\,d\omega_2$$

$$+ (\mathbf{r}_1 \times \boldsymbol{\nabla}_{\mathbf{r}_1})v(\mathbf{r}_1)/kT. \tag{4.54}$$

If $\rho(\mathbf{r})$ changes little over the range of the (spherical) intermolecular forces then the factor $\boldsymbol{\nabla}\rho(\mathbf{r})$ can be taken outside the integral of (4.52) to give

$$\boldsymbol{\nabla}v(\mathbf{r}) + \frac{\boldsymbol{\nabla}\ln \rho(\mathbf{r})}{\rho(\mathbf{r})\kappa} = 0 \tag{4.55}$$

where κ is the compressibility (4.27). If we replace the gradient of the density by that of the pressure, $\boldsymbol{\nabla}p(\mathbf{r}) = (\partial p/\partial \rho)\boldsymbol{\nabla}\rho(\mathbf{r})$, then we obtain the standard hydrostatic equation

$$\rho(\mathbf{r})\boldsymbol{\nabla}v(\mathbf{r}) + \boldsymbol{\nabla}p(\mathbf{r}) = 0. \tag{4.56}$$

Equation (4.53) is the first ($n = 1$) of a hierarchy for the direct correlation functions that parallels the YBG hierarchy (4.33). The general equation is[18]

$$\sum_{i=1}^{n} \boldsymbol{\nabla}_i c(\mathbf{r}_1, \ldots, \mathbf{r}_n) = \int c(\mathbf{r}_1, \ldots, \mathbf{r}_{n+1})\boldsymbol{\nabla}_{n+1}\rho(\mathbf{r}_{n+1})\,d\mathbf{r}_{n+1}. \tag{4.57}$$

Unless there is an external field this set of equations is satisfied trivially.[18]

Many of the results obtained above require the use of the grand canonical ensemble since it is only for an open system that we can envisage arbitrary changes of the numbers of molecules in a system, or of the equilibrium density at any point. In a closed system we should be constrained by the requirement that the total number of molecules is fixed. However it is often convenient to use the simpler equations of the canonical ensemble which can be found by restricting the sums over N in the grand ensemble to their largest term $N = \bar{N}$.

We obtain

$$f^{(N)} = \Xi^{-1} \lambda^N \exp(-\mathcal{H}_N/kT) \tag{4.58}$$

or, in general,

$$f^{(n)} = \Xi^{-1} \frac{\lambda^N}{h^{3(N-n)}(N-n)!} \iint \exp(-\mathcal{H}_N/kT) \, d\mathbf{p}_{n+1} \ldots d\mathbf{p}_N \, d\mathbf{r}_{n+1} \ldots d\mathbf{r}_N \tag{4.59}$$

$$\Xi = \lambda^N Z_N \tag{4.60}$$

$$\rho^{(n)} = \Xi^{-1} \frac{(\lambda/\Lambda^3)^N}{(N-n)!} \int \exp(-(\mathcal{U}_N + \mathcal{V}_N)/kT) \, d\mathbf{r}_{n+1} \ldots d\mathbf{r}_N \tag{4.61}$$

$$= \frac{N!}{(N-n)!} \frac{\displaystyle\int \exp(-(\mathcal{U}_N + \mathcal{V}_N)/kT) \, d\mathbf{r}_{n+1} \ldots d\mathbf{r}_N}{\displaystyle\int \exp(-(\mathcal{U}_N + \mathcal{V}_N)/kT) \, d\mathbf{r}^N}. \tag{4.62}$$

These are the equations of the canonical ensemble, but dressed in the language of the grand ensemble.

We use (4.62) to give a further simple interpretation of the one-body function $c^{(1)}(\mathbf{r})$, introduced in (4.30) and related to $c(\mathbf{r}_1, \mathbf{r}_2)$ by (4.48) and (4.53). Let $\mathcal{U}(\mathbf{r}^N)$ be divided into two terms: $\mathcal{U}(\mathbf{r}^{N-1})$, the energy of mutual interaction of $N-1$ molecules at $\mathbf{r}_2, \ldots, \mathbf{r}_N$, and $u(\mathbf{r}_1; \mathbf{r}^{N-1})$, the increment of energy when an additional molecule is added at $\mathbf{r}_1$, the others retaining their positions; and similarly for $\mathcal{V}$. If we abbreviate $u(\mathbf{r}_1; \mathbf{r}^{N-1})$ to $u(\mathbf{r}_1)$ then

$$\mathcal{U}_N = u(\mathbf{r}_1) + \mathcal{U}_{N-1}, \qquad \mathcal{V}_N = v(\mathbf{r}_1) + \mathcal{V}_{N-1}. \tag{4.63}$$

We now use a double pair of angle brackets to denote the average in a canonical ensemble of N molecules, one of which is fixed at $\mathbf{r}_1$, and a single pair to denote an average in a diminished ensemble of $N-1$

molecules. That is,

$$
\langle\langle\exp(u(\mathbf{r}_1)/kT)\rangle\rangle = \frac{\displaystyle\int \exp(u(\mathbf{r}_1)/kT)\exp(-(\mathcal{U}_N+\mathcal{V}_N)/kT)\,d\mathbf{r}_2\ldots d\mathbf{r}_N}{\displaystyle\int \exp(-(\mathcal{U}_N+\mathcal{V}_N)/kT)\,d\mathbf{r}_2\ldots d\mathbf{r}_N}
$$

$$
= \frac{\exp(-v(\mathbf{r}_1)/kT)\displaystyle\int \exp(-(\mathcal{U}_{N-1}+\mathcal{V}_{N-1})/kT)\,d\mathbf{r}_2\ldots d\mathbf{r}_N}{\displaystyle\int \exp(-(\mathcal{U}_N+\mathcal{V}_N)/kT)\,d\mathbf{r}_2\ldots d\mathbf{r}_N}
$$

$$
= \frac{\displaystyle\int \exp(-(\mathcal{U}_{N-1}+\mathcal{V}_{N-1})/kT)\,d\mathbf{r}_2\ldots d\mathbf{r}_N}{\displaystyle\int \exp(-u(\mathbf{r}_1)/kT)\exp(-(\mathcal{U}_{N-1}+\mathcal{V}_{N-1})/kT)\,d\mathbf{r}_2\ldots d\mathbf{r}_N}
$$

$$
= \langle\exp(-u(\mathbf{r}_1)/kT)\rangle^{-1}. \tag{4.64}
$$

The density at $\mathbf{r}_1$ when no molecule is fixed there, is, from (4.62)

$$
\rho(\mathbf{r}_1) = \frac{N\displaystyle\int \exp(-(\mathcal{U}_N+\mathcal{V}_N)/kT)\,d\mathbf{r}_2\ldots d\mathbf{r}_N}{\displaystyle\int \exp(-(\mathcal{U}_N+\mathcal{V}_N)/kT)\,d\mathbf{r}_1\ldots d\mathbf{r}_N}. \tag{4.65}
$$

Substitution into the second line of (4.64) gives

$$
\langle\langle\exp(u(\mathbf{r}_1)/kT)\rangle\rangle
$$

$$
= \frac{N\exp(-v(\mathbf{r}_1)/kT)}{\rho(\mathbf{r}_1)}\frac{\displaystyle\int \exp(-(\mathcal{U}_{N-1}+\mathcal{V}_{N-1})/kT)\,d\mathbf{r}_2\ldots d\mathbf{r}_N}{\displaystyle\int \exp(-(\mathcal{U}_N+\mathcal{V}_N)/kT)\,d\mathbf{r}_1\ldots d\mathbf{r}_N} \tag{4.66}
$$

$$
= \frac{\exp(-v(\mathbf{r}_1)/kT)}{\Lambda^3\rho(\mathbf{r}_1)}\frac{Z_{N-1}}{Z_N}. \tag{4.67}
$$

Now $kT\ln(Z_{N-1}/Z_N)$ is the change of free energy on going, at fixed V and T, from a system of $(N-1)$ molecules to one of N molecules, (4.14), and so is the chemical potential or $kT\ln\lambda$. Since Z_{N-1} and Z_N are independent of the position $\mathbf{r}_1$, the associated value of λ is also uniform throughout the system. Hence (4.64) and (4.67) can be expressed in the complementary forms

$$\frac{\lambda}{\Lambda^3 \rho(\mathbf{r})} = \exp(v(\mathbf{r})/kT)\langle\langle\exp(u(\mathbf{r})/kT)\rangle\rangle, \tag{4.68}$$

$$\frac{\Lambda^3 \rho(\mathbf{r})}{\lambda} = \exp(-v(\mathbf{r})/kT)\langle\exp(-u(\mathbf{r})/kT)\rangle. \tag{4.69}$$

From (4.30)

$$c^{(1)}(\mathbf{r}) = -\ln\langle\langle\exp(u(\mathbf{r})/kT)\rangle\rangle = \ln\langle\exp(-u(\mathbf{r})/kT)\rangle. \tag{4.70}$$

The pair of equations (4.68) and (4.69) are alternative forms of the *potential distribution theorem*.[19] The first becomes indeterminate if the molecules have hard cores, that is, if there are configurations in which $u(\mathbf{r})$ is positive infinite; the second is well-behaved for all values of $u(\mathbf{r})$. If $v(\mathbf{r})$ is zero (or known) they provide useful links between $\rho(\mathbf{r})$ and the averages of the exponentials of the energy $u(\mathbf{r})$. In taking both of these averages we envisage a molecule fixed at $\mathbf{r}$ and surrounded by $(N-1)$ other molecules. In the first average (double brackets) these $(N-1)$ molecules move subject to their mutual interaction and to their interaction with the fixed molecule at $\mathbf{r}$; in the second average (single brackets) they move as if there were no interaction with the fixed molecule. These averages can often be calculated theoretically, or by computer simulation, even for a non-uniform system, and so the theorem can be yet another useful route to the density profile of a gas–liquid surface.

Another way of looking at this theorem is to say that it provides an answer to the question of what is meant by the chemical potential or activity at a particular point $\mathbf{r}$, in the canonical ensemble. We could define $\lambda(\mathbf{r})$ by the equations, $(v(\mathbf{r}) = 0)$,

$$\lambda(\mathbf{r})/\Lambda^3 \rho(\mathbf{r}) = \langle\langle\exp(u(\mathbf{r})/kT)\rangle\rangle = \langle\exp(-u(\mathbf{r})/kT)\rangle^{-1}. \tag{4.71}$$

The right-hand sides of this equation are, in principle, experimentally accessible quantities (e.g. by computer simulation[19]), and so provide operationally useful definitions of $\lambda(\mathbf{r})$ at a gas–liquid surface or in any other inhomogeneous system. Its constancy for all points $\mathbf{r}$ can then be a matter for experimental verification (see § 5.6). Such checks, like those of the Gibbs adsorption equation (§§ 2.3 and 5.6), are not without value since the facts on which the theoretical structures of both classical and statistical thermodynamics are based are almost entirely facts about homogeneous systems. The extensions of theory to systems which are extremely inhomogeneous even at the scale of length provided by the intermolecular forces is not without hazard, as we have seen already in the attempts to use quasi-thermodynamic arguments (§§ 2.5 and 3.4) and as we shall see again later in this chapter. However, to anticipate later discussion, we may say at once that there are then no known results, experimental or theoretical, which impugn the correctness of any of the equations of this section.

In a homogeneous system there are similar equations for the ratio of λ to the energy density $\phi = U/V$. If $u(\mathbf{r})$ can be written as a sum of two-body, three-body potentials, etc.

$$u(\mathbf{r}) = \sum_{m \geq 2} u_m(\mathbf{r}) \tag{4.72}$$

then[19,20]

$$\phi/\lambda = \sum_{m \geq 2} m^{-1} \langle u_m(\mathbf{r}) \exp(-u(\mathbf{r})/kT) \rangle. \tag{4.73}$$

Instead of (4.63) we can also write,

$$\mathcal{U}_N = u^{(2)}(\mathbf{r}_1, \mathbf{r}_2) + \mathcal{U}_{N-2} \qquad \mathcal{V}_N = v(\mathbf{r}_1) + v(\mathbf{r}_2) + \mathcal{V}_{N-2} \tag{4.74}$$

where $u^{(2)}(\mathbf{r}_1, \mathbf{r}_2)$ is the change of energy on adding a pair of molecules at these positions to a system of $(N-2)$ molecules. For additive pair potentials we can define $u^*(\mathbf{r})$ by

$$u^{(2)}(\mathbf{r}_1, \mathbf{r}_2) = u(\mathbf{r}_{12}) + u^*(\mathbf{r}_1) + u^*(\mathbf{r}_2). \qquad \text{(pp)} \quad (4.75)$$

By arguments that parallel those above we have for a system which may be inhomogeneous

$$\begin{aligned}\frac{\lambda^2}{\Lambda^6 \rho(\mathbf{r}_1)\rho(\mathbf{r}_2)} &= g(\mathbf{r}_1, \mathbf{r}_2) \exp([v(\mathbf{r}_1) + v(\mathbf{r}_2)]/kT) \langle\!\langle \exp(u^{(2)}(\mathbf{r}_1, \mathbf{r}_2)/kT) \rangle\!\rangle \\ &= g(\mathbf{r}_1, \mathbf{r}_2) \exp([v(\mathbf{r}_1) + v(\mathbf{r}_2)]/kT) \langle \exp(-u^{(2)}(\mathbf{r}_1, \mathbf{r}_2)/kT) \rangle^{-1}\end{aligned}$$

$$\tag{4.76}$$

where $g(\mathbf{r}_1, \mathbf{r}_2)$ is the pair distribution function, cf. (4.23),

$$g(\mathbf{r}_1, \mathbf{r}_2) = 1 + h(\mathbf{r}_1, \mathbf{r}_2) = \rho^{(2)}(\mathbf{r}_1, \mathbf{r}_2)/\rho(\mathbf{r}_1)\rho(\mathbf{r}_2) \tag{4.77}$$

and the single brackets are now an average in a diminished ensemble of $(N-2)$ molecules. For a system of pair potentials it is convenient to replace g by the function y

$$y(\mathbf{r}_1, \mathbf{r}_2) = g(\mathbf{r}_1, \mathbf{r}_2) \exp(u(r_{12})/kT \qquad \text{(pp)} \quad (4.78)$$

so that division of (4.76) by (4.68) and (4.69) gives

$$\begin{aligned}y(\mathbf{r}_1, \mathbf{r}_2) &= \frac{\langle\!\langle \exp(u(\mathbf{r}_1)/kT) \rangle\!\rangle \langle\!\langle \exp(u(\mathbf{r}_2)/kT) \rangle\!\rangle}{\langle\!\langle \exp[(u^*(\mathbf{r}_1) + u^*(\mathbf{r}_2))/kT] \rangle\!\rangle} \\ &= \frac{\langle \exp[-(u^*(\mathbf{r}_1) + u^*(\mathbf{r}_2))/kT] \rangle}{\langle \exp(-u(\mathbf{r}_1)/kT) \rangle \langle \exp(-u(\mathbf{r}_2)/kT) \rangle}. \qquad \text{(pp)} \quad (4.79)\end{aligned}$$

If r_{12} is zero the second of these equations becomes

$$y(\mathbf{r}; \mathbf{r}_{12} = 0) = \langle \exp(-2u(\mathbf{r})/kT) \rangle \langle \exp(-u(\mathbf{r})/kT) \rangle^{-2}. \quad \text{(pp)} \quad (4.80)$$

For a hard-sphere potential the averages of the exponentials of $(-2u(\mathbf{r})/kT)$ and of $(-u(\mathbf{r})/kT)$ are the same, and so, from (4.69)

$$\lambda = \Lambda^3 \rho(\mathbf{r}) \exp(v(\mathbf{r})/kT) \, y(\mathbf{r}; r_{12} = 0), \quad \text{(hard-sphere pp)} \quad (4.81)$$

a result first obtained by Hoover and Poirier [21]

The three most useful equations for determining the density at the gas–liquid surface are the first YBG equation (4.34), the integral equation over the direct correlation function (4.52), and the potential distribution theorem (4.69). At a planar surface with $\mathcal{V} \to 0$ these can be re-written in simpler forms. If the direction normal to the surface (the 'height') is that of the z-axis, then the only non-zero gradient is d/dz and we have cylindrical symmetry about this axis. The volume element can then be more usefully expressed in cylindrical or spherical polar coordinates,

$$d\mathbf{r} = 2\pi \, dz \, r \, dr = 2\pi \sin\theta \, d\theta \, r^2 \, dr \qquad (4.82)$$

where θ is the angle of a vector with the z-axis, and where the cylindrical symmetry has allowed an immediate integration over the azimuthal angle. So, from (4.34) we have

$$\frac{d\rho(z_1)}{dz_1} = 2\pi \int_{-\infty}^{\infty} dz_2 \int_{|z_{12}|}^{\infty} dr_{12} r_{12} \rho^{(2)}(r_{12}, z_1, z_2) \frac{d}{dz_2} \left(\frac{u(r_{12})}{kT} \right).$$
$$\text{(pp)} \quad (4.83)$$

In Cartesian coordinates we have

$$\partial r / \partial x = x/r \qquad \partial u / \partial x = (x/r)(du/dr) \text{ etc.} \qquad (4.84)$$

and so

$$\frac{d\rho(z_1)}{dz_1} = \frac{2\pi}{kT} \int_{-\infty}^{\infty} dz_2 z_{12} \int_{|z_{12}|}^{\infty} dr_{12} \rho^{(2)}(r_{12}, z_1, z_2) u'(r_{12}) \quad \text{(pp)} \quad (4.85)$$

Similarly, from (4.52)

$$\frac{d\rho(z_1)}{dz_1} = 2\pi\rho(z_1) \int_{-\infty}^{\infty} dz_2 \int_{|z_{12}|}^{\infty} dr_{12} r_{12} c(r_{12}, z_1, z_2) \frac{d\rho(z_2)}{dz_2}. \quad (4.86)$$

And finally from (4.69)

$$\rho(z_1) = (\lambda/\Lambda^3) \langle \exp(-u(z_1)/kT) \rangle. \qquad (4.87)$$

The first two of these equations (4.85) and (4.86), although they prove to be useful, suffer from a difficulty common in the theory of fluids;

in order to determine the distribution function in which we are interested, in this case the singlet function $\rho(z)$, we need to know first a function of next higher order, in this case the pair functions $\rho^{(2)}(r_{12}, z_1, z_2)$ or $c(r_{12}, z_1, z_2)$. The third equation (4.87) does not exhibit this difficulty explicitly but the average on the right-hand side is a many-body function whose evaluation requires a knowledge of the distribution functions.

4.3 The pressure tensor

The pressure in a homogeneous fluid can be calculated either from the statistical thermodynamic results of the last section or by a direct calculation of the stress across a microscopic area in the interior of the fluid. For an inhomogeneous fluid we use the second route.

We note first that, for a homogeneous fluid, the compressibility and its reciprocal are given in terms of the total and direct correlation function by (4.22) and (4.27). An integration over ρ_u gives us the pressure since the known properties of the perfect gas at $\rho_u = 0$ supply the boundary conditions. The integration requires, however, that we know $h(r_{12})$ and $c(r_{12})$ as functions of density and so this is not a route to the pressure that we shall find useful in discussing the gas–liquid surface.

A more direct route to the pressure of a homogeneous fluid ($\mathcal{V}_N = 0$) is by differentiation of the canonical partition function

$$p = -(\partial F/\partial V)_{N,T} = kT(\partial \ln Z_N/\partial V)_T, \qquad (4.88)$$

a differentiation which is usually made by means of a dimensional argument due independently to Bogoliubov and Green.[22] Consider first the case in which $\mathcal{U}_N$ is composed of pair potentials only, and in which the fluid is confined to a cube of side l, one corner of which is the origin of the coordinate system. Let

$$\mathbf{r}_i = l\mathbf{t}_i \qquad (4.89)$$

where $\mathbf{t}_i$ has components of length between 0 and 1. From (4.7) and (4.31)

$$Z_N = \frac{(l/\Lambda)^{3N}}{N!} \int \exp\left[-\sum_{i<j} u(lt_{ij})/kT \right] dt^N, \qquad \text{(pp)} \quad (4.90)$$

$$\begin{aligned}
\frac{\partial Z_N}{\partial V} &= \frac{1}{3l^2}\left(\frac{\partial \ln Z_N}{\partial l} \right) \\
&= \frac{N}{V} - \frac{(l/\Lambda)^{3N}}{6VZ_N(N-2)!} \int \frac{lt_{12}}{kT} \frac{du(lt_{12})}{d(lt_{12})} e^{-\mathcal{U}_N/kT} dt^N \\
&= \frac{N}{V} - \frac{1}{6} \int \frac{r_{12}}{kT} \frac{du(r_{12})}{dr_{12}} \rho^{(2)}(\mathbf{r}_1, \mathbf{r}_2) d\mathbf{r}_{12}, \qquad \text{(pp)} \quad (4.91)
\end{aligned}$$

or

$$p = kT\rho - \frac{1}{6} \int r_{12} u'(r_{12}) \rho^{(2)}(r_{12}) \, d\mathbf{r}_{12}. \qquad \text{(pp)} \quad (4.92)$$

This is often called the virial equation for the pressure since it can also be derived from the virial theorem of classical mechanics of Clausius.[23] The product $ru'(r)$ is the virial of the pair potential $u(r)$.

If $\mathcal{U}_N$ comprises pair and triplet terms then (4.92) can be extended by adding an integral over the virial of the three-body intermolecular potential,

$$-\frac{1}{18} \int \left(r_{12} \frac{\partial}{\partial r_{12}} + r_{13} \frac{\partial}{\partial r_{13}} + r_{23} \frac{\partial}{\partial r_{23}} \right) u(\mathbf{r}_1, \mathbf{r}_2, \mathbf{r}_3) \, \rho^{(3)}(\mathbf{r}_1, \mathbf{r}_2, \mathbf{r}_3) \, d\mathbf{r}_1 \, d\mathbf{r}_2 \, d\mathbf{r}_3.$$

$$(4.93)$$

The differentiation of F in (4.88)–(4.91) is geometrically equivalent to a small uniform dilation of a cube of homogeneous fluid. There is more than one way of deforming an inhomogeneous fluid, each of which leads to a different expression for the tensor $\mathbf{p}(\mathbf{r})$; indeed, as we shall see, there appears to be no unique definition of this quantity. Its gradient is well defined since in a fluid at equilibrium, cf. (2.82),

$$\nabla \cdot \mathbf{p}(\mathbf{r}) = -\rho(\mathbf{r})\nabla v(\mathbf{r}) \qquad (4.94)$$

so that the gradient vanishes in the absence of the external field, but this restriction is insufficient to determine $\mathbf{p}(\mathbf{r})$ uniquely. There is a symmetry relation[24] which requires that the difference $p_{\alpha\beta}(\mathbf{r}) - p_{\beta\alpha}(\mathbf{r})$ (where $\alpha, \beta = x, y, z$) be the gradient of a third-rank tensor, but this condition adds no new restriction on the form of $\mathbf{p}(\mathbf{r})$. Nevertheless it is worth our while to obtain an expression for this tensor that can be used, for example, in calculating the surface tension through the so-called mechanical definition (2.88). This was done by Kirkwood and Buff[25] but we give first the argument used later by Irving and Kirkwood.[26]

Let $d\mathbf{A}$ be a surface element at point $\mathbf{r}$ within the inhomogeneous fluid, and let $V = 0$. The region of fluid on the side of the surface away from which the vector $d\mathbf{A}$ is pointing is called region 1, and that on the other side, region 2 (Fig. 4.1). The stress across the surface element comprises two terms, a kinetic term from the momentum of the molecules, and a potential term from the intermolecular forces. The first is isotropic and, because of the separability of the translational motion in classical statistical mechanics, has the same value as in a perfect gas, namely $kT\rho(\mathbf{r})\mathbf{1}$, where $\mathbf{1}$ is the unit tensor. It is the second term that introduces the arbitrariness into the definition of $\mathbf{p}(\mathbf{r})$ for there is no unique way of specifying which intermolecular forces contribute to the

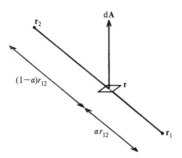

FIG. 4.1

stress across dA. We follow Irving and Kirkwood[26] and choose only those between pairs of molecules at $\mathbf{r}_1$ in region 1, and at $\mathbf{r}_2$ in region 2, for which the vector $\mathbf{r}_2 - \mathbf{r}_1 = \mathbf{r}_{12}$ passes through dA. Such a choice clearly includes all the forces across a macroscopic area A, on integration over dA, but other choices can also satisfy this requirement.

A molecule at $\mathbf{r}_1$ experiences a force $(\mathbf{r}_{12}/r_{12})u'(r_{12})$ from a molecule at $\mathbf{r}_2$. The probability of there being molecules in these positions is proportional to $\rho^{(2)}(\mathbf{r}_1, \mathbf{r}_2)$ which can be re-written $\rho^{(2)}(\mathbf{r} - \alpha\mathbf{r}_{12}, \mathbf{r} + (1-\alpha)\mathbf{r}_{12})$ where α is the ratio of the length $|\mathbf{r} - \mathbf{r}_1|$ to r_{12}. If $\mathbf{r}_{12}$ is fixed, the volume element over which the vector $\mathbf{r}_1 + \alpha\mathbf{r}_{12}$ terminates on dA for α between α and $\alpha + d\alpha$ is $(\mathbf{dA} \cdot \mathbf{r}_{12}) \, d\alpha$. So the average number of pairs of molecules, one in this volume element at $\mathbf{r}_1$ and one in an element $d\mathbf{r}_{12}$ at $\mathbf{r}_2$ is

$$\rho^{(2)}(\mathbf{r} - \alpha\mathbf{r}_{12}, \mathbf{r} + (1-\alpha)\mathbf{r}_{12})(\mathbf{dA} \cdot \mathbf{r}_{12}) \, d\alpha \, d\mathbf{r}_{12}. \tag{4.95}$$

Hence the total force across the surface element is

$$\text{Force} = \mathbf{dA} \cdot \int_{\text{region 2}} d\mathbf{r}_{12} \int_0^1 d\alpha \, \frac{\mathbf{r}_{12}\mathbf{r}_{12}}{r_{12}} \, u'(r_{12})\rho^{(2)}(\mathbf{r} - \alpha\mathbf{r}_{12}, \mathbf{r} + (1-\alpha)\mathbf{r}_{12}).$$

$$\text{(pp)} \quad (4.96)$$

The vector on the right is the product of the vector $\mathbf{dA}$ and the dyadic $\mathbf{r}_{12}\mathbf{r}_{12}$. By substituting $1 - \beta$ for α, and using the result that $\rho^{(2)}(\mathbf{r}_1, \mathbf{r}_2) = \rho^{(2)}(\mathbf{r}_2, \mathbf{r}_1)$ we have that

$$\int_0^1 d\alpha \, \frac{\mathbf{r}_{12}\mathbf{r}_{12}}{r_{12}} \, u'(r_{12})\rho^{(2)}(\mathbf{r} - \alpha\mathbf{r}_{12}, \mathbf{r} + (1-\alpha)\mathbf{r}_{12})$$

$$= \int_0^1 d\beta \, \frac{\mathbf{r}_{12}\mathbf{r}_{12}}{r_{12}} \, u'(r_{12})\rho^{(2)}(\mathbf{r} + \beta\mathbf{r}_{12}, \mathbf{r} - (1-\beta)\mathbf{r}_{12}) \quad \text{(pp)} \quad (4.97)$$

and so (4.96) can be re-written more symmetrically with the integration of $d\mathbf{r}_{12}$ over all space.

$$\text{Force} = \tfrac{1}{2}\, d\mathbf{A} \cdot \int d\mathbf{r}_{12} \int_0^1 d\alpha\, \frac{\mathbf{r}_{12}\mathbf{r}_{12}}{r_{12}}\, u'(r_{12})\rho^{(2)}(\mathbf{r}-\alpha\mathbf{r}_{12},\, \mathbf{r}+(1-\alpha)\mathbf{r}_{12})$$

$$\text{(pp)} \quad (4.98)$$

or

$$\mathbf{p}(\mathbf{r}) = kT\rho(\mathbf{r})\mathbf{1} - \frac{1}{2}\int d\mathbf{r}_{12} \int_0^1 d\alpha\, \frac{\mathbf{r}_{12}\mathbf{r}_{12}}{r_{12}}\, u'(r_{12})\rho^{(2)}(\mathbf{r}-\alpha\mathbf{r}_{12},\, \mathbf{r}+(1-\alpha)\mathbf{r}_{12}).$$

$$\text{(pp)} \quad (4.99)$$

At a planar gas–liquid surface the cylindrical symmetry of the density around the z-axis ensures that $\mathbf{p}$ has only two independent components, the normal component $p_N(z)$ and the tangential component $p_T(z)$, (2.84). We can pick out these terms from the dyadic in (4.99),

$$p_N(z) = kT\rho(z) - \frac{1}{2}\int d\mathbf{r}_{12}\, \frac{z_{12}^2}{r_{12}}\, u'(r_{12})$$

$$\times \int_0^1 d\alpha\, \rho^{(2)}(r_{12},\, z-\alpha z_{12},\, z+(1-\alpha)z_{12}),$$

$$\text{(pp)} \quad (4.100)$$

$$p_T(z) = kT\rho(z) - \frac{1}{4}\int d\mathbf{r}_{12}\, \frac{x_{12}^2 + y_{12}^2}{r_{12}}\, u'(r_{12})$$

$$\times \int_0^1 d\alpha\, \rho^{(2)}(r_{12},\, z-\alpha z_{12},\, z+(1-\alpha)z_{12}).$$

In a homogeneous fluid the pair function is independent of α, and a function only of r_{12}; under this condition either part of (4.100) reduces to (4.92). Differentiation of the expression for $p_N(z)$, and use of the first YBG equation (4.34), shows that $(dp_N(z)/dz)$ is zero, as is required for mechanical stability,[27,28] that is,

$$p_N(z) = p^g = p^l. \tag{4.101}$$

Kirkwood and Buff[25] and Harasima[28] chose a different but equally acceptable set of intermolecular potentials for the calculation of the stress across a surface element $d\mathbf{A}$ at a planar interface. They obtained the same expression for $p_N(z)$ as Irving and Kirkwood but a different one for $p_T(z)$;

$$p_T(z) = kT\rho(z) - \frac{1}{4}\int d\mathbf{r}_{12} u'(r_{12}) \frac{x_{12}^2 + y_{12}^2}{r_{12}}\, \rho^{(2)}(r_{12},\, z,\, z+z_{12}).$$

$$\text{(pp)} \quad (4.102)$$

Schofield and Henderson[24] have recently re-analysed the problem and have concluded that there are infinitely many ways of writing $\mathbf{p}(\mathbf{r})$ which can be distinguished by different choices of a contour joining two interacting molecules. The expression of Irving and Kirkwood (4.100) corresponds to a straight line from the molecule at $\mathbf{r}_1$ to that at $\mathbf{r}_2$, as shown in Fig. 4.1. Harasima's expression (4.102) is obtained by choosing an unsymmetric contour that goes parallel to the planar surface from $\mathbf{r}_1$ to (x_2, y_2, z_1), and then along the normal to $\mathbf{r}_2$. Any other choice, for plane or curved surfaces, satisfies all known restrictions on $\mathbf{p}$. Nevertheless the choice of Irving and Kirkwood is the most natural, and the one generally made. Measurable properties of the system are invariant to the choice of tensor.

4.4 The virial route to the surface tension

The first attempt to express the surface tension in terms of the intermolecular forces was that of Laplace (1.9). His result was based on a static view of matter and restricted to a sharp profile at the gas–liquid surface. We have seen in Chapter 3 how van der Waals and his successors removed both restrictions and how their work has led to a rich vein of quasi-thermodynamic results. In following this line, however, the direct relation of the surface tension to the intermolecular forces has been lost from sight; now is the time to resume the search for this relation, unrestricted by Laplacian simplifications. There was no advance along this line between Laplace's derivation of (1.9) in 1806 and a paper by Fowler[29] in 1937, since Rayleigh's work in this direction was no more than a re-statement of Laplace's result. We return to Fowler's result later in this section but turn first to a more general expression for σ that can be obtained by differentiation of F or Ω, or directly from $\mathbf{p}$.

The surface tension is the derivative of both F and Ω with respect to the area of the liquid surface, (2.7) and (2.16). Both differentiations can be made by the dimensional argument used in the last section to obtain the pressure; this was first done independently by Buff,[30] McLellan,[31] and Harasima[32] for F in the canonical ensemble, and later by Buff[27] for Ω in the grand ensemble. We give only the latter derivation which follows at once[2] from (4.44). The change of coordinates needed to increase the area at fixed volume is the change of x_i and y_i to $(1+\xi)x_i$ and $(1+\xi)y_i$, and z_i to $(1-2\xi)z_i$, where $\xi \ll 1$;

$$\delta u(r_{12}) = \xi\left(x_1\frac{d}{dx_1} + y_1\frac{d}{dy_1} - 2z_1\frac{d}{dz_1} + x_2\frac{d}{dx_2} + y_2\frac{d}{dy_2} - 2z_2\frac{d}{dz_2}\right)u(r_{12})$$

$$= \frac{\xi}{r_{12}}(x_{12}^2 + y_{12}^2 - 2z_{12}^2)\frac{du(r_{12})}{dr_{12}} = \xi\left(r_{12} - \frac{3z_{12}^2}{r_{12}}\right)u'(r_{12}).$$

$$(4.103)$$

Since $\delta A = 2\xi A$, and the 'free' integration over $d\mathbf{r}_1$ can be written as an integration of $A\,dz_1$, we obtain from (4.44)

$$\sigma = \left(\frac{\partial\Omega}{\partial A}\right)_{V,\,T,\,\mu} = \frac{1}{4}\int_{-\infty}^{\infty} dz_1 \int d\mathbf{r}_{12}\left(r_{12} - \frac{3z_{12}^2}{r_{12}}\right)u'(r_{12})\rho^{(2)}(\mathbf{r}_1, \mathbf{r}_2).$$

(pp) (4.104)

Because of the cylindrical symmetry about the z-axis the term in brackets in the integrand of (4.104) can equally well be written as $2(x_{12}^2 - z_{12}^2)/r_{12}$ or $2(y_{12}^2 - z_{12}^2)/r_{12}$. Since u is a function only of r^2 a third form of (4.104) is

$$\sigma = \frac{1}{2}\int_{-\infty}^{\infty} dz_1 \int d\mathbf{r}_{12}\left(x_{12}^2\frac{\partial^2 u_{12}}{\partial z_{12}^2} - z_{12}^2\frac{\partial^2 u_{12}}{\partial x_{12}^2}\right)\rho^{(2)}(\mathbf{r}_1, \mathbf{r}_2).$$

(pp) (4.105)

The same result (4.104) can be obtained, and indeed was so first obtained by Kirkwood and Buff,[25] from the components of the pressure tensor. Substitution of (4.100) into the so-called mechanical definition of σ (2.88) yields

$$\sigma = \frac{1}{4}\int_{-\infty}^{\infty} dz_1 \int d\mathbf{r}_{12}\left(r_{12} - \frac{3z_{12}^2}{r_{12}}\right)u'(r_{12})$$

$$\times \int_0^1 d\alpha\,\rho^{(2)}(r_{12}, z_1 - \alpha z_{12}, z_1 + (1 - \alpha)z_{12}). \quad \text{(pp)} \quad (4.106)$$

Now the effect of integration over α is simply to translate the vector $\mathbf{r}_{12}$ in the z-direction by a distance z_{12}. If this operation is coupled with integration over all z_1 then we can take these integrations before that over $d\mathbf{r}_{12}$ and obtain at once

$$\int_{-\infty}^{\infty} dz_1 \int_0^1 d\alpha\,\rho^{(2)}(r_{12}, z_1 - \alpha z_{12}, z_1 + (1 - \alpha)z_{12})$$

$$= \int_{-\infty}^{\infty} dz_1\rho^{(2)}(r_{12}, z_1, z_2). \quad (4.107)$$

Substitution in (4.106) yields (4.104). Harasima's expression for $p_T(z)$, (4.102), leads also to (4.104). The value of the surface tension of a plane surface, that is, the zeroth moment of the difference $p_N(z) - p_T(z)$, (2.88), is invariant to the choice of contour joining the interacting molecules.[24]

The first moment of the difference $p_N(z) - p_T(z)$ was used by Tolman as a definition of the surface of tension, (2.89), but this integral is dependent on the choice of contour joining the interacting molecules. This was noted first by Harasima,[28] but the implication, that z_s of (2.89) is indeterminate by an amount comparable with the range of $u(r)$, has generally been ignored. If the fluid is confined to a rectangular box and has a plane surface, and if the limits of integration in (2.89) are taken as the walls of the box, then the integral is invariant to choice of contour (as long as this lies wholly within the fluid), but the integrand is now non-zero near the top and bottom walls as well as near the interface, and so does not measure the height of the surface of tension.[24] Similarly the force and torque measured in the thought-experiment described in § 2.2 are the zeroth and first moments of $p_N(z) - p_T(z)$ from wall-to-wall, and so are not purely properties of the liquid–gas surface. The complications due to the walls can be avoided in the measurement of σ by means of the auxiliary pistons used in § 2.2, but cannot be avoided in attempts to define z_s without ambiguity. The indeterminacy of z_s of (2.89) leads to difficulties when we consider the spherical surface, as we shall see in §§ 4.8 and 5.7.

The virial expression for the surface tension (4.104)–(4.106) has been extended in various ways. For a c-component mixture of spherical molecules, we sum the right-hand side of (4.104) over the $\frac{1}{2}c(c+1)$ pairs of interactions.[1] Carey, Scriven, and Davis[33] have extended the expressions for the pressure tensor to mixtures. If $\mathcal{U}_N$ contains multi-body potentials then σ can be written as a sum of virial terms,[34,35] each of the same general form as (4.104). If $\mathcal{U}_N$ is pair-wise additive, but if the pair potentials depend also on the orientations ω_i and ω_j of the ij-pair, then σ comprises two terms σ_r and σ_ω which can be written as integrals over the angle-averages of the derivatives $(\partial u/\partial r)$ and $(\partial u/\partial \theta)$ respectively, where θ is the angle between $\mathbf{r}_{12}$ and the z-axis.[36]

These virial equations relate σ to the pair (or higher) intermolecular potential and to the corresponding distribution functions that describe the density and correlation of density in the surface region. The first YBG equation (4.85) relates the density $\rho(z_1)$ to $\rho^{(2)}(r_{12}, z_1, z_2)$ and the pair potential, and the integral equation (4.86) relates $\rho'(z_1)$ and $\rho'(z_2)$ to the direct correlation function $c(r_{12}, z_1, z_2)$. What we lack, however, is a practicable and exact route to any of these densities or distribution functions from a knowledge of $u(r)$ alone. Further progress requires approximations for ρ, $\rho^{(2)}$ or c, as we shall see in the following chapters. Here we consider only the simplest approximation, namely that $\rho(z)$ is a constant ρ^l on the liquid side of a dividing surface, and $\rho^g \approx 0$ on the other side;

$$\rho(z) = \rho^l \quad (z \geq 0), \qquad \rho(z) = 0 \quad (z < 0). \tag{4.108}$$

We can then write

$$\rho^{(2)}(r_{12}, z_1, z_2) = \rho(z_1)\rho(z_2)g(r_{12}) \tag{4.109}$$

where $g(r_{12})$ is the radial distribution function of the homogeneous liquid. Substitution into (4.104) and a change to cylindrical coordinates (4.82) gives

$$\sigma = \tfrac{1}{2}\pi(\rho^{l})^{2} \int_{0}^{\infty} dr_{12} \int\!\!\int_{-\infty}^{\infty} dz_{1}\, dz_{2}(r_{12}^{2}-3z_{12}^{2})u'(r_{12})g(r_{12}). \quad \text{(pp)} \quad (4.110)$$

In view of the condition on the density (4.108) we can express the integration over z_{1} and z_{2}, at fixed r_{12}, as

$$\int_{0}^{r_{12}} dz_{1} \int_{-z_{1}}^{r_{12}} dz_{12} + \int_{r_{12}}^{\infty} dz_{1} \int_{-r_{12}}^{r_{12}} dz_{12}. \qquad (4.111)$$

The second integral vanishes; that is, the whole contribution to σ comes from the molecules within the range of $u(r_{12})$ of the sharp surface. So σ reduces to the single integral

$$\sigma = \frac{\pi}{8}(\rho^{l})^{2} \int_{0}^{\infty} dr_{12}r_{12}^{4}u'(r_{12})g(r_{12}). \qquad (4.112)$$

This was the expression for σ obtained by a different route, by Fowler[29] in 1937. By making the drastic Laplacian approximation of a sharp liquid surface he replaced the unknown function $\rho^{(2)}(r_{12}, z_{1}, z_{2})$ in the interface by the simpler function $(\rho^{l})^{2}g(r_{12})$ in the homogeneous liquid. The latter can now be determined by X-ray or neutron diffraction or by accurate and well-tested (but not exact) theoretical methods.[37]

If we go further and assume an extreme form of the mean-field approximation, namely that the liquid is homogeneous even on the scale of length of $u(r)$, then we can put $g(r) = 1$ in (4.112). An integration by parts then returns us to Laplace's equation for σ, (1.9),

$$\sigma = -\tfrac{1}{8}(\rho^{l})^{2} \int r_{12}u(r_{12})\, d\mathbf{r}_{12}. \qquad \text{(pp)} \quad (4.113)$$

Another crude approximation for $g(r)$, but one which is more realistic than that of Laplace and which is exact at low densities, is $g(r) = \exp(-u(r)/kT)$. If we take for $u(r)$ a Lennard–Jones $(n, \tfrac{1}{2}n)$ potential $(n > 8)$, with a minimum of $-\varepsilon$ at $r = r_{m}$, then (4.112) can be integrated to give

$$\sigma = \tfrac{1}{2}\pi(\rho^{l})^{2}kTr_{m}^{4}(\varepsilon/kT)^{4/n}I_{4/n}^{n}(\varepsilon/kT) \qquad \text{(pp)} \quad (4.114)$$

where $I_{p}^{n}(x)$ is a solution of the confluent hypergeometric equation.[38] Its asymptotic form at low temperature yields

$$\sigma = \frac{1}{n}(\rho^{l})^{2}\varepsilon r_{m}^{4}\left(\frac{\pi\varepsilon}{kT}\right)^{\frac{3}{2}}e^{\varepsilon/kT}. \qquad \text{(pp)} \quad (4.115)$$

Thus the surface tension rises rapidly as the temperature falls, a more realistic result than that of the Laplace approximation (4.113).

The equation equivalent to (4.112) for δ, the separation of the equimolar surface and the surface of tension defined via (2.89) and the pressure tensor of Irving and Kirkwood, is[25]

$$\delta = z_e - z_s = \frac{4}{15} \frac{\displaystyle\int_0^\infty r^5 u'(r) g(r)\,dr}{\displaystyle\int_0^\infty r^4 u'(r) g(r)\,dr}. \qquad \text{(pp)} \quad (4.116)$$

With the same drastic approximation as above (and $n > 10$)

$$\delta = \tfrac{1}{3} r_m \left(\frac{\varepsilon}{kT}\right)^{1/n} \frac{I_{5/n}^n(\varepsilon/kT)}{I_{4/n}^n(\varepsilon/kT)} \simeq \tfrac{1}{3} r_m \quad \text{as} \quad T \to 0. \quad \text{(pp)} \quad (4.117)$$

That is, the surface of tension (so defined) lies about one-third of a molecular diameter on the liquid side of the equimolar dividing surface, and their limiting separation is, in this approximation, independent of the index of the Lennard–Jones potential. Harasima's pressure tensor gives a value of δ smaller by about a factor of $\tfrac{3}{4}$.[24]

4.5 Functionals of the distribution functions

The classical mechanics of the first three sections of Chapter 1, the classical thermodynamics of Chapter 2, and the statistical mechanics of the first four sections of this chapter all have one feature in common—the discussion is strictly confined to equilibrium states of matter that can, at least in principle, be studied in the laboratory. The first breach in this position of self-restraint came when J. Thomson (1871) and van der Waals (1873) suggested that Andrews's experiments on the 'continuity of state' made it reasonable to discuss the properties of fluids whose densities were between those of the orthobaric gas and liquid. These states played a key role in the development of the theories of surface tension of Rayleigh (1892) and, more important, of van der Waals (1893). The later quasi-thermodynamic work of Chapter 3 shows how fruitful it has been to postulate the existence of thermodynamic functions such as the free energy for values of their arguments other than those that describe the state of equilibrium.

Similar generalizations came more slowly in equilibrium statistical mechanics, but have proved equally fruitful. In this section we introduce functionals of arbitrary distribution functions which have the two properties, first, that they are at extrema when the distribution functions are those of the equilibrium state, and secondly, that these extremal values are the equilibrium values of a thermodynamic potential such as F or Ω. There is a clear analogy between these functionals and the thermodynamic functions of arbitrary argument of Chapter 3. Such functionals were apparently first used by Lee and Yang[39] and by Green;[40] their application to the statistical mechanics of non-uniform systems in a way that is useful for the purposes of this and later Chapters was made first by Morita and Hiroike,[41] De Dominicis,[42] Stillinger and Buff,[43] and Lebowitz and Percus.[44] All but the last of these papers use graphical expansions in powers of the density, and so their results are not immediately applicable to the liquid state. The treatment below derives more closely from that of Mermin,[45] as developed by Ebner and Saam[46,47] and reviewed by Evans.[2] All these authors have been concerned, in one paper or another, with distinctively quantal systems (the electron gas, liquid helium, or liquid metals) and, possibly for this reason, have chosen to work with a set of arbitrary but complete distribution functions $\hat{f}^{(N)}$, where the 'hat' shows that f does not necessarily have its equilibrium form. Here we are concerned only with classical systems in which the kinetic part of the Hamiltonian is separable from the potential part (4.3). In this section we shall not need to discuss states in which the velocities depart from the equilibrium or Maxwellian distribution and so can confine the arbitrariness of $\hat{f}^{(N)}$ to its configurational part $\hat{\rho}^{(N)}$.

Let us therefore consider in the canonical ensemble a distribution function $\hat{\rho}^{(N, N)}$ which is normalized (cf. 4.2)

$$\int \hat{\rho}^{(N, N)}(\mathbf{r}^N) \, d\mathbf{r}^N = N! \tag{4.118}$$

but which is otherwise almost arbitrary.† We assume for the moment that $\mathcal{V} = 0$ and define the functional $F[\hat{\rho}^{(N, N)}(\mathbf{r}^N)]$, or more shortly $F[\hat{\rho}]$, by

$$F[\hat{\rho}] = \frac{1}{N!} \int \hat{\rho}^{(N, N)}(\mathcal{U}_N + kT \ln(\Lambda^{3N} \hat{\rho}^{(N, N)})) \, d\mathbf{r}^N \tag{4.119}$$

The import of this functional is seen by considering two simple special cases. The first is $\mathcal{U}_N = 0$, when $\hat{\rho}^{(N, N)}$ can be factorized into the product of singlet densities:

$$\hat{\rho}^{(N, N)}(\mathbf{r}^N) = (N!/N^N) \prod_{i=1}^{N} \hat{\rho}(\mathbf{r}_i) \tag{4.120}$$

†We restrict the class of $\hat{\rho}^{(N,N)}(\mathbf{r}^N)$, which we abbreviate $\hat{\rho}$, as follows. From any equilibrium singlet density $\rho(\mathbf{r})$ that corresponds to a fixed external field $v(\mathbf{r})$, we can generate an arbitrary set of $\hat{\rho}(\mathbf{r})$, each of which is the equilibrium density that corresponds to a different external field, say $\hat{v}(\mathbf{r})$, but which is a non-equilibrium distribution with respect to the original field $v(\mathbf{r})$. To each $\hat{\rho}(\mathbf{r}) = \rho(\mathbf{r})[\hat{v}(\mathbf{r})]$ there corresponds a unique[2] N-body function $\hat{\rho} \equiv \hat{\rho}^{(N, N)}(\mathbf{r}^N) = \rho^{(N, N)}(\mathbf{r}^N)[\hat{v}(\mathbf{r})]$. Hence, when we write $F[\hat{\rho}]$, and later $\Omega[\hat{\rho}]$, we are implying a restriction on $\hat{\rho}$ that means we can equally take F and Ω to be functionals of the singlet density, $F[\hat{\rho}(\mathbf{r})]$ and $\Omega[\hat{\rho}(\mathbf{r})]$, and we shall exploit this property later, e.g. in (4.130) below. A fortiori, $F[\hat{\rho} = \rho]$ or $F[\rho]$ is a functional of the singlet density $\rho(\mathbf{r})$.

and the normalization (4.118) becomes

$$\int \rho(\mathbf{r}_i)\, d\mathbf{r}_i = N. \tag{4.121}$$

This result requires that

$$\int \prod_i \hat{\rho}(\mathbf{r}_i) \ln\left(\prod_i \hat{\rho}(\mathbf{r}_i)\right) d\mathbf{r}^N = N^N \int \hat{\rho}(\mathbf{r}) \ln \hat{\rho}(\mathbf{r})\, d\mathbf{r} \tag{4.122}$$

so that for the perfect gas (but with $\hat{\rho}(\mathbf{r})$ not necessarily equal to its uniform equilibrium value ρ) we have

$$F[\hat{\rho}] = kT \int \hat{\rho}(\mathbf{r})(\ln(\Lambda^3 \hat{\rho}(\mathbf{r})) - 1)\, d\mathbf{r}. \tag{4.123}$$

The second special case of (4.119) is that when $\hat{\rho}^{(N,\,N)}$ has its equilibrium value $\rho^{(N,\,N)}$, cf. (4.61),

$$\rho^{(N,N)} = \Lambda^{-3N} Z_N^{-1} \exp(-\mathcal{U}_N/kT). \tag{4.124}$$

Substitution in (4.119) gives

$$F[\rho] = -kT \ln Z_N \tag{4.125}$$

where Z_N is the partition function and $F[\rho]$ the free energy of a system with $\mathcal{V} = 0$.

We now define a second functional by

$$\Omega[\hat{\rho}; \mathcal{V}] = F[\hat{\rho}] + \int \hat{\rho}(\mathbf{r})(v(\mathbf{r}) - kT \ln \lambda)\, d\mathbf{r} \tag{4.126}$$

$$= F[\hat{\rho}] - N\mu + \int v(\mathbf{r})\hat{\rho}(\mathbf{r})\, d\mathbf{r}. \tag{4.127}$$

It follows at once that

$$\Omega[\rho; \mathcal{V}] = F - N\mu = \Omega \tag{4.128}$$

where

$$F = F[\rho] + \int v(\mathbf{r})\rho(\mathbf{r})\, d\mathbf{r} \tag{4.129}$$

is the total free energy, that is, the sum of the 'intrinsic' free energy $F[\rho]$ and a term that comes from the external field $\mathcal{V}$. Thus replacing the arbitrary distribution function by the equilibrium one reduces both $F[\hat{\rho}]$ and $\Omega[\hat{\rho}]$ to the equilibrium or thermodynamic functions F and Ω of § 4.2. Moreover it can readily be shown that $\Omega[\hat{\rho} = \rho; \mathcal{V}]$ is a minimum of the functional; the proof[2] is an example of the Gibbs inequality.[48] Hence, differentiating at fixed $v(\mathbf{r})$, we have

$$(\delta\Omega[\hat{\rho}; \mathcal{V}]/\delta\hat{\rho}(\mathbf{r}))_\rho = 0 \tag{4.130}$$

or, from (4.126)

$$(\delta F[\hat{\rho}]/\delta\hat{\rho}(\mathbf{r}))_\rho + v(\mathbf{r}) - kT \ln \lambda = 0. \qquad (4.131)$$

The functional derivative of the intrinsic free energy $F[\rho]$ is the intrinsic chemical potential. It is related to the total chemical potential $\mu = kT \ln \lambda$ by (4.131) which can therefore be written

$$\mu = \mu_{int}(\mathbf{r}) + v(\mathbf{r}) \qquad (4.132)$$

an equation which has already been used in § 4.2. It is related also to the singlet direct correlation function $c^{(1)}(\mathbf{r})$ of (4.30) by

$$(\delta F[\hat{\rho}]/\delta\hat{\rho}(\mathbf{r}))_\rho = \mu_{int}(\mathbf{r}) = kT \ln(\Lambda^3\rho(\mathbf{r})) - kTc^{(1)}(\mathbf{r}). \qquad (4.133)$$

By similar arguments,[2] and use of either (4.42) or (4.48), we have also

$$\left(\frac{\delta^2 F[\hat{\rho}]}{\delta\hat{\rho}(\mathbf{r}_1)\,\delta\hat{\rho}(\mathbf{r}_2)}\right)_\rho = -\left(\frac{\delta v(\mathbf{r}_1)}{\delta\rho(\mathbf{r}_2)}\right)$$
$$= \frac{kT\delta(\mathbf{r}_1 - \mathbf{r}_2)}{\rho(\mathbf{r}_1)} - kTc(\mathbf{r}_1, \mathbf{r}_2). \qquad (4.134)$$

Thus in a perfect gas, by differentiation of (4.123),

$$(\delta F[\hat{\rho}]/\delta\hat{\rho}(\mathbf{r}))_\rho = kT \ln(\Lambda^3\rho(\mathbf{r})) \quad \text{or} \quad c^{(1)}(\mathbf{r}) = 0. \qquad (4.135)$$

In general $c^{(1)}(\mathbf{r})$ and so $\mu_{int}(\mathbf{r})$ are not solely functions of the state of the system at $\mathbf{r}$ but depend also on the states of neighbouring parts; see (4.70).

We now introduce a density $\hat{\rho}(\mathbf{r}; \alpha)$ which changes linearly with α from zero to $\hat{\rho}(\mathbf{r})$ as α goes from zero to unity,

$$\hat{\rho}(\mathbf{r}; \alpha) = \alpha\hat{\rho}(\mathbf{r}). \qquad (4.136)$$

We can integrate (4.133) along this path to give

$$F[\hat{\rho}] = kT \int_0^1 d\alpha \int d\mathbf{r}\, \hat{\rho}(\mathbf{r})[\ln(\Lambda^3\hat{\rho}(\mathbf{r})) - 1 - c^{(1)}(\mathbf{r}; \alpha\hat{\rho})] \qquad (4.137)$$

which is checked most easily by differentiation to regain (4.133), noting that

$$\int \frac{\delta c^{(1)}(\mathbf{r}_1; \alpha\hat{\rho})}{\delta\hat{\rho}(\mathbf{r}_2)} \hat{\rho}(\mathbf{r}_2)\, d\mathbf{r}_2 = \alpha \frac{d}{d\alpha} c^{(1)}(\mathbf{r}_1; \alpha\hat{\rho}). \qquad (4.138)$$

Since $c^{(1)}(\mathbf{r}_1)$ is itself the functional integral of the pair correlation function $c(\mathbf{r}_1, \mathbf{r}_2)$ with respect to $\hat{\rho}(\mathbf{r}_2)$, (4.48), we can integrate (4.137) again to obtain

$$F[\rho] = kT \int d\mathbf{r}\rho(\mathbf{r})(\ln(\Lambda^3\rho(\mathbf{r})) - 1)$$

$$- kT \int_0^1 d\alpha \int_0^\alpha d\beta \iint d\mathbf{r}_1 \, d\mathbf{r}_2 \rho(\mathbf{r}_1)\rho(\mathbf{r}_2)c(\mathbf{r}_1, \mathbf{r}_2; \beta\rho). \quad (4.139)$$

The integration is over a triangular area on the α, β-plane and we can change the order of integration to give

$$\int_0^1 d\alpha \int_0^\alpha d\beta \, c(\beta\rho) = \int_0^1 d\beta \int_\beta^1 d\alpha \, c(\beta\rho) = \int_0^1 d\beta(1 - \beta)c(\beta\rho).$$

$$(4.140)$$

So, changing the symbol β back to α and adding the external potential we have for the total free energy

$$F = kT \int \rho(\mathbf{r}_1) \, d\mathbf{r}_1 \left[\ln(\Lambda^3\rho(\mathbf{r}_1)) - 1 + v(\mathbf{r}_1)/kT \right.$$

$$\left. - \int_0^1 (1 - \alpha) \, d\alpha \int d\mathbf{r}_2\rho(\mathbf{r}_2)c(\mathbf{r}_1, \mathbf{r}_2; \alpha\rho) \right]. \quad (4.141)$$

The grand potential is $F - NkT \ln \lambda$ and, after some manipulation, can be written

$$\Omega = kT \int \rho(\mathbf{r}_1) \, d\mathbf{r}_1 \left[\int_0^1 \alpha \, d\alpha \int d\mathbf{r}_2\rho(\mathbf{r}_2)c(\mathbf{r}_1, \mathbf{r}_2; \alpha\rho) - 1 \right]. \quad (4.142)$$

The functional derivative $\delta v(\mathbf{r}_1)/\delta\rho(\mathbf{r}_2)$ of (4.42) can also be integrated along a path similar to (4.136) from a uniform density ρ_u to the density $\rho(\mathbf{r})$ in the presence of the field $v(\mathbf{r})$, to give[46,47]

$$kT \ln\left(\frac{\rho(\mathbf{r}_1)}{\rho_u}\right) + v(\mathbf{r}_1) = kT \int_0^1 d\alpha \int d\mathbf{r}_2 c(\mathbf{r}_1, \mathbf{r}_2; \alpha\rho)(\rho(\mathbf{r}_2) - \rho_u). \quad (4.143)$$

Finally, from (4.126) and (4.45), we have

$$\delta F[\rho]/\delta u(r_{12}) = \tfrac{1}{2}\rho^{(2)}(\mathbf{r}_1, \mathbf{r}_2) \qquad \text{(pp)} \quad (4.144)$$

and this equation can be integrated from a reference potential $u_0(r_{12})$ to the potential of interest $u(r_{12})$ along a path of α from zero to unity;

$$u(r_{12}; \alpha) = \alpha u(r_{12}) + (1 - \alpha)u_0(r_{12}). \qquad (4.145)$$

The result is

$$F[\rho] = F_0[\rho] + \frac{1}{2}\int_0^1 d\alpha \iint d\mathbf{r}_1 \, d\mathbf{r}_2[u(r_{12}) - u_0(r_{12})]\rho^{(2)}(\mathbf{r}_1, \mathbf{r}_2; \alpha)$$

$$\text{(pp)} \quad (4.146)$$

where $\rho^{(2)}(\mathbf{r}_1, \mathbf{r}_2; \alpha)$ is the pair distribution function when the potential is $u(r_{12}; \alpha)$ but the density $\rho(\mathbf{r})$. Similarly $F_0[\rho]$ is the 'free energy' for a system of potential $u_0(r_{12})$ but density $\rho(\mathbf{r})$.

The first two of these results (4.141) and (4.142) are of little use unless we know or can approximate the function $c(\mathbf{r}_1, \mathbf{r}_2; \alpha\rho)$ for all values of α. They were obtained first by Stillinger and Buff[43] by a cluster expansion and by Lebowitz and Percus[44] by using functional integration, but we owe to Saam and Ebner[47] the comment that since $F[\hat{\rho}]$ is a unique functional of $\hat{\rho}(\mathbf{r})$ then the values of F and Ω calculated in this way are independent of the path in $\hat{\rho}$-space (4.136). The third result (4.146), although restricted to pair potentials, is a useful starting point for the development of perturbation theories of both bulk liquids and of the gas–liquid surface.

4.6 The surface tension from the direct correlation function

We now turn to the justification and improvement, by the methods of statistical mechanics, of the quasi-thermodynamic calculations of Chapter 3, where the surface tension was calculated from the gradient of the density along the normal to the surface. If this gradient approaches zero then the density of any extensive thermodynamic function at a point $\mathbf{r}$ approaches that found in a uniform fluid of density $\rho(\mathbf{r})$ and of the same temperature and chemical potential. As in the early part of Chapter 3, we ignore for the moment the problem that such a fluid may have no stable existence, but may be metastable or even unstable with respect to its separation into two or more phases. If the gradient is everywhere small but not zero then we can expand a generalized density, such as the functional $\psi[\hat{\rho}(\mathbf{r})]$, in the gradient and higher derivatives of the number density $\hat{\rho}(\mathbf{r})$. Such a series may not converge but, because of the short range of the intermolecular potential, we assume that it has at least an asymptotic validity. It converges only if all moments of $c(r_{12})$ exist, as for example for a fluid (not at its critical point) in which the pair potential decays exponentially (or faster) with distance.[49] It diverges for potentials that decay as r^{-6}.

Let $\psi[\hat{\rho}(\mathbf{r})]$ denote the local density of the functional $F[\hat{\rho}]$ of (4.119); that is,[2,50]

$$F[\hat{\rho}] = \int d\mathbf{r} \, \psi[\hat{\rho}(\mathbf{r})] \tag{4.147}$$

and

$$\psi[\hat{\rho}(\mathbf{r})] = \psi^{(0)}(\hat{\rho}(\mathbf{r})) + \sum_\alpha \psi_\alpha^{(1)}(\hat{\rho}(\mathbf{r}))\nabla_\alpha\hat{\rho}(\mathbf{r})$$

$$+ \sum_\alpha \sum_\beta [\psi_{\alpha\beta}^{(2a)}(\hat{\rho}(\mathbf{r}))\nabla_\alpha\hat{\rho}(\mathbf{r})\nabla_\beta\hat{\rho}(\mathbf{r}) + \psi_{\alpha\beta}^{(2b)}(\hat{\rho}(\mathbf{r}))\nabla_\alpha\nabla_\beta\hat{\rho}(\mathbf{r})]$$

$$(\alpha, \beta = x, y, z) \tag{4.148}$$

where $\psi^{(0)}$, $\psi^{(1)}$ etc. are functions (not functionals) of $\hat{\rho}(\mathbf{r})$ which are still to be determined. Since $\psi[\hat{\rho}(\mathbf{r})]$ must be rotationally invariant the first-order terms vanish and the second-order can be simplified;[2]

$$F[\hat{\rho}] = \int d\mathbf{r}[\psi^{(0)}(\hat{\rho}(\mathbf{r})) + \psi^{(2)}(\hat{\rho}(\mathbf{r}))\,|\nabla\hat{\rho}(\mathbf{r})|^2 + O(\nabla^4)]. \qquad (4.149)$$

The chemical potential of the equilibrium distribution $\rho(\mathbf{r})$ follows from (4.131),

$$\mu[\rho(\mathbf{r})] = v(\mathbf{r}) + \mu^{(0)}(\rho(\mathbf{r})) - (d\psi^{(2)}(\rho(\mathbf{r}))/d\rho(\mathbf{r}))\,|\nabla\rho(\mathbf{r})|^2$$
$$- 2\psi^{(2)}(\rho(\mathbf{r}))\nabla^2\rho(\mathbf{r}) + \dots. \qquad (4.150)$$

where $\mu^{(0)} = d\psi^{(0)}(\rho(\mathbf{r}))/d\rho(\mathbf{r})$ is the chemical potential of a uniform fluid of density $\rho(\mathbf{r})$ in the absence of any external potential. The chemical potential $\mu[\rho(\mathbf{r})] = \mu$ is constant throughout the system, and so (4.150) is, in principle, an equation which could be solved for the equilibrium density $\rho(\mathbf{r})$.

A second expansion of $F[\hat{\rho}]$ can be made if $\hat{\rho}(\mathbf{r})$ is everywhere close to an average uniform value ρ_u;

$$\Delta\hat{\rho}(\mathbf{r}) \equiv \hat{\rho}(\mathbf{r}) - \rho_u, \qquad \Delta\hat{\rho}(\mathbf{r}) \ll \rho_u, \qquad (4.151)$$

$$F[\hat{\rho}] = F[\rho_u] + \int \left(\frac{\delta F[\hat{\rho}]}{\delta\hat{\rho}(\mathbf{r}_1)}\right)_{\rho_u} \Delta\hat{\rho}(\mathbf{r}_1)\,d\mathbf{r}_1$$
$$+ \frac{1}{2}\iint \left(\frac{\delta^2 F[\hat{\rho}]}{\delta\hat{\rho}(\mathbf{r}_1)\delta\hat{\rho}(\mathbf{r}_2)}\right)_{\rho_u} \Delta\hat{\rho}(\mathbf{r}_1)\Delta\hat{\rho}(\mathbf{r}_2)\,d\mathbf{r}_1\,d\mathbf{r}_2. \qquad (4.152)$$

The functional derivatives are those of (4.133) and (4.134); so

$$F[\hat{\rho}] = F[\rho_u] + \int \mu(\rho_u)\Delta\hat{\rho}(\mathbf{r}_1)\,d\mathbf{r}_1$$
$$+ \tfrac{1}{2}kT \iint (\rho_u^{-1}\delta(\mathbf{r}_{12}) - c(r_{12}, \rho_u))\Delta\hat{\rho}(\mathbf{r}_1)\Delta\hat{\rho}(\mathbf{r}_2)\,d\mathbf{r}_1\,d\mathbf{r}_2. \qquad (4.153)$$

Another route to this equation was obtained by Green[51] who observed that the quadratic form of the distribution of density fluctuations in the last term may be approximated by a generalized Gaussian. A theorem[52] on multivariate normal distributions then leads to (4.153).

The direct correlation function of a uniform fluid is of short range and so we can expand $\Delta\hat{\rho}(\mathbf{r}_2)$ about $\Delta\hat{\rho}(\mathbf{r}_1)$ in a Taylor series:

$$\Delta\hat{\rho}(\mathbf{r}_2) = \Delta\hat{\rho}(\mathbf{r}_1) + (\mathbf{r}_{12}\cdot\nabla)\hat{\rho}(\mathbf{r}_1) + \tfrac{1}{2}(\mathbf{r}_{12}\cdot\nabla)^2\hat{\rho}(\mathbf{r}_1) + \dots \qquad (4.154)$$

where it is understood in the last term that $\mathbf{r}_{12}\cdot\nabla$ operates twice on $\hat{\rho}(\mathbf{r}_1)$.

Equation (4.153) can be written

$$F[\hat{\rho}] = F[\rho_u] + \int d\mathbf{r}_1 \mu(\rho_u) \Delta\hat{\rho}(\mathbf{r}_1)$$

$$+ \tfrac{1}{2}kT \int d\mathbf{r}_1 \Delta\hat{\rho}(\mathbf{r}_1)^2 \left[\rho_u^{-1} - \int d\mathbf{r}_{12} c(r_{12}; \rho_u) \right]$$

$$- \tfrac{1}{4}kT \int d\mathbf{r}_1 \Delta\hat{\rho}(\mathbf{r}_1) \int d\mathbf{r}_2 c(r_{12}; \rho_u)(\mathbf{r}_{12} \cdot \boldsymbol{\nabla})^2 \Delta\hat{\rho}(\mathbf{r}_1). \quad (4.155)$$

The term of first-order in $(\mathbf{r}_{12} \cdot \boldsymbol{\nabla})$ has been omitted since it vanishes on integration over $d\mathbf{r}_1$. The Cartesian terms in $(\mathbf{r}_{12} \cdot \boldsymbol{\nabla})^2$ of odd order, such as $(\partial^2/\partial x_1 \, \partial y_1)$, also vanish. The remaining terms can be integrated by parts,

$$\int \Delta\hat{\rho}(\mathbf{r}_1) \frac{\partial^2 \Delta\hat{\rho}(\mathbf{r}_1)}{\partial x_1^2} \, dx_1 = - \int \left(\frac{\partial \Delta\hat{\rho}(\mathbf{r}_1)}{\partial x_1} \right)^2 dx_1 \quad (4.156)$$

and we note that

$$\int x_{12}^2 c(r_{12}; \rho_u) \, d\mathbf{r}_{12} = \frac{1}{3} \int r_{12}^2 c(r_{12}; \rho_u) \, d\mathbf{r}_{12}. \quad (4.157)$$

Hence

$$F[\hat{\rho}] = F[\rho_u] + \int d\mathbf{r}_1 \mu(\rho_u) \Delta\hat{\rho}(\mathbf{r}_1)$$

$$+ \tfrac{1}{2}kT \int d\mathbf{r}_1 \Delta\hat{\rho}(\mathbf{r}_1)^2 \left(\rho_u^{-1} - \int d\mathbf{r}_{12} c(r_{12}; \rho_u) \right)$$

$$+ \tfrac{1}{12}kT \int d\mathbf{r}_1 |\boldsymbol{\nabla}\Delta\rho(\mathbf{r}_1)|^2 \int d\mathbf{r}_{12} r_{12}^2 c(r_{12}; \rho_u). \quad (4.158)$$

We compare this with the expansion of (4.149) about ρ_u;

$$F[\hat{\rho}] = \int d\mathbf{r} \, [\psi^{(0)}(\rho_u) + (d\psi^{(0)}(\rho_u)/d\rho_u)/d\rho_u)\Delta\hat{\rho}(\mathbf{r})$$

$$+ \tfrac{1}{2}(d^2\psi^{(0)}(\rho_u)/d\rho_u^2)\Delta\hat{\rho}(\mathbf{r})^2 + \ldots + \psi^{(2)}(\rho_u) \, |\boldsymbol{\nabla}\Delta\hat{\rho}(\mathbf{r})|^2 + \ldots +]$$

$$(4.159)$$

with the consequent results

$$d\psi^{(0)}(\rho_u)/d\rho_u = \mu(\rho_u),$$

$$d^2\psi^{(0)}(\rho_u)/d\rho_u^2 = \rho_u^{-2}\kappa^{-1} = kT\left(\rho_u^{-1} - \int c(r_{12}; \rho_u) \, d\mathbf{r}_{12} \right),$$

$$\psi^{(2)}(\rho_u) = \tfrac{1}{12}kT \int r_{12}^2 c(r_{12}; \rho_u) \, d\mathbf{r}_{12} \quad (4.160)$$

where κ is the isothermal compressibility. The first two equations have been obtained already; the first is the thermodynamic identity (2.95) and the second the compressibility equation of state (4.27). The third equation[53] is the result we need, for it is the coefficient of the square of the gradient which plays so important a role in the quasi-thermodynamic theory of van der Waals described in the last chapter. We can therefore write (4.159) as

$$F[\hat{\rho}] = \int d\mathbf{r}[\psi(\rho_u) + \mu(\rho_u)\Delta\rho(\mathbf{r}) + \tfrac{1}{2}\rho_u^{-2}\kappa^{-1}\Delta\rho(\mathbf{r})^2 + \ldots$$
$$+ \tfrac{1}{2}m(\rho_u)\,|\boldsymbol{\nabla}\rho(\mathbf{r})|^2 + \ldots]$$
$$= \int d\mathbf{r}[\psi^{(0)}(\rho(\mathbf{r})) + \tfrac{1}{2}m(\rho_u)\,|\boldsymbol{\nabla}\rho(\mathbf{r})|^2 + \ldots] \qquad (4.161)$$

where $\psi^0(\rho(\mathbf{r}))$ is the free-energy density of a homogeneous fluid of density $\rho(\mathbf{r})$ and where

$$m(\rho) = \tfrac{1}{6}kT \int r^2 c(r;\rho)\,d\mathbf{r}. \qquad (4.162)$$

Equation (3.11) for the surface tension follows from (4.162) by the arguments already given in Chapter 3. That is,

$$\sigma = \tfrac{1}{6}kT \int_{-\infty}^{\infty} dz_1\rho'(z_1)^2 \int d\mathbf{r}_2 r_{12}^2 c(r_{12};\rho(z_1)) \qquad (4.163)$$

where $c(r_{12})$ is an isotropic function of r_{12}, and is the correlation function appropriate to a uniform fluid of density $\rho(z_1)$. The expression for $m(\rho)$, (4.162), can be compared with that of Rayleigh and van der Waals,

$$m = -\frac{1}{6} \int_{r>d} r^2 u(r)\,d\mathbf{r} \qquad (4.164)$$

where the integral is understood to be taken only over the negative or attractive part of the intermolecular potential $r > d$. The similarity of (4.162) and (4.164) is obvious, and becomes more striking if we introduce the Percus–Yevick approximation[55]

$$c(r) = g(r)[1 - e^{u(r)/kT}]. \qquad (4.165)$$

On linearizing this equation and on introducing the mean-field approximation $g(r) = 1$ we obtain the mean-spherical approximation,

$$kTc(r) \approx -u(r), \qquad (4.166)$$

a substitution which converts (4.162) into (4.164). Hence we expect $m(\rho)$ of (4.162) to be similar in size to m of (4.164), but note however that it is not a constant but a function of both temperature and density. This difference is conceptually crucial in the critical region (§ 9.2) and numerically important in the calculation of surface tension at lower temperatures (§§ 3.2, 5.6, and 7.4).

The results above, (4.161) and (4.162), are the best justification that can be given for quasi-thermodynamic theories, but their derivation reveals some of the fundamental weaknesses in the whole structure discussed in § 3.5. We have had to pass over in silence any further discussion of the meaning to be attached to $\psi(\rho_u)$, when $\rho^g < \rho_u < \rho^l$, and of the convergence of the expansions of $F[\hat{\rho}]$ if the higher moments of $c(r_{12})$ do not exist. It is therefore extremely valuable that there are other lines of argument which lead to an equation for σ that is similar to that of the van der Waals theory but which is free from these approximations and uncertainties. This equation was first obtained by Yvon and reported by him in a paper given at a meeting in Brussels in 1948, but not formally published. It was derived independently by Triezenberg and Zwanzig[56] in 1972 from a calculation of the change of the grand potential arising from an increase in surface area caused by a fluctuation in density, and by Lovett and others[54] in 1973 from the pressure difference across a curved surface. The last derivation follows most directly from the equations so far obtained.

Consider a flat gas–liquid surface lying in the x, y-plane and take the origin of the coordinate system to lie on a convenient dividing surface, e.g. the equimolar surface, $\Gamma_\rho = 0$. Let $v(\mathbf{r})$ be an external potential which deforms the flat surface into a spherical one with radius of curvature R (Fig. 4.2). The potential can be arbitrarily weak since we shall be interested only in the limiting behaviour $R^{-1} \to 0$. If $\Delta\rho(\mathbf{r})$ is the change of density on applying the potential

$$\Delta\rho(\mathbf{r}) = \rho(\mathbf{r}; R) - \rho(\mathbf{r}; R = \infty) \tag{4.167}$$

then geometrical considerations and a Taylor expansion give

$$\Delta\rho(\mathbf{r}) = \tfrac{1}{2}(x^2 + y^2)R^{-1}\rho'(z) + O(R^{-2}). \tag{4.168}$$

The potential needed to bring about this change in density can be found from the second Yvon equation (4.42)

$$v(\mathbf{r}_1) = \int \left(\frac{\delta v(\mathbf{r}_1)}{\delta\rho(\mathbf{r}_2)}\right)_{R=\infty} \Delta\rho(\mathbf{r}_2)\, d\mathbf{r}_2 \tag{4.169}$$

$$= \tfrac{1}{2}kT \int \left(c(\mathbf{r}_1, \mathbf{r}_2) - \frac{\delta(\mathbf{r}_{12})}{\rho(\mathbf{r}_1)}\right) \frac{x_2^2 + y_2^2}{R}\, \rho'(z_2)\, d\mathbf{r}_2. \tag{4.170}$$

The difference of pressure between two points on the z-axis is found by integrating the force $v'(z)$ along this path. Since $v(z)$ vanishes except

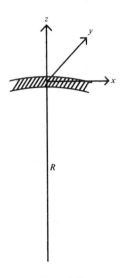

FIG. 4.2

where $\rho'(z)$ is non-zero we can write this difference[54]

$$p^1 - p^g = -\int_{-\infty}^{\infty} \rho(z_1)v'(z_1)\,dz_1 = \int_{-\infty}^{\infty} v(z_1)\rho'(z_1)\,dz_1. \quad (4.171)$$

The integral is found by substitution of v from (4.170); the delta-function contributes nothing since it is zero when x_2 or y_2 is non-zero. Hence

$$p^1 - p^g = \tfrac{1}{2}kT\int_{-\infty}^{\infty} dz_1\rho'(z_1)\int d\mathbf{r}_2\rho'(z_2)\left(\frac{x_2^2 + y_2^2}{R}\right)c(\mathbf{r}_1, \mathbf{r}_2). \quad (4.172)$$

From thermodynamic arguments, § 2.1, the pressure difference is $2\sigma/R$ hence

$$\sigma = \tfrac{1}{4}kT\int_{-\infty}^{\infty} dz_1\rho'(z_1)\int d\mathbf{r}_2\rho'(z_2)(x_2^2 + y_2^2)c(\mathbf{r}_1, \mathbf{r}_2). \quad (4.173)$$

This is the exact result to which (3.11) and (4.163) are approximations. Like most exact results it has its disadvantages; first, we rarely know anything about the direct correlation function in the surface layer $c(\mathbf{r}_1, \mathbf{r}_2)$, and secondly, the equation introduces the gradient of the profile $\rho'(z)$ but provides no recipe for determining it. We note however that $\rho'(z)$ and $c(\mathbf{r}_1, \mathbf{r}_2)$ are also linked by the integral equation (4.52), with $v(\mathbf{r}) = 0$. The primitive van der Waals theory overcomes both problems; the coefficient m is a constant determined solely by $u(r)$, and the integral over $\rho'(z)$ can be transformed into an integral over the free-energy density, (3.13). The more correct expression for m, (4.162), introduces only the reasonably accessible isotropic function $c(r; \rho)$, not the inaccessible $c(\mathbf{r}_1, \mathbf{r}_2)$.

If we assume that $c(\mathbf{r}_1, \mathbf{r}_2)$ is, indeed, isotropic then (4.173) simplifies to

$$\sigma = \tfrac{1}{4}kT \int_{-\infty}^{\infty} dz_1 \rho'(z_1) \int d\mathbf{r}_2 \rho'(z_2)(r_{12}^2 - z_{12}^2)c(\mathbf{r}_{12}; \rho(z_1)) \quad (4.174)$$

where c is the isotropic correlation function appropriate to a uniform fluid of density $\rho(z_1)$. If $d\mathbf{r}_2$ is expressed in cylindrical coordinates as $2\pi r_{12} dr_{12} dz_{12}$, and if $\rho'(z_2)$ is expanded about $\rho'(z_1)$ as a Taylor series, we obtain[57]

$$\sigma = \tfrac{1}{2}\pi kT \int_{-\infty}^{\infty} dz_1 \int_{0}^{\infty} dr_{12} r_{12} c(r_{12}; \rho(z_1)) \int_{-r_{12}}^{r_{12}} dz_{12}(r_{12}^2 - z_{12}^2)$$

$$\times [\rho'(z_1)^2 + z_{12}\rho'(z_1)\rho''(z_1) + \ldots]$$

$$= \frac{2\pi}{3} kT \int_{-\infty}^{\infty} dz_1 \rho'(z_1)^2 \int_{0}^{\infty} dr_{12} r_{12}^4 c(r_{12}; \rho(z_1))$$

$$+ \text{higher order terms,} \quad (4.175)$$

which is (4.163). As before, the introduction of a gradient expansion turns a 'non-local' expression (4.173) into a 'local' expression, (4.163) or (4.175).

We now have two apparently exact expressions for the surface tension; (4.104) and (4.173). The first is restricted, in its usual form, to fluids composed of molecules which interact only with pair potentials; the second is free of this restriction. Their forms are reminiscent of the two equations of state for a homogeneous fluid, the pressure equation (4.92) and the compressibility equation (4.27), but there seems to be no real link to them. The equivalence of (4.92) and (4.27) for a system of pair potentials was demonstrated nearly thirty years ago,[58] but it is only recently that (4.104) and (4.173) have been shown to be equivalent. The proof is the subject of the next section.

4.7 Equivalence of the two expressions for the surface tension

It has not been easy to establish that the routes to σ through the virial of the pair potential and through the direct correlation function lead to the same result; that is, that σ of (4.104), and its trivial transformation (4.105), is the same as σ of (4.173). They are the same in the critical region,[56] and in the low density limit[59] when

$$\rho^{(2)}(\mathbf{r}_1, \mathbf{r}_2) = \rho(\mathbf{r}_1)\rho(\mathbf{r}_2)\exp(-u(r_{12})/kT). \qquad \text{(pp)} \quad (4.176)$$

We shall see in § 5.6 a proof of their equivalence for a particular model. The first published attempt at a general proof was that of Jhon, Desai, and Dahler[60] which led to an inequality; σ from (4.104) is equal to or larger than σ from (4.173). Their methods were taken further by Schofield[61] who first established the equality of these two results for a system of pair potentials, and whose treatment we follow here. The extension to multibody potentials was made independently by Schofield[62] and by Grant and Desai,[35] and the extension to molecular fluids by the latter.[63]

Their methods borrow results from the theory of the dynamical properties of inhomogeneous fluids. It should be possible to reformulate the proof without this use of time-dependent functions but since this has not yet been done we must first make a digression to collect the auxiliary results we need. Naturally these results involve only ensemble averages that are stationary with respect to time.

A fluid in which the distribution of velocities is statistically independent but which does not necessarily follow the Maxwell–Boltzmann distribution has a momentum density $\mathbf{J}(\mathbf{r})$ defined by

$$\mathbf{J}(\mathbf{r}) = \sum_{i=1}^{N} \mathbf{p}_i \, \delta(\mathbf{r} - \mathbf{r}_i) \quad \text{with} \quad \hat{\rho}(\mathbf{r}) = \sum_{i=1}^{N} \delta(\mathbf{r} - \mathbf{r}_i). \qquad (4.177)$$

(The microscopic density of (4.177) becomes the smooth function $\hat{\rho}$ of § 4.5 on averaging.) The change of momentum density with time is the gradient of the instantaneous value of the stress tensor. The pressure tensor of § 4.3 is the ensemble average of the negative of this stress tensor in a system at equilibrium in the absence of an external field. Hence, cf. (2.82) and (4.94),

$$\langle \dot{\mathbf{J}}(\mathbf{r}) \rangle = \rho(\mathbf{r})\nabla v(\mathbf{r}) - \nabla \cdot \mathbf{p}(\mathbf{r}) = 0. \qquad (4.178)$$

The starting point of the proof is two sum-rules of Jhon *et al.* These are identities, obtained from different ways of evaluating time derivatives, which are consequences of the stationary values of ensemble averages in a system at equilibrium. Thus if the average of the time derivative of the correlation function $\langle \mathbf{J}(\mathbf{r}_1; t)\mathbf{J}(\mathbf{r}_2; t) \rangle$ is to be zero at all times, then

$$\langle \dot{\mathbf{J}}(\mathbf{r}_1)\dot{\mathbf{J}}(\mathbf{r}_2) \rangle = -\langle \mathbf{J}(\mathbf{r}_1)\ddot{\mathbf{J}}(\mathbf{r}_2) \rangle. \qquad (4.179)$$

The correlation of $\dot{\mathbf{J}}(\mathbf{r}_1)$ and $\hat{\rho}(\mathbf{r}_2)$ can be expressed[64] by means of the Poisson bracket of $\mathbf{J}(\mathbf{r}_1)$ and $\hat{\rho}(\mathbf{r}_2)$. If α denotes one of the Cartesian components, then

$$\langle \dot{j}^{\alpha}(\mathbf{r}_1)\hat{\rho}(\mathbf{r}_2)\rangle = -\langle J^{\alpha}(\mathbf{r}_1)\dot{\hat{\rho}}(\mathbf{r}_2)\rangle \tag{4.180}$$

$$= -kT\left\langle \sum_i \left[\frac{\partial J^{\alpha}(\mathbf{r}_1)}{\partial p_i^{\alpha}}\frac{\partial \hat{\rho}(\mathbf{r}_2)}{\partial r_i^{\alpha}} - \frac{\partial J^{\alpha}(\mathbf{r}_1)}{\partial r_i^{\alpha}}\frac{\partial \hat{\rho}(\mathbf{r}_2)}{\partial p_i^{\alpha}}\right]\right\rangle \tag{4.181}$$

$$= kT\left\langle \sum_i \delta(\mathbf{r}_1-\mathbf{r}_i)\nabla_2^{\alpha}\,\delta(\mathbf{r}_2-\mathbf{r}_i)\right\rangle \tag{4.182}$$

$$= -kT\rho(\mathbf{r}_1)\nabla_1^{\alpha}\,\delta(\mathbf{r}_1-\mathbf{r}_2). \tag{4.183}$$

The last two steps follow from (4.177) and the properties of the delta-function (Appendix 2). Equations (4.179) and (4.183) are the two sum-rules needed below.

We shall see in § 4.9 that a weak external field is needed to stabilize the position and form of the interface in an infinite system. The rate of change of $J^{\alpha}(\mathbf{r})$ is determined, via the equations of motion, by the gradient of this external field $v(\mathbf{r})$ and that of the intermolecular field on a molecule at $\mathbf{r}$, $u(\mathbf{r})$ of (4.63).

$$\dot{j}^{\alpha}(\mathbf{r}_1) = \sum_i \dot{p}_i^{\alpha}\,\delta(\mathbf{r}_1-\mathbf{r}_i) + \sum_i p_i^{\alpha}\frac{\partial}{\partial t}\,\delta(\mathbf{r}_1-\mathbf{r}_i) \tag{4.184}$$

$$= -\sum_i \delta(\mathbf{r}_1-\mathbf{r}_i)\frac{\partial}{\partial r_i^{\alpha}}[u(\mathbf{r}_i)+v(\mathbf{r}_i)] + m^{-1}\sum_i p_i^{\alpha}\mathbf{p}_i\cdot\mathbf{\nabla}_i\,\delta(\mathbf{r}_1-\mathbf{r}_i). \tag{4.185}$$

The first term is

$$-\sum_{i\neq j}\sum \delta(\mathbf{r}_1-\mathbf{r}_i)\frac{r_i^{\alpha}-r_j^{\alpha}}{r_{ij}}\,u'(r_{ij}) - \rho(\mathbf{r}_1)\frac{\partial v(\mathbf{r}_1)}{\partial r_1^{\alpha}}$$

$$= \int \frac{r_2^{\alpha}-r_1^{\alpha}}{r_{12}}\,u'(r_{12})\sum_{i\neq j}\sum \delta(\mathbf{r}_1-\mathbf{r}_i)\,\delta(\mathbf{r}_1+\mathbf{r}_2-\mathbf{r}_j)\,d\mathbf{r}_2 - \rho(\mathbf{r}_1)\frac{\partial v(\mathbf{r}_1)}{\partial r_1^{\alpha}}. \tag{pp} \quad (4.186)$$

The sum of the second term in (4.185) and the first in (4.186) is just the α-component of the gradient of the stress tensor. Hence, cf. (4.94),

$$\dot{j}^{\alpha}(\mathbf{r}) = -(\mathbf{\nabla}\cdot\hat{\mathbf{p}}(\mathbf{r}))^{\alpha} - \hat{\rho}(\mathbf{r})\nabla^{\alpha}v(\mathbf{r}) \tag{4.187}$$

where $\hat{\mathbf{p}}$ is the negative of the stress tensor, or the instantaneous value of the pressure tensor. Since $\langle \dot{j}^{\alpha}(\mathbf{r})\rangle$ is zero this can be re-written

$$\dot{j}^{\alpha}(\mathbf{r}) = -(\mathbf{\nabla}\cdot\delta\hat{\mathbf{p}}(\mathbf{r}))^{\alpha} - \delta\hat{\rho}(\mathbf{r})\nabla^{\alpha}v(\mathbf{r}), \tag{4.188}$$

where

$$\delta\hat{\mathbf{p}}(\mathbf{r}) = \hat{\mathbf{p}}(\mathbf{r}) - \mathbf{p}(\mathbf{r}),$$
$$\delta\hat{\rho}(\mathbf{r}) = \hat{\rho}(\mathbf{r}) - \rho(\mathbf{r}) = \sum_i \delta(\mathbf{r}-\mathbf{r}_i) - \rho(\mathbf{r}). \tag{4.189}$$

The final result we need is an expression for $\langle J^\alpha(\mathbf{r}_1)\ddot{J}^\alpha(\mathbf{r}_2)\rangle$ in terms of derivatives of $u(r_{12})$. From the Poisson bracket

$$\langle J^\alpha(\mathbf{r}_1)\ddot{J}^\alpha(\mathbf{r}_2)\rangle = kT\left\langle \sum_i \sum_\beta \left[\frac{\partial}{\partial p_i^\beta} J^\alpha(\mathbf{r}_1) \frac{\partial}{\partial r_i^\beta} \dot{J}^\alpha(\mathbf{r}_2) - \frac{\partial}{\partial r_i^\beta} J^\alpha(\mathbf{r}_1) \frac{\partial}{\partial p_i^\beta} \dot{J}^\alpha(\mathbf{r}_2)\right]\right\rangle$$
(4.190)

where the sum over β is over the three Cartesian coordinates. The first term vanishes unless $\beta = \alpha$, (4.177), when it becomes, from (4.184),

$$kT\left\langle \sum_j \delta(\mathbf{r}_1 - \mathbf{r}_j)\left[-\sum_i \frac{\partial^2(u+v)_i}{\partial r_i^\alpha \partial r_j^\alpha} \delta(\mathbf{r}_2 - \mathbf{r}_i)\right.\right.$$
$$\left.\left.+\frac{\partial}{\partial r_2^\alpha}\left(\sum_\gamma m^{-1} p_i^\alpha p_j^\gamma \frac{\partial}{\partial r_2^\gamma} + \nabla_j^\alpha(u+v)_i\right)\delta(\mathbf{r}_2 - \mathbf{r}_j)\right]\right\rangle. \quad (4.191)$$

The second term in this expression contains the product $\delta(\mathbf{r}_1 - \mathbf{r}_j)\,\delta(\mathbf{r}_2 - \mathbf{r}_j)$ whose ensemble average is $\rho(\mathbf{r}_1)\,\delta(\mathbf{r}_1 - \mathbf{r}_2)$. Such a local term will be seen later to contribute nothing to the surface tension. The second term in the Poisson bracket (4.190) contains the same factor. Thus

$$\langle J^\alpha(\mathbf{r}_1)\ddot{J}^\alpha(\mathbf{r}_2)\rangle = -kT\left\langle \sum_{i\neq j}\sum \frac{\partial^2(u+v)_i}{\partial r_i^\alpha \partial r_j^\alpha}\delta(\mathbf{r}_1 - \mathbf{r}_j)\delta(\mathbf{r}_2 - \mathbf{r}_i)\right\rangle + \text{local terms}$$
$$= -kT\rho^{(2)}(\mathbf{r}_1, \mathbf{r}_2)\frac{\partial^2 u(\mathbf{r}_{12})}{\partial r_1^\alpha \partial r_2^\alpha} + \text{local terms.} \qquad \text{(pp)} \quad (4.192)$$

Here the ensemble average of these delta functions is a formal definition of the equilibrium pair density $\rho^{(2)}$, cf. (4.177).

With these preliminary results we can now convert the virial expression for σ into that in terms of the direct correlation function. We start with the variant (4.105), use (4.192) and (4.179), and then the identity

$$\partial^2 u_{12}/\partial z_{12}^2 = -\partial^2 u_{12}/\partial z_1 \, \partial z_2 \qquad (4.193)$$

to show that σ can be expressed

$$kT\sigma = -\frac{1}{2}\int_{-\infty}^\infty dz_1 \int d\mathbf{r}_2[x_{12}^2\langle \dot{J}^z(\mathbf{r}_1)\dot{J}^z(\mathbf{r}_2)\rangle - z_{12}^2\langle \dot{J}^x(\mathbf{r}_1)\dot{J}^x(\mathbf{r}_2)\rangle]. \quad (4.194)$$

It is the presence of the factors x_{12}^2 and z_{12}^2 which ensure the absence of any contribution from the configurational local terms; the local kinetic terms cancel. Into (4.194) we substitute for $\dot{J}^x$ and $\dot{J}^z$ from (4.188). The products in (4.194) generate four terms. The first $(\boldsymbol{\nabla}_1 \cdot \delta\hat{\mathbf{p}}(\mathbf{r}_1)\boldsymbol{\nabla}_2 \cdot \delta\hat{\mathbf{p}}(\mathbf{r}_2))$ vanishes on integration by parts since $\hat{p}_{xz}(\mathbf{r}) = \hat{p}_{zx}(\mathbf{r})$. The two cross-terms of the form

$$(\boldsymbol{\nabla}_1\langle\delta\hat{\mathbf{p}}(\mathbf{r}_1)\,\delta\hat{\rho}(\mathbf{r}_2)\rangle\boldsymbol{\nabla}_2 u(r_{12}))^\alpha$$

can be simplified by replacing the gradient of the pressure with j by (4.188) to reduce each of them to the form of the fourth term but with opposite sign, viz.

$$-\nabla_1^\alpha v(\mathbf{r}_1)\nabla_2^\alpha v(\mathbf{r}_2)\langle\delta\rho(\mathbf{r}_1)\delta\rho(\mathbf{r}_2)\rangle$$

where the ensemble average here is the density–density correlation function, that is, the right-hand side of (4.40). We now invoke the cylindrical symmetry of the system and replace the variables $\mathbf{r}_1$ and $\mathbf{r}_2$ by z_1 and z_2 and the plane vector $\mathbf{s}_{12}$, whose components are x_{12} and y_{12}. Equation (4.194) becomes

$$kT\sigma = \frac{1}{4}\int_{-\infty}^{\infty} dz_1\,dz_2 v'(z_1)v'(z_2)\int d\mathbf{s}_{12}s_{12}^2 H(s_{12};z_1,z_2) \quad \text{(pp)} \quad (4.195)$$

where

$$H(s_{12};z_1,z_2) = \rho^{(2)}(\mathbf{s}_{12};z_1,z_2) - \rho(z_1)\rho(z_2) + \rho(z_1)\,\delta\mathbf{s}_{12}\,\delta z_{12}. \quad (4.196)$$

The inverse of this function is, from the second Yvon equation (4.42)

$$C(s_{12};z_1,z_2) = -c(s_{12};z_1,z_2) + [\rho(z_1)]^{-1}\delta\mathbf{s}_{12}\,\delta z_{12} \quad (4.197)$$

where, cf. (4.43), we have, with the abbreviation H_{12} and C_{12} for these functions

$$\int_{-\infty}^{\infty} dz_3\int d\mathbf{s}_3 H_{13}C_{32} = \delta\mathbf{s}_{12}\delta z_{12}, \quad (4.198)$$

and so

$$\int d\mathbf{s}_{12}s_{12}^2\int_{-\infty}^{\infty} dz_3\int d\mathbf{s}_3 H_{13}C_{32} = 0. \quad (4.199)$$

Moreover, from (4.197)

$$H(s_{12};z_1,z_2) = \int_{-\infty}^{\infty} dz_3\,dz_4 \iint d\mathbf{s}_3\,d\mathbf{s}_4 H_{13}C_{34}H_{42}, \quad (4.200)$$

and so, from the last three equations

$$\int d\mathbf{s}_{12}s_{12}^2 H_{12} = -\int_{-\infty}^{\infty} dz_3\,dz_4\int d\mathbf{s}_{13}H_{13}\int d\mathbf{s}_{34}s_{34}^2 C_{34}\int d\mathbf{s}_{42}H_{42}. \quad (4.201)$$

Finally the gradient of the stabilizing potential $v'(\mathbf{r})$ is related to the density gradient by (4.51) which can be rewritten

$$kT\rho'(z_1) = \int_{-\infty}^{\infty} dz_2\int d\mathbf{s}_2 v'(z_2)H_{12}. \quad (4.202)$$

Substitution of (4.201) and (4.202) into (4.195) yields (4.174), which is the result sought. Extensions of the proof to multibody and non-central potentials require further complications but no new point of principle.

4.8 The spherical surface

The equations for the pressure tensor and the surface tension can be cast into forms that are appropriate for a spherical interface by using quasi-thermodynamic (and therefore dubious) arguments.

Consider a drop of phase α, immersed in phase β, and described by polar coordinates r, θ, ϕ. As in § 2.4 the system under discussion is bounded by two concentric spheres, one of radius R^α, which lies in phase α, and one of radius R^β in phase β. An arbitrary dividing surface of radius R separates the two phases, $R^\alpha < R < R^\beta$. As for a planar surface, the symmetry of the system requires that the pressure tensor has only two independent components, normal and tangential;

$$\mathbf{p}(\mathbf{r}) = p_N(r)[\mathbf{e}_r\mathbf{e}_r] + p_T(r)[\mathbf{e}_\theta\mathbf{e}_\theta + \mathbf{e}_\phi\mathbf{e}_\phi] \qquad (4.203)$$

where $\mathbf{e}_r$, $\mathbf{e}_\theta$, and $\mathbf{e}_\phi$ are orthogonal unit vectors, and r is the distance from the centre.

Consider now a flat radial strip, stretching from R^α to R^β, of angular width $d\theta$, and lying in a plane of constant ϕ (Fig. 4.3). The force acting on one side of this strip, and the moment of the force about the centre of the drop, can each be expressed in two ways: first, as an integral over $p_T(r)$, and secondly, as the result of the pressure p^α acting on the part of the strip between R^α and R_s, of p^β on the part between R_s and R^β, and of

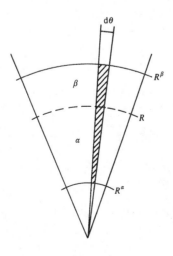

FIG. 4.3

the surface tension $\sigma[R_s] \equiv \sigma_s$ acting at R_s, where R_s is the surface of tension. The two ways of expressing the force and its moment must be equivalent if σ_s and R_s are to carry their usual mechanical interpretations. The two equations that describe the equivalence can be written in a common form, with $n = 1$ for the force, and $n = 2$ for the moment,

$$\int_{R^\alpha}^{R_s} r^n \, dr \, p^\alpha + \int_{R_s}^{R^\beta} r^n \, dr \, p^\beta - R_s^n \sigma_s = \int_{R^\alpha}^{R^\beta} r^n \, dr \, p_T(r) \qquad (n = 1, 2).$$

(4.204)

It is useful to define a pressure $p^{\alpha, \beta}(r; R)$, for any dividing surface R, where, cf. (3.1), $p^{\alpha, \beta}$ is equal to p^α for $r < R$ and p^β for $r > R$. With the aid of this function, (4.204) yields the two solutions

$$\sigma_s = \int_0^\infty \left(\frac{r}{R_s}\right) dr[p^{\alpha, \beta}(r; R_s) - p_T(r)] \qquad (4.205)$$

$$= \int_0^\infty \left(\frac{r}{R_s}\right)^2 dr[p^{\alpha, \beta}(r; R_s) - p_T(r)] \qquad (4.206)$$

or

$$R_s = \frac{\displaystyle\int_0^\infty r^2 \, dr[p^{\alpha, \beta}(r; R_s) - p_T(r)]}{\displaystyle\int_0^\infty r \, dr[p^{\alpha, \beta}(r; R_s) - p_T(r)]}. \qquad (4.207)$$

Further expressions for σ can be obtained by exploiting the more general condition of hydrostatic equilibrium $\nabla \cdot \mathbf{p} = 0$, (2.82) or (4.94). Substitution of (4.203) into this equation gives a result that can be written

$$\frac{d}{dr}[r^n p_N(r)] = r^{n-1}[(n-2)p_N(r) + 2p_T(r)]. \qquad (4.208)$$

This equation, a generalization of three due to Buff,[65] $n = 0, 2, 3$, holds for all values of n. By integration

$$(R^\beta)^n p^\beta - (R^\alpha)^n p^\alpha = \int_{R^\alpha}^{R^\beta} r^{n-1} \, dr[(n-2)p_N(r) + 2p_T(r)]. \qquad (4.209)$$

Similarly by integration of $(nr^{n-1}p^{\alpha,\beta})$ defined with respect to an arbitrary dividing surface, we have

$$(R^\beta)^n p^\beta - (R^\alpha)^n p^\alpha + R^n(p^\alpha - p^\beta) = \int_{R^\alpha}^{R^\beta} nr^{n-1} \, dr \, p^{\alpha,\beta}(r; R). \qquad (4.210)$$

By subtraction of these equations, and use of the generalized Laplace equation (2.57),

$$R^n(p^\alpha - p^\beta) = 2R^{n-1}\sigma[R] + R^n[\mathrm{d}\sigma/\mathrm{d}R]$$

$$= \int_0^\infty r^{n-1}\,\mathrm{d}r\{n[p^{\alpha,\beta}(r; R) - p_N(r)] + 2[p_N(r) - p_T(r)]\}. \tag{4.211}$$

This equation has a 'mechanical' or 'model' solution (in the sense of § 2.2), namely

$$p_N(r) = p^{\alpha,\beta}(r; R), \qquad p_N(r) - p_T(r) = \sigma_s\delta(r - R_s). \tag{4.212}$$

The formal derivative $[\mathrm{d}\sigma/\mathrm{d}R]$ vanishes at R_s, (2.59), so

$$\sigma_s = \frac{1}{2}\int_0^\infty \left(\frac{r}{R_s}\right)^{n-1}\,\mathrm{d}r\{n[p^{\alpha,\beta}(r; R_s) - p_N(r)] + 2[p_N(r) - p_T(r)]\}. \tag{4.213}$$

Equation (4.205) is this equation with $n = 2$; other cases of interest are $n = 0$ and $n = 1$;

$$\sigma_s = \int_0^\infty \left(\frac{R_s}{r}\right)\,\mathrm{d}r[p_N(r) - p_T(r)] \tag{4.214}$$

$$= \frac{1}{2}\int_0^\infty \mathrm{d}r[p^{\alpha,\beta}(r; R_s) - p_T(r)] + \frac{1}{2}\int_0^\infty \mathrm{d}r[p_N(r) - p_T(r)]. \tag{4.215}$$

The general result (4.211) can be written ($n = 2$)

$$p^\alpha - p^\beta = \frac{2}{R^3}\int_0^\infty r^2\,\mathrm{d}r[p^{\alpha,\beta}(r; R) - p_T(r)]$$

$$- \frac{2}{R^3}\int_0^\infty r(r - R)\,\mathrm{d}r[p^{\alpha,\beta}(r; R) - p_T(r)] \tag{4.216}$$

and the two terms identified with $2\sigma/R$ and $[\mathrm{d}\sigma/\mathrm{d}R]$. Differentiation of the expression for $2\sigma/R$ yields the second term since the derivative of $p^{\alpha,\beta}(r; R)$ is $(p^\alpha - p^\beta)\delta(r - R)$. The identification appears to be arbitrary since $(p^\alpha - p^\beta)$ does not determine σ and $[\mathrm{d}\sigma/\mathrm{d}R]$ separately but only the function $[\mathrm{d}(R^2\sigma)/\mathrm{d}R]$, cf. (2.75). It appears, therefore, as if an arbitrary term of the form (a/R^2) could be added to σ and subtracted from $[\mathrm{d}\sigma/\mathrm{d}R]$, but this term must be zero if $\sigma[R]$ is to reduce to (4.206) on putting $R = R_s$, and if $[\mathrm{d}\sigma/\mathrm{d}R]_{R_s}$ is to be zero. Thus

$$\sigma[R] = \int_0^\infty \left(\frac{r}{R}\right)^2\,\mathrm{d}r[p^{\alpha,\beta}(r; R) - p_T(r)]. \tag{4.217}$$

Subtraction of (4.206) from (4.211) with $n = 3$, $R = R_s$, yields the identity,

$$\int_0^\infty r^2 \, dr[p^{\alpha, \beta}(r; R_s) - p_N(r)] = 0 \qquad (4.218)$$

and so we have also[65, 66]

$$\sigma_s = \int_0^\infty \left(\frac{r}{R_s}\right)^2 dr[p_N(r) - p_T(r)], \qquad (4.219)$$

another equation for R_s,

$$R_s^3 = \frac{\int_0^\infty r^2 \, dr[p_N(r) - p_T(r)]}{\int_0^\infty r^{-1} \, dr[p_N(r) - p_T(r)]}, \qquad (4.220)$$

and an alternative form[65] for $d\sigma/dR$,

$$\left[\frac{d\sigma}{dR}\right] = \frac{1}{R^3} \int_0^\infty r^2 \, dr[p^{\alpha, \beta}(r; R) - p_N(r)]. \qquad (4.221)$$

We saw in §2.4 that thermodynamic arguments showed that $\sigma[R]$ has a minimum at $R = R_s$, and that its value is independent of the formal choice of R only to terms of the order of $1/R$. To one lower order, the equations above are all equivalent and can be written

$$\sigma_s = \int_0^\infty dr[p_N(r) - p_T(r)],$$

$$\sigma_s R_s = \int_0^\infty r \, dr[p_N(r) - p_T(r)]. \qquad (4.222)$$

So far the arguments of this section have been mechanical or quasi-thermodynamic in the sense of §2.5. They lead to statistical mechanical results when we insert into these equations one of the expressions of §4.3 for the pressure tensor. Ono and Kondo[66] choose the form due to Irving and Kirkwood, (4.99), and extract $p_T(r)$ from (4.203)

$$p_T(r_1) = kT\rho(r_1) - \frac{1}{2} \int d\mathbf{r}_{12} \int_0^1 d\alpha \frac{(\mathbf{r}_{12} \cdot (\mathbf{e}_\phi)_1)^2}{r_{12}} u'(r_{12})\rho^{(2)}(|\mathbf{r}_1 - \alpha\mathbf{r}_{12}|, \mathbf{r}_{12}).$$

$$\text{(pp)} \quad (4.223)$$

Substitution in (4.217) yields, after some algebra,

$$\sigma[R] = -kT\Gamma(R) + \frac{1}{2} \int_0^\infty \left(\frac{r_1}{R}\right)^2 dr_1 \int d\mathbf{r}_2 \frac{(\mathbf{r}_{12} \cdot (\mathbf{e}_\phi)_1)^2}{r_{12}} u'(r_{12})$$
$$\times [\rho^{(2)}(\mathbf{r}_{12}, r_1) - (\rho^{(2)})^{\alpha, \beta}(\mathbf{r}_{12}, r_1; R)], \quad \text{(pp)} \quad (4.224)$$

where $\Gamma(R)$ is the adsorption with respect to the surface R,

$$\Gamma(R) = \int_0^\infty \left(\frac{r}{R}\right)^2 dr[\rho(r) - \rho^{\alpha,\beta}(r;R)], \qquad (4.225)$$

and where $\rho^{\alpha,\beta}$ and $(\rho^{(2)})^{\alpha,\beta}$ are defined in the same way as $p^{\alpha,\beta}$. The scalar product $\mathbf{r}_{12} \cdot (\mathbf{e}_\phi)_1$ is $r_2 \sin\theta_2 \sin(\phi_2 - \phi_1)$. Although formally valid for arbitrary R, (4.224) is useful only if $R \approx R_s$. An expression for R_s can be obtained by substituting (4.223) into (4.207).

We saw in § 4.4 that the arbitrariness of the pressure tensor at a plane surface did not affect the value of σ but did affect that of z_s, defined by the usual mechanical definition, (2.89). It might therefore be expected that σ_s obtained from any of the above equations, e.g. (4.219) or (4.222), is invariant with choice of pressure tensor, but that R_s is not. All that we can be sure of, however, is that the *planar limit* of (4.219) is invariant, as are any expressions that are equal to $p^\alpha - p^\beta$, such as the sum of the two terms in (4.216), but not the separate terms, and as are expressions derived directly from the hydrostatic condition of equilibrium, such as (4.208).

We now face a difficulty since the thermodynamic argument that leads to (2.74) relates the ratio σ_s/σ_∞ to the separation $\delta = z_e - z_s$, which is the planar limit of $R_e - R_s$. If z_s and R_s are not well-defined then there is an inconsistency in some of the definitions or arguments. One way of analysing the difficulty would be to obtain an expression for σ_s in terms of the direct correlation function; that is, to extend (4.173) to spherical surfaces. We do this below, but find that the extension itself is not without a degree of arbitrariness, at least for a diffuse interface. We shall show, however, in § 5.7 that, for a particular molecular model, the extension proposed below yields values of σ_s and R_s which are consistent with the thermodynamic argument that leads to (2.74). We therefore believe that the surface of tension introduced into the thermodynamic argument of § 2.4 (that is, via Laplace's equation) is well-defined, but it is not the same as the ill-defined quantity given by the mechanical or quasi-thermodynamic arguments that lead to (2.89) and (4.207) etc. There could, of course, be other solutions to the problem—the pressure tensor may have a unique form, notwithstanding the arguments of Harasima and of Schofield and Henderson, or the surface of tension may be well-defined for a curved surface but have no well-defined planar limit, etc.—but we think these solutions less likely.

The second route to the surface tension can be found through the direct correlation function by an adaptation of the argument of Lovett *et al.* used in § 4.6. Consider a portion of a curved interface near the z-axis of a Cartesian coordinate system with its origin at the centre of curvature. As before, apply a weak external potential which increases the radius of curvature of each dividing surface by a small increment ΔR. It is impossible to specify a deformation that bends a thick interface uniformly in such a way that the local environment of each point is unchanged,[67] and so there is an imprecision about the form of the expression for $\Delta\rho(\mathbf{r})$, the consequent change of density. To a good approximation, which is exact for a sharp surface, we may choose the equimolar dividing surface, R_e, as the mean radius about which the surface is bent, so that

$$\Delta\rho(\mathbf{r}) \approx -\frac{1}{2}\left(\frac{x^2+y^2}{r^2}\right)\rho'(r)\Delta R \approx \tfrac{1}{2}(x^2+y^2)\rho'(r)\Delta(R_e^{-1}). \quad (4.226)$$

The potential needed to effect this change is, cf. (4.170),

$$v(\mathbf{r}_1) = \tfrac{1}{2}kT \int \left(c(\mathbf{r}_1, \mathbf{r}_2) - \frac{\delta(\mathbf{r}_{12})}{\rho(\mathbf{r}_1)}\right)(x_2^2 + y_2^2)\rho'(r_2)\Delta(R_e^{-1})\,d\mathbf{r}_2. \quad (4.227)$$

Hence, by a repetition of the previous argument

$$\frac{1}{2}\frac{d(p^\alpha - p^\beta)}{d(R_e^{-1})} = \frac{\pi kT}{8}\int_0^\infty dr_1 r_1^{-3}\rho'(r_1)\int_0^\infty dr_2 r_2\rho'(r_2)$$

$$\times \int_{\|\mathbf{r}_1\|-\|\mathbf{r}_2\|}^\infty dr_{12} r_{12} c(r_{12}, r_1, r_2) f(r_1, r_2, r_{12}) \quad (4.228)$$

where

$$f(r_1, r_2, r_{12}) = (r_1 + r_2 + r_{12})(r_1 + r_2 - r_{12})(r_1 - r_2 + r_{12})(-r_1 + r_2 + r_{12}).$$
$$(4.229)$$

The derivative on the left of (4.228) is the real change of $p^\alpha - p^\beta$ with real change of curvature; it can be expressed

$$\frac{1}{2}\frac{d(p^\alpha - p^\beta)}{d(R_e^{-1})} = \frac{1}{2}\left(\frac{R_e}{R_s}\right)^2\left(\frac{dR_e}{dR_s}\right)\frac{d}{d(R_s^{-1})}\left(\frac{2\sigma_s}{R_s}\right). \quad (4.230)$$

To an adequate approximation R_e and R_s move together as the surface is bent, and so (dR_e/dR_s) is unity. Since we expect from (2.74) that σ_s varies with R_s according to the equation,

$$\sigma_s = \sigma_\infty\left(1 - \frac{2\delta}{R_s}\right), \qquad \delta = R_e - R_s, \quad (4.231)$$

then (4.230) can be written

$$\frac{1}{2}\frac{d(p^\alpha - p^\beta)}{d(R_e^{-1})} = \left(1 + \frac{2\delta}{R_s}\right)\left(\sigma_s - R_s\frac{d\sigma_s}{dR_s}\right) = \sigma_s + O\left(\frac{\delta}{R_s}\right)^2. $$
$$(4.232)$$

Hence σ_s is given by the right-hand side of (4.228). We test this conclusion in § 5.7.

4.9 Density fluctuations and their correlation

Local fluctuations of density occur at all points even in a macroscopically homogeneous fluid, but are sufficiently small in magnitude and in the range of their correlations for them properly to be neglected in many calculations in equilibrium statistical mechanics. Fluids near critical points are obvious exceptions and, as was mentioned in § 3.5, so is the fluid at the gas–liquid surface. We saw there that the van der Waals theory was justifiable only if certain constraints were imposed on the fluctuations at the interface. We now use results established earlier in this chapter to study these fluctuations in more detail.

In § 4.2 we set out the statistical mechanics of three-dimensional systems of arbitrary inhomogeneity in the presence of an arbitrary external field; we now specialize these equations to a system with a square interface in the x,y-plane of area $A = L^2$, whose equimolar dividing surface $\zeta(x, y)$ has its mean height at $z = 0$, and whose mean planarity is maintained by a gravitational potential $v(z) = mgz$, where m is the mass of a molecule. The positions of a pair of points $\mathbf{r}_1$ and $\mathbf{r}_2$ can, as before, be represented by their heights z_1 and z_2 and the vectors $\mathbf{s}_1$ and $\mathbf{s}_2$, which are the projections of $\mathbf{r}_1$ and $\mathbf{r}_2$ onto the x, y-plane.

A simple but natural way of discussing fluctuations of density at the surface is to suppose[68] that there is a spectrum of capillary waves imposed on an intrinsic or *bare* profile (denoted by a subscript b) given by

$$\begin{aligned}
\rho_b(z) &= \rho^l \qquad z < \langle \zeta \rangle = 0 \\
\rho_b(z) &= \rho^g \qquad z > \langle \zeta \rangle = 0
\end{aligned} \qquad (4.233)$$

where ρ^l and ρ^g are the densities of liquid and gas and $\langle \zeta \rangle$ is the mean height of the dividing surface $\zeta(\mathbf{s})$. Capillary waves of all numbers q cause ζ to oscillate in a way that can be represented

$$\zeta(\mathbf{s}) = \sum_q \alpha(q) e^{i\mathbf{q}\cdot\mathbf{s}}, \qquad \alpha(0) = 0. \qquad (4.234)$$

The discussion of the liquid surface in terms of such capillary waves by Mandelstam (1913) and Born and von Karman (1913) is reviewed by Frenkel.[69] Such waves were later used to analyse the surface of liquid helium by Atkins and others,[70] but the development here follows that of Buff, Lovett, and Stillinger[68] and later similar work.[17,71] These treatments are not restricted to a bare profile that is necessarily a step-function, as in (4.233), although that is a convenient simplification to keep in mind. Buff *et al.* calculated the reversible work w needed to create an interface with surface $\zeta(\mathbf{s})$. It comprises two terms—work against the gravitational field, and the work needed to create an interface with the area of the deformed surface and a tension σ_b.

$$w = \iint_A ds \left[\int_0^{\zeta(s)} (\rho^l - \rho^g)mgz \, dz + \sigma_b(1 + \zeta_x^2 + \zeta_y^2)^{\frac{1}{2}} \right] \quad (4.235)$$

where ζ_x is the x-derivative of $\zeta(s)$. The fluctuations are expected to be small in amplitude or of long wavelength so the root can be expanded and terms of order higher than ζ_x^2 and ζ_y^2 discarded. Substitution of (4.234) into (4.235) gives

$$w = \sigma_b A + \frac{1}{2} \sum_{q_1} \sum_{q_2} \alpha(q_1)\alpha(q_2) \int_A ds \exp(i(\mathbf{q}_1 + \mathbf{q}_2) \cdot \mathbf{s})[(\rho^l - \rho^g)mg - \sigma_b \mathbf{q}_1 \cdot \mathbf{q}_2].$$
$$(4.236)$$

With periodic boundary conditions, $\zeta(L, y) = \zeta(0, y)$ and $\zeta(x, L) = \zeta(x, 0)$, the allowed values of q_1 and q_2 are products of $(2\pi/L)$ with all positive, zero, and negative integers. All terms with $\mathbf{q}_2 \neq -\mathbf{q}_1$ vanish on integration over s, and so

$$w = \sigma_b A \left\{ 1 + \frac{1}{2} \sum_q \alpha(q)\alpha(-q)[2a^{-2} + q^2] \right\} \quad (4.237)$$

where the capillary length a is given by (1.24); that is,

$$a^2 = 2\sigma_b/mg(\rho^l - \rho^g). \quad (4.238)$$

We recall that this length is of the order of 1 mm, e.g. 1·41 mm for argon at its triple point if we identify σ_b with the observed surface tension.

The mean square fluctuation $\overline{\zeta^2}$ for a given set of amplitudes is

$$\overline{\zeta^2} = A^{-1} \int \zeta^2(\mathbf{s}) \, d\mathbf{s} = \sum_{q>0} \alpha(q)\alpha(-q). \quad (4.239)$$

The probability of a given amplitude $\alpha(q)$ is proportional to the Boltzmann factor $\exp(-w/kT)$, which has a Gaussian form (4.237). Hence[72] the mean-square amplitude is determined by $(2a^{-2} + q^2)$, or

$$\langle \zeta^2 \rangle = \sum_{q>0} \langle \alpha(q)\alpha(-q) \rangle = (kT/\sigma_b A) \sum_{q>0} (2a^{-2} + q^2)^{-1} \quad (4.240)$$

where the angle brackets denote a canonical average. The sum starts at $(2\pi/L)$ the lowest value of $|q|$, and has an upper limit beyond which it is meaningless to discuss fluctuations in terms of a set of continuous waves. Buff et al. took this limit to be fixed by the mean thickness $\langle \zeta^2 \rangle^{\frac{1}{2}}$; a simpler but similar choice is a length l which is of the order of the diameter of a molecule, and so of the mean molecular separation in a dense liquid.

Thus

$$q_{min} = 2\pi/L \qquad q_{max} = 2\pi/l \qquad (4.241)$$

and the integration (summation) of (4.240) gives

$$\langle \zeta^2 \rangle = \frac{kT}{4\pi\sigma_b} \ln\left[\frac{1 + 2(\pi a/l)^2}{1 + 2(\pi a/L)^2}\right]. \qquad (4.242)$$

Thus if we first fix g and then take the thermodynamic limit of $A = L^2 = \infty$, then (4.242) shows that the thickness of a bare surface with capillary waves restrained by gravity and surface tension is given by

$$\langle \zeta^2 \rangle = \frac{kT}{4\pi\sigma_b} \ln[1 + 2(\pi a/l)^2]. \qquad (4.243)$$

The ratio $(\pi a/l)$ is about 10^7 in the earth's gravitational field and increases as g diminishes, so in the thermodynamic limit the thickness diverges, as g goes to zero, as $(-\ln g)^{\frac{1}{2}}$.

Similarly if we first fix the size of the system, $A = L^2$, and then let g become zero, we find that the thickness is proportional to $(\ln A)^{\frac{1}{2}}$, and so again diverges in the thermodynamic limit. These are the weak divergences whose consequences for the van der Waals theory are discussed below.

The weakness of a divergence as $(\ln A)^{\frac{1}{2}}$ is seen by calculating $\langle \zeta^2 \rangle^{\frac{1}{2}}$ from (4.242) in the limit $g = 0$, with an arbitrary, but reasonable choice of $l = 5$ Å, and with the replacement of σ_b by the observed tension. For argon at its triple point we find

$$\langle \zeta^2 \rangle^{\frac{1}{2}} = 4 \cdot 47 \text{ Å} \quad \text{for} \quad L = 1 \text{ mm} \qquad A = 10^{-6} \text{ m}^2,$$

$$\langle \zeta^2 \rangle^{\frac{1}{2}} = 5 \cdot 43 \text{ Å} \quad \text{for} \quad L = 1 \text{ m} \qquad A = 1 \text{ m}^2.$$

In the earth's gravitational field these thicknesses are 4·46 Å and 4·74 Å, respectively; the latter being indistinguishable from that of the limit $A = \infty$.

A fuller analysis of the correlation of fluctuations was made by Wertheim[17] in terms of the two-dimensional Fourier transform of the direct and total correlation functions;

$$\tilde{c}(q; z_1, z_2) = \int c(s_{12}; z_1, z_2)\exp(i\mathbf{q} \cdot \mathbf{s}_{12}) \, d\mathbf{s}_{12}$$

$$= (2\pi/q) \int c(s_{12}; z_1, z_2)\sin(qs_{12}) \, ds_{12} \qquad (4.244)$$

of which the limit at zero wavenumber is

$$\tilde{c}(z_1, z_2) = 2\pi \int c(s_{12}; z_1, z_2)s_{12} \, ds_{12}. \qquad (4.245)$$

The Fourier transform of the total correlation function $h(s_{12}; z_1, z_2)$ is defined similarly, and in terms of these two functions we may define $C(q; z_1, z_2)$ and $H(q; z_1, z_2)$,

$$C(q; z_1, z_2) = \delta(z_1 - z_2)/\rho(z_1) - \tilde{c}(q; z_1, z_2)$$

$$H(q; z_1, z_2) = \rho(z_1)\delta(z_1 - z_2) + \rho(z_1)\rho(z_2)\tilde{h}(q; z_1, z_2). \tag{4.246}$$

It follows from the two Yvon equations (4.41) and (4.42) that

$$C(z_1, z_2) = -\frac{1}{kT} \int \frac{\delta v(\mathbf{r}_1)}{\delta \rho(\mathbf{r}_2)} \, d\mathbf{s}_{12}$$

$$H(z_1, z_2) = -kT \int \frac{\delta \rho(\mathbf{r}_1)}{\delta v(\mathbf{r}_2)} \, d\mathbf{s}_{12} \tag{4.247}$$

where again $C(z_1, z_2)$ and $H(z_1, z_2)$ denote the limit of $q = 0$. The latter is therefore the integral over the plane of the density–density correlation function, (4.40), used in (4.196)–(4.202);

$$H(z_1, z_2) = 2\pi \int H(s_{12}; z_1, z_2)s_{12} \, ds_{12}. \tag{4.248}$$

Integration of (4.43) over the area, s_3, gives

$$\int C(z_1, z_3)H(z_3, z_2) \, dz_3 = \delta(z_1 - z_2), \tag{4.249}$$

a relation which holds[17] for all q;

$$\int C(q; z_1, z_3)H(q; z_3, z_2) \, dz_3 = \delta(z_1 - z_2). \tag{4.250}$$

From (4.52) with $v(\mathbf{r}) = mgz$,

$$\frac{mg}{kT} + \frac{\rho'(z_1)}{\rho(z_1)} = \int d\mathbf{s}_{12} \int dz_2 c(s_{12}; z_1, z_2)\rho'(z_2) \tag{4.251}$$

or, bringing $\rho'(z_1)$ within the integral, cf. (4.39), and carrying out the integration over $\mathbf{s}_{12}$,

$$mg = -kT \int C(z_1, z_2)\rho'(z_2) \, dz_2. \tag{4.252}$$

Similarly from (4.51)

$$kT\rho'(z_1) = -mg \int H(z_1, z_2) \, dz_2. \tag{4.253}$$

The capillary-wave argument above leads to a divergence of the surface thickness as $(-\ln g)^{\frac{1}{2}}$, and any divergence of $\rho'(z)^{-1}$ will be no stronger. Hence (4.253) implies that the integral of $H(z_1, z_2)$ over z_2 diverges at least as rapidly as g^{-1}. This result puts more precisely the preliminary conclusion drawn in § 4.2 that (4.51) implies that $\rho'(z)$ vanishes in the absence of an external field.

Wertheim analysed (4.252) and (4.253) further by noting that the form of the first, with $g = 0$, is that of a non-negative continuous symmetric matrix $C(z_1, z_2)$, of which $\rho'(z_2)$ is the eigenfunction of zero eigenvalue. Such a matrix can be expanded in a spectral form (or parametric representation[73])

$$C(z_1, z_2) = \sum_q \lambda_q \varepsilon_q^*(z_1) \varepsilon_q(z_2) \qquad (4.254)$$

where λ_q are the eigenvalues and $\varepsilon_q(z)$ an orthonormal set of eigenfunctions. Since $H(z_1, z_2)$ is the inverse of $C(z_1, z_2)$, (4.249), it follows that[73]

$$H(z_1, z_2) = \sum_q \lambda_q^{-1} \varepsilon_q^*(z_1) \varepsilon_q(z_2). \qquad (4.255)$$

In a homogeneous fluid $\varepsilon_q = e^{iqz}$, and λ_q and λ_q^{-1} have, respectively, densities of $[\rho^{-1} - \tilde{c}(q)]$ and $[\rho + \rho^2 \tilde{h}(q)]$. This can be seen by substituting these expressions into (4.254) and (4.255), and integrating over q to give,

$$\begin{aligned} C(z_{12}) &= \delta(z_{12})/\rho - c(z_{12}), \\ H(z_{12}) &= \rho\delta(z_{12}) + \rho^2 h(z_{12}). \end{aligned} \qquad (4.256)$$

Substitution in (4.249) then gives

$$h(z_{12}) = c(z_{12}) + \rho \int c(z_{13}) h(z_{23}) \, dz_3 \qquad (z_{12} \neq 0) \qquad (4.257)$$

which is the one-dimensional analogue of (4.25).

At a liquid surface some of the eigenvalues λ_q go to zero linearly with g, and, according to Wertheim's arguments, none goes faster than this. It follows from (4.255) that $H(z_1, z_2)$ has a strong divergence as g^{-1}. Since we have seen already that its integral over z_2, (4.253), has the same divergence, it further follows that this divergence arises from strong correlations in the x, y-plane. This is the most important conclusion of Wertheim's analysis.

An approximation for $H(z_1, z_2)$ can be obtained by assuming that the sum in (4.255) is dominated by the smallest eigenvalue that goes to zero linearly with g, which we can represent by $\lambda_0 \equiv (mg/kT)n_0$. From (4.253)

$$\rho'(z_1) \approx -n_0^{-1} \varepsilon_0^*(z_1) \int \varepsilon_0(z_2) \, dz_2 \equiv -n_0^{-1} \varepsilon_0^*(z_1) E_0 \qquad (4.258)$$

and on integrating over z_1

$$\rho^l - \rho^g \approx n_0^{-1} |E_0|^2. \tag{4.259}$$

So from (4.258)

$$\rho'(z_1)\rho'(z_2) \approx H(z_1, z_2)(mg/kT)(\rho^l - \rho^g),$$

or

$$H(z_1, z_2) \approx (kT/mg)(\rho^l - \rho^g)^{-1}\rho'(z_1)\rho'(z_2). \tag{4.260}$$

an approximation which shows explicitly the dependence on g^{-1}. If the same analysis is made to the order of q^2 then there is an extra term in σq^2, so that we can write[74]

$$H(q; z_1, z_2) \approx (kT/\sigma)\rho'(z_1)\rho'(z_2)[2a^{-2} + q^2]^{-1} \tag{4.261}$$

where σ is now the surface tension given by (4.163). Since $aq_{min} \approx 1$, it follows that the dominant contribution to $H(q; z_1, z_2)$ comes from the fluctuations of longest wavelength, $q \approx q_{min}$.

Thus if the interface is imagined to be a taut membrane with an average energy kT in each of its modes of transverse vibration, then the mean-square displacement of the interface from its equilibrium position is proportional to the logarithm of its area, $\ln A$, and so diverges in the macroscopic limit $A \to \infty$. When the restoring force in the transverse displacements is not due to surface tension alone, but has a gravitational component as well, the mean-square displacement no longer diverges as $A \to \infty$, but its infinite-A limit now depends on the gravitational acceleration g, and diverges proportionally to $-\ln g$ as $g \to 0$. That the interface is not well localized in the absence of a stabilizing field such as that of gravity is seen also in the two-dimensional Ising model,[75] and in lattice-gas models generally, where it manifests itself in three dimensions as the phenomenon of interface 'roughening'.[76]

Because of these great excursions from its equilibrium position made by an interface of infinite area in the absence of a stabilizing field, the mean density $\rho(z)$ at any height z is just some average of that of the two bulk phases, with no z-dependence, and the interface appears to be infinitely thick. Alternatively, if A or $1/g$ is finite, then so is the apparent thickness of the interface, but that thickness, being now dependent on A or g, is not intrinsic to the phase equilibrium itself. Certainly, that smeared-out profile without a well-defined thickness is not the profile calculated by the van der Waals theory of Chapter 3, or from the approximated solutions of the equations of this chapter that are described in Chapter 7.

It was shown by Weeks[71] that fluctuations in the number of molecules in any vertical column of fluid, as molecules enter or leave it from or to adjacent columns, are responsible for fluctuations in the location of the $\Gamma_\rho = 0$ dividing surface in that column; and that these fluctuations are correlated over large horizontal distances, and thus manifest themselves as the divergences predicted by the capillary-wave theory. But Weeks also observes that if the columns are taken to be of cross-sectional area ξ^2 (where ξ, as in § 3.5, is the range of the correlations), then the mean difference in the heights of the $\Gamma_\rho = 0$ dividing surfaces in contiguous columns is of the order ξ. He takes that mean difference to be an estimate of the fluctuations in the height within a single column, hence of the local or intrinsic width of the interface. That this width is then of order ξ is in agreement with the van der Waals theory. By this view, the density profiles of the van der Waals theory are those of a hypothetical fluid in which fluctuations in density over distances greater than ξ, in directions parallel to the plane of the interface, have been suppressed[71] (cf. § 3.5).

The pessimistic conclusion that in the absence of gravity there is no definable interfacial structure and thickness, unless one limits the area, has been challenged by Evans.[77] He analyses the generalized van der Waals approximation, that is, (4.161)–(4.163), and shows that in the absence of gravity the density profile cannot be located in space but is well-determined, of finite thickness, and found by taking the limit $g \to 0$ of the profile in the presence of gravity. The interface is stable against density fluctuations and, in particular, against capillary-wave fluctuations. He shows, moreover, that these fluctuations are contained within the theory, and that the horizontal density–density correlation contains the same q^{-2} divergence as (4.261), when $g \to 0$. Thus for the van der Waals approximation, Evans shows that divergent long-range horizontal correlations do not imply a divergent thickness. This approximation cannot therefore be used as a model for the bare interface to which capillary waves are added since it already includes such fluctuations. It is, however, still not clear if his conclusions are applicable in general, or at least to a wider range of approximations. We return to some of these points and to Evans's theory in § 9.7.

The internal coherence of the statistical mechanical treatment, and the numerical agreement of the calculated values of σ with those obtained by computer simulation and by experiment (Chapters 5–7), suggest strongly that the intrinsic profiles, correlation functions, and surface tension of the first eight sections of this chapter do not become entirely figments of the imagination as $g \to 0$. Even if the profile is washed out by a $\ln g$ divergence of $\langle \zeta \rangle^2$, integrals over the profile can still have well-defined limits; for example,[2] the divergence merely adds a non-divergent term in $g \ln g$ to the usual expression for the surface tension. Even if Evans's analysis cannot be extended beyond the van der Waals approximation, the following argument might suggest a way of salvaging the g-independent results of this chapter.

The divergence of $\langle \zeta^2 \rangle$ is due to the low-wave-number modes. In two dimensions (one-dimensional interface), at $g = 0$ and $L = \infty$, the contribution to $\langle \zeta^2 \rangle$ made by all capillary waves of wave number greater than any prescribed q is $kT/\pi\sigma q$. In three dimensions, again at $g = 0$ and $L = \infty$, it is $(kT/2\pi\sigma)\ln(q_{max}/q)$, with $q_{max} = 2\pi/l$, from (4.240) and (4.241). In both cases the divergence in $\langle \zeta^2 \rangle$ comes from $q \to 0$. (The elimination of the high-q divergence by imposing a q_{max} is natural and necessary in $d = 3$; it is the low-q divergence which is the essential result of the theory.) In systems of more than three dimensions the divergences are absent, capillary waves can legitimately be ignored and mean-field or van der Waals approximations are essentially correct. That three is the 'borderline' dimensionality is shown by the presence of logarithmic singularities. The simple translation of one phase in the other is a $q = 0$ mode that is in many ways typical of those which cause the divergence of $\langle \zeta^2 \rangle$. For the moment, consider only it, and ignore those capillary waves that change the shape of the interface. At $g = 0$, one phase is spherical and floats in the other. The interfacial profile washes out because of the wandering of the sphere. The $\langle \rho \rangle$ we calculate from equilibrium statistical mechanics is uniform, and is an average of the densities ρ^l and ρ^g of the two phases with weights equal to their volume fractions. But that smearing of the density does not mean there is not an intrinsic profile—there is one, and in the present picture it is a step function—merely that the profile must be defined relative to some instantaneous position of the sphere. We could measure that profile experimentally by correcting for any small effect that the sphere's slow wandering might have on the observations. As the sphere wanders, it continually reconstructs itself, exchanging molecules with the other phase and within itself; all such ways of composing the sphere are included in the statistical-mechanical calculation.

Now, the capillary waves that lead to the divergent $\langle \zeta^2 \rangle$ are of the same character, and the divergence is due to just the same kind of smearing of the density. Such smearing is not inconsistent with an intrinsic profile of finite width, as we saw in the example of the wandering sphere. Thus, we are encouraged to believe that there will someday be a successful definition of an intrinsic profile, though necessarily defined relative to the momentary location of some dividing surface. Matters would be easier if there were a natural, low-q cutoff; but alas, there is none.

4.10 Local thermodynamic functions

We conclude this chapter with a brief assessment of what is meant by a *local thermodynamic function*. These entities were introduced first in §§ 2.3 and 2.5, and have been used repeatedly in the succeeding chapters.

We have for example, assumed throughout this chapter that it is meaning-ful to write $\rho(\mathbf{r})$ for the equilibrium density at a point $\mathbf{r}$. In a canonical ensemble the formal definition of this function follows from (4.177)

$$\rho(\mathbf{r}) = \left\langle \sum_{i=1}^{N} \delta(\mathbf{r}-\mathbf{r}_i) \right\rangle \qquad (4.262)$$

since this is the limit of the ratio $\langle \delta N \rangle / \delta V$, as δV tends to zero, where δN is the number of molecules with centres in a volume δV that contains the point $\mathbf{r}$. We saw in the last section that there is doubt about the existence of a value for such a function that is independent of both A and g, but if we suppress these doubts, then $\rho(\mathbf{r})$ is a well-defined local thermodynamic function. We can define similarly local densities of momentum and kinetic energy. In a classical system the latter leads to an unambiguous definition of a local temperature $T(\mathbf{r})$

$$3mkT(\mathbf{r})\rho(\mathbf{r}) = \left\langle \sum_{i=1}^{N} p_i^2 \delta(\mathbf{r}-\mathbf{r}_i) \right\rangle \qquad (4.263)$$

where $\mathbf{p}_i$ is the linear momentum of molecule i.

All these functions pertain to properties of single molecules. Difficul-ties arise, however, with the definition of local functions that are averages over the properties of pairs or larger groups of molecules. The simplest such function is, perhaps, the energy density $\phi(\mathbf{r})$, defined as the limit of $\langle \delta U \rangle / \delta V$, but the value of this limit depends on how δU is calculated. Molecules within δV interact with those outside it, and it is an arbitrary decision on how much of that energy of interaction should be included in δU. Thus, in a system of pair intermolecular potentials, half the energy $u(r_{ij})$ could be assigned to point $\mathbf{r}_i$ and half to $\mathbf{r}_j$; this is the definition of Rao and Levesque and of Ladd and Woodcock.[78] It would be equally satisfactory, and no less arbitrary, to assign the whole of $u(r_{ij})$ to the mid-point of the vector $\mathbf{r}_i - \mathbf{r}_j$; the two definitions would lead to different functions $\phi(\mathbf{r})$ in a system, such as a gas–liquid surface, with in-homogeneities on the scale of the range of $u(r)$. To be satisfactory, any definition must lead to invariant values for any measurable property of the system. The most obvious condition for $\phi(\mathbf{r})$ is that of the constancy of the integral

$$\int_V \phi(\mathbf{r}) \, d\mathbf{r} = U \qquad (4.264)$$

but, for a system with a planar gas–liquid interface and an equimolar dividing surface at $z = z_e$, it must also lead to an invariant value of the surface energy $U^s(z_e)$ of § 2.3,

$$U^s(z_e) = \int_{-\infty}^{z_e} [\phi(z) - \phi^l] \, dz + \int_{z_e}^{\infty} [\phi(z) - \phi^g] \, dz. \qquad (4.265)$$

The local entropy density $\eta(\mathbf{r})$ and free-energy density $\psi(\mathbf{r})$ have similar ambiguities. The free-energy density of Chapter 3 and § 4.6 is not an exact expression for a system with arbitrary inhomogeneities but a limiting form to which we expect all consistent definitions of ψ to reduce when the scale length of the inhomogeneity becomes long compared with the range of the intermolecular potential.

The local pressure tensor $\mathbf{p}(\mathbf{r})$ suffers from ambiguities similar to those of the energy density since it can be expressed in terms of the virial of the intermolecular potential (§ 4.3), but again there are the constraints that any consistent definition must lead to invariant values for observable quantities. In a fluid with a gradient only in the z-direction mechanical stability requires that $p_{\mathrm{N}}(z)$ is a constant, and equal to p^{l} and p^{g}. The zeroth-moment of the difference $p_{\mathrm{N}}(z) - p_{\mathrm{T}}(z)$, which is σ, is likewise invariant, but we have seen that the first moment is apparently not, and have discussed in § 4.8 the difficulties that follow if this moment is identified with the planar limit of the surface of tension that enters into the thermodynamic discussion of the spherical drop.

With the chemical potential or activity, $\lambda(\mathbf{r})$, we are again on firmer ground. There have been several proposed definitions of this function,[79] some of which depend on simplified views of the molecular structure in an interface.[80] All these may usefully be supplanted, however, by either of those based on the potential distribution theorem of § 4.2. This definition has the two advantages; first, that $\lambda(\mathbf{r})$ so defined is demonstrably constant in an inhomogeneous system at equilibrium, and second, that $\lambda(\mathbf{r})$ then admits of operationally unambiguous and practicable rules for its measurement. Examples of this second virtue will be given in Chapter 5.

These observations can be summarized, as follows.

(1) Local thermodynamic functions can be defined unambiguously for 'single-molecule' properties, e.g. $\rho(\mathbf{r})$, $T(\mathbf{r})$.

(2) They can be defined locally and unambiguously for functions which are, in fact, constant, throughout a system at equilibrium, e.g. $\lambda(\mathbf{r})$, and any component of $\mathbf{p}(\mathbf{r})$ whose constancy is required for mechanical equilibrium.

(3) Other local functions such as $\phi(\mathbf{r})$ and $\eta(\mathbf{r})$, and therefore $\psi(\mathbf{r})$, and the (other) components of $\mathbf{p}(\mathbf{r})$ cannot be defined unambiguously, but are required to satisfy certain constraints, typically the invariance of some of the lower moments of the function.

Notes and references

1. Ono, S. and Kondo, S. *Encyclopedia of Physics* (ed. S. Flügge) Vol. 10, p. 134. Springer, Berlin (1960). The review of M. S. Jhon, J. S. Dahler, and R.

C. Desai (*Adv. chem. Phys.* **46,** 279 (1981)) is concerned primarily with the dynamics of the interface.

2. Evans, R. *Adv. Phys.* **28,** 143 (1979). A later review by J. K. Percus, to be published in *Studies on Statistical Mechanics* (ed. E. W. Montroll and J. L. Lebowitz), North-Holland, Amsterdam, covers much of the same material.

3. Distribution functions were used by Maxwell and Boltzmann for discussion of the transport properties of gases. They were first used extensively in the equilibrium theory of dense fluids by Ornstein, L. S. and Zernike, F. *Proc. Roy. Acad. Sci., Amsterdam* **17,** 793 (1914), reprinted in *Equilibrium Theory of Classical Fluids* (ed. H. L. Frisch and J. L. Lebowitz) p. III-3, Benjamin, New York (1964).

4. See, for example, Münster, A. *Statistical Thermodynamics,* Vol. 1, Chapter 4. Springer, Berlin (1969).

5. Introduced by Gibbs, J. W. in *Elementary Principles in Statistical Mechanics,* Chapter 15, University Press, Yale (1902); reprinted by Dover Publications in 1960.

6. Ornstein and Zernike, ref. 3.

7. See, for example, Münster, ref. 4, Vol. 1, Chapter 5, and Ramshaw, J. D. *Mol. Phys.* **41,** 219 (1980).

8. This result is apparently due to N. Wiener, see Mayer, J. E. and Montroll, E. W. *J. chem. Phys.* **9,** 2 (1941).

9. Yvon, J. *Le Théorie Statistique des Fluides et l'Equation d'Etat, Act. Sci. et Indust.* No. 203, Hermann, Paris, (1935); Born, M. and Green, H. S. *Proc. Roy. Soc. A.* **188,** 10 (1946), reprinted in Born M. and Green, H. S. *A General Kinetic Theory of Liquids,* Cambridge University Press (1949). The equation was obtained independently also by Bogoliubov, N. N. *J. Phys. URSS* **10,** 256, 265 (1946); these papers in English are excerpts from a treatise in Russian of which an English translation appears in *Studies in Statistical Mechanics* (ed. J. de Boer and G. E. Uhlenbeck) Vol. 1, Part A, North-Holland, Amsterdam (1962). An equivalent but not identical hierarchy was obtained by Kirkwood, J. G. *J. chem. Phys.* **3,** 330 (1935). The abbreviation BBGKY is sometimes used for these equations, but we prefer the shorter form, YBG.

10. Tarazona, P. and Evans, R. *Chem. Phys. Lett.* **97,** 279 (1983).

11. Integral equations for the distribution functions of homogeneous fluids are reviewed in detail by Watts, R. O. *Statistical Mechanics* (ed. K. Singer) Vol. 1, Chapter 1, Specialist Periodical Rep., Chem. Soc., London (1973), and in most recent monographs on liquids.

12. Functional derivatives and integrals were introduced by V. Volterra, whose account of them is still one of the most readable, *Theory of Functionals* (trans. M. Long), Blackie, London (1929); a later edition was published by Dover Publications in 1959. According to Bogoliubov (ref. 9), who used this technique, the first application in statistical mechanics was that of M. A. Leontovich in 1935. For brief accounts of the use of functional differentiation in the theory of liquids, see Hansen, J. P. and McDonald, I. R. *Theory of Simple Liquids,* p. 68, Academic Press, London (1976) and Münster, ref. 4 Vol. 1, Appendix 7.

13. Yvon, J. *Nuovo Cimento (Suppl.)* **9,** 144 (1958).

14. Percus, J. K. in Frisch and Lebowitz, ref. 3, p. II-33.

15. Triezenberg, D. G. Thesis, University of Maryland, 1973.

16. Lovett, R. A., Mou, C. Y. and Buff, F. P. *J. chem. Phys.* **65,** 570 (1976).
17. Wertheim, M. S. *J. chem. Phys.* **65,** 2377 (1976).
18. Kayser, R. F. and Raveché, H. J. *Phys. Rev.* B **22,** 424 (1980).
19. The second form of this theorem, (4.69), was obtained by Widom, B. *J. chem. Phys.* **39,** 2808 (1963), and Jackson, J. L. and Klein, L. S. *Phys. Fluids* **7,** 228 (1964). A special case, restricted to a system of hard spheres, and in a form appropriate for Monte Carlo simulation, was first put forward by Byckling, E. *Physica* **27,** 1030 (1961). More extensive computer calculations for homogeneous fluids have been made by Adams, D. J. *Mol. Phys.* **28,** 1241 (1974); Romano, S. and Singer, K. *Mol. Phys.* **37,** 1765 (1979); and Powles, J. G. *Mol. Phys.* **41,** 715 (1980). The first proof of the constancy of λ of (4.69) in an inhomogeneous system is in Widom, B. *J. stat. Phys.* **19,** 563 (1978). The first form of the theorem, (4.68), was obtained more recently by de Oliveira, M. J., personal communication (1979); Robledo, A. and Varea, C. *J. stat. Phys.*, **26,** 513 (1981) have derived the second form by a variational principle from the grand potential of a non-uniform fluid as a functional of the density.
20. Snider, N. S. *J. chem. Phys.* **55,** 1481 (1971).
21. Hoover, W. G. and Poirier, J. C. *J. chem. Phys.* **37,** 1041 (1962).
22. Bogoliubov, ref. 9; Green, H. S. *Proc. Roy. Soc.* A. **189,** 103 (1947).
23. Clausius, R. *Pogg. Ann. Phys.* **141,** 124 (1870); English translation in *Phil. Mag.* **40,** 122 (1870), reprinted in Brush, S. G. *Kinetic Theory*, Vol. 1, p. 172, Pergamon Press, Oxford (1965). For the derivation of (4.92) from the virial theorem, see e.g. Münster, ref. 4, p. 324.
24. Schofield, P. and Henderson, J. R. *Proc. Roy. Soc.* A **379,** 231 (1982).
25. Kirkwood, J. G. and Buff, F. P. *J. chem. Phys.* **17,** 338 (1949).
26. Irving, J. H. and Kirkwood, J. G. *J. chem. Phys.* **18,** 817 (1950) (Appendix); see also Ono and Kondo, ref. 1.
27. Buff, F. P. *J. chem. Phys.* **23,** 419 (1955).
28. Harasima, A. *Adv. chem. Phys.* **1,** 203 (1958), and Ono and Kondo, ref. 1.
29. Fowler, R. H. *Proc. Roy. Soc.* A **159,** 229 (1937).
30. Buff, F. P. *Zeit. Elektrochem.* **56,** 311 (1952).
31. McLellan, A. G. *Proc. Roy. Soc.* A. **213,** 274 (1952), **217,** 92 (1953).
32. Harasima, A. *J. phys. Soc. Japan* **8,** 343 (1953).
33. Carey, B. S., Scriven, L. E. and Davis, H. T. *J. chem. Phys.* **69,** 5040 (1978).
34. Toxvaerd, S. *Prog. Surf. Sci.* **3,** 189 (1973); *Statistical Mechanics* (ed. K. Singer) Vol. 2, Chapter 4, Specialist Periodical Rep., Chemical Society, London (1975); Present, R. D., Shih, C. C. and Uang, Y.-H. *Phys. Rev.* A **14,** 863 (1976).
35. Grant, M. and Desai, R. C. *J. chem. Phys.* **72,** 1482 (1980).
36. Gray, C. G. and Gubbins, K. E. *Mol. Phys.* **30,** 179 (1975); Davis, H. T. *J. chem. Phys.* **62,** 3412 (1975).
37. See, for example, Münster, ref. 4, Hansen and McDonald, ref. 12 or Barker, J. A. and Henderson, D. *Rev. mod. Phys.* **48,** 587 (1976).
38. Rowlinson, J. S. *Mol. Phys.* **6,** 75 (1963) (Appendix).
39. Lee, T. D. and Yang, C. N. *Phys. Rev.* **117,** 22 (1960).
40. Green, M. S. *J. chem. Phys.* **33,** 1403 (1960).
41. Morita, T. and Hiroike, K. *Prog. theor. Phys.* **25,** 537 (1961).
42. De Dominicis, C. *J. math. Phys.* **3,** 983 (1962).
43. Stillinger, F. H. and Buff, F. P. *J. chem. Phys.* **37,** 1 (1962).
44. Lebowitz, J. L. and Percus, J. K. *J. math. Phys.* **4,** 116 (1963).
45. Mermin, N. D. *Phys. Rev.* A **137,** 1441 (1965); see also Hohenberg, P. and Kohn, W. *Phys. Rev.* B **136,** 864 (1964).

46. Ebner, C. and Saam, W. F. *Phys. Rev.* B **12**, 923 (1975); Ebner, C., Saam, W. F. and Stroud, D. *Phys. Rev.* A **14**, 2264 (1976).
47. Saam, W. F. and Ebner, C. *Phys. Rev.* A **15**, 2566 (1977).
48. Gibbs, ref. 5, Chapter 11; for a more recent account see Hansen and McDonald, ref. 12, p. 149.
49. Evans, R. personal communication (1981).
50. Yang, A. J. M., Fleming, P. D. and Gibbs, J. H. *J. chem. Phys.* **64**, 3732 (1976); **67**, 74 (1977).
51. Green, M. S. *J. math. Phys.* **9**, 875 (1968).
52. Cramér, H. *Mathematical Methods of Statistics*, p. 310 *et seq.* University Press, Princeton (1946).
53. First obtained explicitly in this form by Yang, Fleming, and Gibbs, ref. 50, and by Bongiorno, V., Scriven, L. E. and Davis, H. T. *J. Coll. Interf. Sci.* **57**, 462 (1976). Ornstein and Zernike, ref. 3, had earlier found that the second moment of $c(r)$ enters into related equations for density fluctuations and the compressibility near the critical point. This work was quoted as a justification of (4.162) by Lovett, *et al.* ref. 54. For related equations in terms of the second moment of the total correlation function $h(r_{12}; \rho_u)$ see Lebowitz and Percus, ref. 44, and B. Widom, in *Phase Transitions and Critical Phenomena*, (ed. C. Domb and M. S. Green) Vol. 2, Chapter 3, Academic Press, London (1972).
54. Lovett, R., DeHaven, P. W., Vieceli, J. J. and Buff, F. P. *J. chem. Phys.* **58**, 1880 (1973).
55. Percus, J. K. and Yevick, G. T. *Phys. Rev.* **110**, 1 (1958). This and the mean-spherical approximation are discussed in recent monographs on the molecular theory of fluids.
56. Triezenberg, D. G. and Zwanzig, R. *Phys. Rev. Lett.* **28**, 1183 (1972).
57. Henderson, J. R. *Mol. Phys.* **39**, 709 (1980).
58. Rushbrooke, G. S. and Scoins, H. I. *Proc. Roy. Soc.* A. **216**, 203 (1953); see also Schofield, P. *Proc. Phys. Soc.* **88**, 149 (1966) (Appendix).
59. Lekner, J. and Henderson, J. R. *Mol. Phys.* **34**, 333 (1977).
60. Jhon, M. S., Desai, R. C. and Dahler, J. S. *Chem. Phys. Lett.* **56**, 151 (1978); *J. chem. Phys.* **70**, 5228 (1979).
61. Schofield, P. *Chem. Phys. Lett.* **62**, 413 (1979). We are indebted to Mr Schofield for explanatory notes on this paper.
62. Schofield, P. personal communication (1979).
63. Grant, M. and Desai, R. C. *Mol. Phys.* **43**, 1035 (1981).
64. Kubo, R. *Rep. Prog. Phys.* **29**(i), 255 (1966), eqn (5.20); McQuarrie, D. A. *Statistical Mechanics* p. 509, Harper and Row, New York (1976).
65. Buff, F. P. *J. chem. Phys.* **23**, 419 (1955).
66. Ono and Kondo, ref. 1, § 37. In § 15 they ascribe (4.219) to Bakker.
67. Landau, L. D. and Lifshitz, E. M. *Theory of Elasticity*, 2nd edn, p. 75. Pergamon Press, Oxford (1970).
68. Buff, F. P., Lovett, R. A. and Stillinger, F. H. *Phys. Rev. Lett.* **15**, 621 (1965).
69. Frenkel, J. *Kinetic Theory of Liquids*, Chapter 6. Oxford University Press (1946). Mandelstam, L. *Ann. Phys.* **41**, 609 (1913), obtained equation (4.240).
70. Atkins, K. R. *Canad. J. Phys.* **31**, 1165 (1953); Widom, A. *Phys. Rev.* A **1**, 216 (1970); Cole, M. W. *Phys. Rev.* A **1**, 1838 (1970).
71. Weeks, J. D. *J. chem. Phys.* **67**, 3106 (1977); Davis, H. T. *J. chem. Phys.* **67**, 3636 (1977); Kalos, M. H., Percus, J. K. and Rao, M. *J. stat. Phys.* **17**, 111 (1977); Abraham, F. F. *Chem. Phys. Lett.* **58**, 259 (1978); *Phys. Rep.* **53**, 93

(1979); Henderson, J. R. and Lekner, J. *Mol. Phys.* **36,** 781 (1978). Long-range correlations in the interface were earlier discussed by Hart, E. W. *J. chem. Phys.* **34,** 1471 (1961). A two-dimensional lattice model that is analogous to Weeks's continuum model has been analysed by Abraham, D. B. *Phys. Rev. Lett.* **47,** 545 (1981).

72. See, for example, Landau, L. D. and Lifshitz, E. M. *Statistical Physics,* 2nd edn. pp. 343–5. Pergamon Press, Oxford (1969).

73. See, for example, Bellman, R. *Introduction to Matrix Analysis,* 2nd edn., pp. 37, 94. McGraw-Hill, New York (1970).

74. See Evans, ref. 2, and Kalos *et al.*, ref. 71.

75. Abraham, D. B. and Reed, P. *Phys. Rev. Lett.* **33,** 377 (1974); Abraham. D. B. and Issigoni, M. *J. Phys. A* **12,** L 125 (1979); van Beijeren, H. *Commun. math. Phys.* **40,** 1 (1975); Bricmont, J., Lebowitz, J. L., Pfister, C. E. and Olivieri, E. *Commun. math. Phys.* **66,** 1 (1979).

76. Weeks, J. D. Gilmer, G. H. and Leamy, H. J. *Phys. Rev. Lett.* **31,** 549 (1973). See the reviews of H. Müller-Krumbhaar, in *Proc. First European Conf. on Crystal Growth (ECCG-1) Zürich, September 1976* (ed. E. Kaldis and H. J. Scheel) p. 116, North-Holland, Amsterdam (1977), and Weeks, J. D. and Gilmer, G. H. *Adv. chem. Phys.* **40,** 157 (1979).

77. Evans, R. *Mol. Phys.* **42,** 1169 (1981). His analysis follows the earlier work of Wertheim, ref. 17, and of Yang *et al.*, ref. 50.

78. Rao, M. and Levesque, D. *J. chem. Phys.* **65,** 3233 (1976); Ladd, A. J. C. and Woodcock, L. V. *Mol. Phys.* **36,** 611 (1978).

79. Hill, T. L. *J. chem. Phys.* **30,** 1521 (1959); Stillinger and Buff, ref. 43.

80. Hill, T. L. *J. chem. Phys.* **19,** 261 (1951); Plesner, I. W. and Platz, O. *J. chem. Phys.* **48,** 5361 (1968); Bongiorno, Scriven, and Davis, ref. 53.

Additions to References

Ref. 2   The review by J. K. Percus is in *Studies in Statistical Mechanics,* Vol. 8, p. 31 (1982). For a more recent review see Evans, R. in *Liquides aux Interfaces/Liquids at Interfaces* (ed. J. Charvolin, J. F. Joanny, and J. Zinn-Justin) Les Houches, Session XLVIII, 1988, Elsevier, Amsterdam (1989).

Ref. 9   Yvon's paper of 1935 has been reprinted in Yvon, J. *Oeuvre Scientifique,* Vol. 1, p. 37, Comm. de l'Energie Atomique, Paris (1985).

Ref. 10   For a review of orientation at interfaces see Gubbins, K. E. in *Fluid Interfacial Phenomena,* Chap. 10 (ed. C. A. Croxton) Wiley, Chichester (1986).

§ 4.5 and ref. 40–47   Density functional theory has been developed further and applied particularly to fluids near walls. Four recent important papers are Tarazona, P. *Phys. Rev. A* **31,** 2672 (1985); **32,** 3148 (1985); Meister, T. F. and Kroll, D. M. *Phys. Rev. A* **31,** 4055 (1985); Curtin, W. A. and Ashcroft, N. W. *Phys. Rev. A* **32,** 2909 (1985).

§ 4.8 and ref. 65   For a review of recent work on the spherical surface, see Henderson, J. R. in *Fluid Interfacial Phenomena,* Chap. 12 (ed. C. A. Croxton) Wiley, Chichester (1986).

§ 4.9 and ref. 68–77   Capillary-wave theory continues to generate controversy. See Bedeaux, D. and Weeks, J. D. *J. chem. Phys.* **82,** 972 (1985); Ciach, A. *Phys. Rev. A* **36,** 3990 (1987); Requardt, M. and Wagner, H. J. *J. Phys. C.*, in press (1988).

5

MODEL FLUIDS IN THE
MEAN-FIELD APPROXIMATION

5.1 Introduction: mean-field theory of a homogeneous fluid of attracting hard spheres

In § 1.6 we saw how important was the role played in the historical development of ideas about the constitution of fluids generally, and that of the liquid–gas interface particularly, by the mean-molecular-field (or mean-field) approximation. We asserted in § 3.5 (anticipating the development in the present chapter) that this approximation contains the assumption that within the range of the attractive forces of every molecule there is always found the same number of neighbouring molecules. This means that the potential energy of attraction felt by the molecules is assumed to be a constant, not varying from molecule to molecule at any one time or with time for any one molecule.

The purpose of this chapter is to develop that idea as a calculational tool, first for a homogeneous fluid and then for an interface. The homogeneous fluid is treated in the present section, and then the interface between phases in the model of attracting hard spheres is treated in § 5.2, the interfaces in two lattice–gas models in §§ 5.3–5.4, and in two versions of the penetrable-sphere model in §§ 5.5–5.7. The results of the model calculations illustrate many of the ideas of the van der Waals theory described in Chapter 3 and of the general statistical-mechanical theory of Chapter 4.

By (4.69), the ratio of the density ρ to the activity $\zeta = \lambda/\Lambda^3$ in a fluid, when there is no external potential $(v(\mathbf{r}) = 0)$, is

$$\rho/\zeta = \langle \exp(-u/kT) \rangle \tag{5.1}$$

where u is the potential energy of interaction of a test particle with the molecules of the fluid (in a typical equilibrium configuration unaffected by the test particle), and where the average is taken uniformly over the whole volume. If the fluid is inhomogeneous, ρ and $\langle \exp(-u/kT) \rangle$ depend on the position $\mathbf{r}$; if the fluid is uniform, these are, too. In any case, ζ or λ, as remarked after (4.67), is uniform. In the remainder of this section we take the fluid to be homogeneous.

The first defining characteristic of the mean-field approximation is that it replaces all or part of the average of the exponential in (5.1) by an exponential of the form $\exp(-u_{\mathrm{mf}}/kT)$ with some temperature-independent u_{mf} ('mf' for mean field). The second part of the mean-field approximation is a prescription for calculating u_{mf}, which will be given later (e.g. in (5.6) below). There is no unique way of decomposing $\langle\exp(-u/kT)\rangle$ and identifying a factor to be approximated by $\exp(-u_{mf}/kT)$. Some procedures will be more successful—more accurate, more convenient, or more illuminating—than others, and the choices that are made will clearly vary with the model systems that are treated. It is now well established that to describe dense fluids successfully it is important to treat accurately the effects of the strong, very-short-ranged repulsive forces, while it is often sufficient to treat the weaker and longer-ranged attractive component of the intermolecular interaction by the mean-field approximation. We shall here apply that idea to the model of attracting hard spheres, which will allow us to illustrate the use of the mean-field approximation in deriving a simple but useful theory of dense fluids.

For attracting hard spheres (5.1) may be written

$$\rho/\zeta = \omega\langle\exp(-u_{attr}/kT)\rangle \qquad (5.2)$$

where ω is the probability that the hard-sphere test molecule will fit into the fluid at an arbitrary point without overlapping any of the hard-sphere molecules that are already there, and u_{attr} is the attractive component of the intermolecular pair potential. The decomposition (5.2) is exact for the model fluid. The characteristic first step in the mean-field approximation for this model is

$$\langle\exp(-u_{attr}/kT)\rangle \simeq \exp(-u_{mf}/kT) \qquad (5.3)$$

(to be followed shortly by a prescription for calculating the supposedly temperature-independent u_{mf}). We are thereby assuming that, wherever the test particle is placed, if its hard core fits among those of the molecules that are already there it then feels the constant potential energy $u_{attr} = u_{mf}$. It is as though the fluid consisted of hard spheres immersed in a uniform background potential of strength u_{mf}. The essence of the mean-field approximation (5.3) is to have ignored the fluctuations—spatial and temporal variations—in that background potential.

If the attractive forces were infinitely weak [$O(1/N)$ in a fluid of N molecules] and of infinitely long range, so that all N molecules contributed equally to the attractive potential energy felt by the test particle independently of its position, then (5.3), and the picture of hard spheres in a uniform background potential, would be exact for this model. That is the reason, as we saw in Chapter 1, that weak, long-ranged forces were thought to be necessary for such a mean-field theory to be accurate. But the mean-field theory may be accurate (albeit not exact) for other reasons. In a dense fluid almost every molecule and almost every empty space is surrounded by many molecules, so that while the attractive potential energy felt by each molecule of the fluid, or by a test particle, is great, being the sum of contributions from many neighbouring molecules, the net attractive force need not be, because each attractive interaction is compensated by nearly equal attractions in the opposite direction. The attractive background potential is therefore relatively gradient-free, i.e., uniform. It is therefore really because the liquid is of high density, with each molecule having many neighbours, that the mean-field approximation yields a useful picture, not because the attractions are weak and long ranged. To be sure, the attractions are weaker and longer ranged than the repulsions; but they are not weak compared with kT, nor are they much longer-ranged than the sphere diameters themselves, or than the typical distance between neighbouring molecules. In lattice models, analogously, the mean-field theory becomes accurate in the limit of high coordination number.[1] Such a lattice model of interacting magnetic moments, treated in mean-field approximation, yields the Weiss mean-molecular-field theory of magnets, which has given its name to this whole class of approximations.[2]

The background potential, being uniform in this approximation, exerts no forces, so the equilibrium configuration of the molecules is the same as that in a fluid of hard spheres without attraction, at the same density. The probability ω that a test particle will fit at an arbitrary point is then also the same as in such a hard-sphere reference fluid. Then setting $u_{attr} = 0$ in (5.2),

$$\omega = \rho/\zeta_{hs}(\rho) \qquad (5.4)$$

where $\zeta_{hs}(\rho)$, a function of ρ alone, is the activity of the hard-sphere reference fluid, in which $u_{attr} = 0$, at the same ρ as in the fluid of interest. From (5.1)–(5.4), the activity of our model fluid, in mean-field approximation, is then

$$\zeta(\rho, T) = \zeta_{hs}(\rho)\exp(u_{mf}/kT). \qquad (5.5)$$

The second essential part of the mean-field approximation is the prescription for calculating u_{mf}. Here u_{mf} is the value of u_{attr} that is felt by a test particle at any point at which its hard core fits among those of the molecules of the fluid. The prescription of the mean-field approximation is now that u_{mf} be identified with the hypothetical u_{attr} that would be felt by the test particle if the molecules of the fluid were distributed around it with the local macroscopic density, but otherwise at random. If the fluid were inhomogeneous, with density $\rho(\mathbf{r})$ at $\mathbf{r}$, the u_{mf} so calculated would be $\mathbf{r}$-dependent, and, assuming pair forces alone, would be given by

$$u_{mf}(\mathbf{r}_1) = \int_{r_{12}>b} u_{attr}(r_{12})\rho(\mathbf{r}_2)\, d\mathbf{r}_2 \qquad (5.6)$$

where $u_{attr}(r_{12})$ is the attractive part of the pair potential at distance r_{12}, where $d\mathbf{r}_2$ is an element of volume at $\mathbf{r}_2$, and where the integration is over all space excluding a sphere of radius b about $\mathbf{r}_1$, where b is the diameter of the hard-sphere molecule. We shall require the general (5.6) in the next section; but for now, since we are assuming the fluid to be homogeneous, (5.6) is simply

$$u_{mf} = -2a\rho \tag{5.7}$$

with

$$a = -\frac{1}{2} \int_{r>b} u_{attr}(r) \, d\mathbf{r}. \tag{5.8}$$

The background potential is thus proportional to the density. From (5.5) and (5.7),

$$\zeta(\rho, T) = \zeta_{hs}(\rho)\exp(-2a\rho/kT). \tag{5.9}$$

Since each molecule is immersed in the background potential $-2a\rho$, the total potential energy U of the fluid of N molecules in the volume V is

$$U = \tfrac{1}{2}N(-2a\rho) = -a\rho^2 V. \tag{5.10}$$

The factor $\tfrac{1}{2}$ is needed so as not to count each interacting pair twice, once when one member of the pair is the central molecule and the other contributes to the background potential felt by the first, and a second time when their roles are reversed. Equation (5.10) may also be obtained more formally from (4.73) with the sum restricted to its first term, $m = 2$, since the total interaction energy is the sum of the energies of interacting pairs. If the u_{attr} felt by a test particle is uniformly u_{mf}, that relation requires $U/V = \tfrac{1}{2}\zeta\omega u_{mf} \exp(-u_{mf}/kT)$, which, with (5.4), (5.5), and (5.7), is just (5.10).

Equations (5.9) and (5.10) are consistent with the thermodynamic identity (Appendix 1)

$$\left[\frac{\partial \ln \zeta}{\partial(1/kT)}\right]_\rho = \left[\frac{\partial(U/V)}{\partial\rho}\right]_T. \tag{5.11}$$

With this identity, either of (5.9) or (5.10) could have been inferred from the other. Indeed, with ρ/ζ given by (4.69) and U/V by (4.73), it follows from (5.11) that the only thermodynamically consistent u_{mf} that is uniform and temperature-independent is one which is proportional to the density[3], as in (5.7). This shows that the two aspects of the mean-field approximation in a homogeneous fluid, (5.3) and (5.7), are not independent. It is important always, in any application of the ideas of the mean-field theory, to verify its thermodynamic consistency in this way.[4]

From the further thermodynamic identity (Appendix 1)

$$\zeta(\partial p/\partial \zeta)_T = \rho kT, \tag{5.12}$$

and (5.9), we obtain for the pressure of the model fluid

$$p = p_{hs} - a\rho^2, \tag{5.13}$$

where p_{hs} is the pressure of the hard-sphere fluid without attractions at the same ρ and T. (p_{hs}/T is a function of ρ alone.) We could have added an arbitrary function of T alone to the right-hand side of (5.13), and (5.13) would still have satisfied (5.12) with (5.9); but since p and p_{hs} must both vanish in the limit $\rho \to 0$ at fixed T, the added function of T could only be 0.

If p_{hs} of the three-dimensional hard-sphere fluid were approximated by p_{hs} of the one-dimensional fluid of hard rods, $p_{hs} = \rho kT/(1 - b\rho)$, where b is the rod length (sphere diameter), (5.13) would be recognized as the van der Waals equation of state, (1.34), with the parameter a calculable from the intermolecular forces by (5.8). With an accurate three-dimensional p_{hs}, (5.13) is a reasonable equation of state of a liquid.[5]

5.2 Liquid–gas interface in the model of attracting hard spheres

We continue here to treat the model of attracting hard spheres in mean-field approximation, but we no longer assume the fluid to be homogeneous. Taking the more general (5.6) instead of (5.7), and using the inhomogeneous form of (5.1), we shall find a functional equation for the density profile of the liquid–gas interface.

Now ρ in (5.1) is $\rho(\mathbf{r})$ and $\langle \exp(-u/kT) \rangle$ depends on the position $\mathbf{r}$ at which the average is evaluated, although ζ, the activity in the two-phase fluid, is still uniform. The decomposition (5.2) is still exact, with ω and $\langle \exp(-u_{attr}/kT) \rangle$ now $\mathbf{r}$-dependent but otherwise defined as before. We again adopt (5.3) as the characteristic mean-field approximation, with u_{mf} now $\mathbf{r}$-dependent and given by (5.6) as a functional of $\rho(\mathbf{r})$. If the gradient of $\rho(\mathbf{r})$ were so small that ρ changed negligibly over distances comparable with the sphere diameter or with the distance between neighbouring spheres, then on the scale of such distances the fluid would appear to be a uniform, equilibrium fluid of hard spheres of density equal to the local density $\rho(\mathbf{r})$ of the inhomogeneous fluid; so ω would again be given by (5.4), now with $\rho = \rho(\mathbf{r})$. We shall assume that to be so; but we must recognize that we are thereby making a second approximation[6]—a small-gradient assumption—beyond the mean-field approximation (5.3).

From (5.2)–(5.4) and (5.6),

$$\zeta_{hs}[\rho(\mathbf{r}_1)] \exp\left[\frac{1}{kT} \int_{r_{12}>b} u_{attr}(r_{12})\rho(\mathbf{r}_2) \, d\mathbf{r}_2 \right] = \zeta, \tag{5.14}$$

a functional equation which must be satisfied by any equilibrium $\rho(\mathbf{r})$ in the model fluid, in this approximation. It has been assumed that there are no external forces acting on the body of the fluid. By (4.69), if there were an external $v(\mathbf{r})$ there would be an additional factor $\exp[-v(\mathbf{r})/kT]$ on the right-hand side of (5.1); hence also of $\exp[-v(\mathbf{r}_1)/kT]$ on the right-hand side of (5.14), if $\zeta_{hs}(\rho)$ were understood still to mean the activity of the uniform, equilibrium hard-sphere fluid of density ρ with no external forces.

It is convenient to rewrite (5.14) as

$$-\int_{r_{12}>b} u_{attr}(r_{12})[\rho(\mathbf{r}_2)-\rho(\mathbf{r}_1)]\,d\mathbf{r}_2 = -2a\rho(\mathbf{r}_1) + kT\ln\{\zeta_{hs}[\rho(\mathbf{r}_1)]/\zeta\} \quad (5.15)$$

with a still defined by (5.8). When the fluid is homogeneous $\rho(\mathbf{r})$ is independent of $\mathbf{r}$, the left-hand side of (5.15) vanishes, and (5.15) reduces to (5.9).

The uniform value of the chemical potential in the two-phase system is

$$\mu = kT\ln\lambda = kT\ln(\Lambda^3\zeta), \quad (5.16)$$

while, according to (5.9), the chemical potential $\mu(\rho, T)$ of the equilibrium homogeneous fluid at the density ρ and temperature T is

$$\mu(\rho, T) = kT\ln[\Lambda^3\zeta(\rho, T)] = -2a\rho + kT\ln[\Lambda^3\zeta_{hs}(\rho)]. \quad (5.17)$$

Then from (5.15)–(5.17), the functional equation that must be satisfied by $\rho(\mathbf{r})$ takes the form

$$-\int_{r_{12}>b} u_{attr}(r_{12})[\rho(\mathbf{r}_2)-\rho(\mathbf{r}_1)]\,d\mathbf{r}_2 = \mu[\rho(\mathbf{r}_1), T] - \mu. \quad (5.18)$$

In any of the forms (5.14), (5.15), or (5.18), the functional equation is non-local, in that it expresses $\rho(\mathbf{r}_1)$ in terms of the density $\rho(\mathbf{r}_2)$ at all points $\mathbf{r}_2$ in the fluid outside a sphere of radius b centred at $\mathbf{r}_1$. When the gradients are small—as we have already assumed in obtaining (5.14)—we may expand $\rho(\mathbf{r}_2)$ about $\mathbf{r}_1$ and truncate the expansion at some finite order. The result is a differential equation for $\rho(\mathbf{r}_1)$ which is of order equal to that of the last term retained in the expansion, and which is then local, for it expresses $\rho(\mathbf{r}_1)$ only in terms of its derivatives of finite order evaluated at $\mathbf{r}_1$. If we truncate after the term of second order (recognizing that the term of first order vanishes by symmetry), and at the same time specialize the discussion to allow non-uniformity only in the z-direction, perpendicular to the plane of the interface, (5.18) becomes

$$m\,d^2\rho(z)/dz^2 = \mu[\rho(z), T] - \mu \quad (5.19)$$

where

$$m = -\frac{1}{2} \int_{r>b} z^2 u_{attr}(r) \, d\mathbf{r} = -\frac{1}{6} \int_{r>b} r^2 u_{attr}(r) \, d\mathbf{r}, \qquad (5.20)$$

proportional to the second moment of the attractive component of the pair potential.

Equation (5.19) is (3.7) of the van der Waals theory—$\mu(\rho, T)$ is the $M(\rho)$ of Chapter 3—and (5.20) is (1.38) or (4.164). The function $\mu(\rho, T)$, or $M(\rho)$, as a function of ρ for fixed T, given explicitly by (5.17), has the form anticipated in Fig. 3.4. It would be that of the van der Waals equation of state if $\zeta_{hs}(\rho)$ were taken to be that for hard spheres (hard rods) in one dimension,

$$b\zeta_{hs}(\rho) = [b\rho/(1-b\rho)] \exp[b\rho/(1-b\rho)]. \qquad (5.21)$$

With the three-dimensional $\zeta_{hs}(\rho)$ there are important quantitative differences from (5.21), but the associated $\mu(\rho, T)$ or $M(\rho)$ from (5.17) remains qualitatively the same and is still like that in Fig. 3.4. We know from Chapter 3 that (5.19) then yields a density profile $\rho(z)$ of the expected form, like that shown as (c) in Fig. 3.3.

The liquid–gas surface tension in the model, with the approximations that led to (5.19)–(5.20), is given by (3.11)–(3.13) with m independent of ρ. For example, from (3.13),

$$\sigma = (2m)^{\frac{1}{2}} \int_{\rho^g}^{\rho^l} [-W(\rho)]^{\frac{1}{2}} \, d\rho \qquad (5.22)$$

where ρ^g and ρ^l are the gas and liquid densities obtained from the equal-areas construction, as in Fig. 3.4, applied to the $\mu(\rho, T)$, or $M(\rho)$, in (5.17); and where $W(\rho)$ is the potential, like that in Fig. 3.2, that is related to $M(\rho)$ by $-dW(\rho)/d\rho = M(\rho) - \mu$.

The small-gradient approximation, which is contained in (5.14) and (5.18) and which was made again in the subsequent reduction to (5.19), can be quantitatively accurate only near the critical point, where, as we shall see in Chapter 9, the interface is diffuse and the gradients are indeed small. The small-gradient approximation may not greatly affect the qualitative character of the results, but the mean-field approximation, (5.3), does. The latter ignores fluctuations in u_{attr}, which are particularly significant near the critical point, where the density is typically only about one third of that in the liquid near the triple point. We defer to Chapter 9 a discussion of density profiles and surface tensions near critical points.

5.3   Lattice–gas model: one component

The lattice-gas model[7] of liquid–vapour equilibrium has been particularly useful because of its simplicity, and because it can be related by exact transcriptions to the much-studied Ising model of a ferromagnet (§ 9.7).

The volume V that contains the fluid is imagined divided into V/v_0 cells each of volume v_0. We now redefine the density ρ to be Nv_0/V, so that it is dimensionless, and similarly take ζ to be $\lambda v_0/\Lambda^3$. Each cell has c neighbours (c = coordination number). The energy of interaction of the molecules is the sum of the energies of interaction of pairs: when the two molecules of the pair are in the same cell that energy is $+\infty$; when they are in neighbouring cells it is $-\varepsilon(\varepsilon > 0)$; and when they are neither in the same cell nor in neighbouring cells it is 0. The cells do not have walls that impede the motions of the molecules; those motions are affected only by the intermolecular interactions, as in any fluid. The cells serve merely as a system of coordinates, one in which distances are discrete rather than continuous. Except that the intermolecular separations on which they depend are now measured in discrete units, the pair interactions in the lattice gas are much like those in the model of attracting hard spheres: infinite repulsion at short distances, finite attraction at intermediate distances, and u_{attr} of short enough range to be integrable, as in (5.8).

We again have (5.2), but now ω is the probability that a cell is empty, and is given simply and exactly by

$$\omega = 1 - \rho. \tag{5.23}$$

The mean-field approximation is again (5.3), with u_{mf} calculable from the discrete analogue of (5.6) if the fluid is inhomogeneous, or more simply from (5.7) and the discrete analogue of (5.8) if the fluid is uniform. For the uniform fluid we have from (5.8) for this model

$$a = \tfrac{1}{2}c\varepsilon v_0 \tag{5.24}$$

so from (5.2), (5.3), (5.7), (5.8), (5.23), and (5.24),

$$\zeta(\rho, T) = [\rho/(1-\rho)] \exp(-c\rho\varepsilon/kT). \tag{5.25}$$

The mean-field approximation for the uniform fluid has thus amounted to the assumption that a test particle feels the same potential energy, $-c\varepsilon\rho$, in every empty cell; it is again the picture of hard particles in a uniform background potential of depth proportional to the density. The only difference between (5.25) and (5.9) is that in the lattice-gas model $\zeta_{\text{hs}}(\rho)$ is simply and explicitly $\rho/(1-\rho)$ whatever the structure or dimensionality of the lattice of cells, whereas in the continuum model $\zeta_{\text{hs}}(\rho)$ is given by (5.21) in one dimension, and is known only numerically or approximately in two and three dimensions. Equations (5.10) and (5.13), with a now given by (5.24), and with

$$p_{\text{hs}}v_0/kT = -\ln(1-\rho) \tag{5.26}$$

hold also for the lattice gas.

Equation (5.25), still for the homogeneous fluid, may be rewritten

$$\ln[\zeta(\rho, T)\exp(\tfrac{1}{2}c\varepsilon/kT)] = \ln[\rho/(1-\rho)] - (c\varepsilon/kT)(\rho - \tfrac{1}{2}). \qquad (5.27)$$

The right-hand side is an odd function of $\rho - \tfrac{1}{2}$; it is plotted against ρ for each of three different, fixed $c\theta$ in Fig. 5.1, where $\theta = \varepsilon/kT$. The critical temperature T^c of the model fluid in this mean-field approximation is such that $c\theta = 4$. (In the model of a ferromagnet to which this lattice gas is equivalent,[7] the ordinate in Fig. 5.1 is proportional to the magnetic field and $\rho - \tfrac{1}{2}$ is proportional to the magnetization per lattice site. Figure 5.1 then depicts the magnetic field-magnetization isotherms of the ferromagnet in the Weiss mean-molecular-field approximation.[2])

The equal-areas rule

$$\int_{\rho^g}^{\rho^1} \ln[\zeta(\rho, T)/\zeta]\,d\rho = 0 \qquad (5.28)$$

and equilibrium conditions

$$\zeta(\rho^g, T) = \zeta(\rho^1, T) = \zeta \qquad (5.29)$$

determine the densities ρ^g and ρ^1 of the coexisting gas and liquid phases and the common value ζ of their activity. From (5.27), and Fig. 5.1, they imply

$$\rho^g + \rho^1 = 1 \qquad \zeta = \exp(-\tfrac{1}{2}c\varepsilon/kT). \qquad (5.30)$$

These are exact for the model.[7] They reflect the 'hole–particle' symmetry of the lattice gas, according to which the liquid and vapour phases are

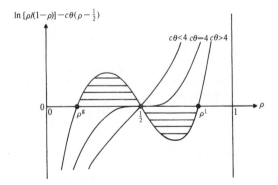

FIG. 5.1. $\ln[\rho/(1-\rho)] - c\theta(\rho - \tfrac{1}{2})$ plotted against ρ for each of three different, fixed $c\theta$. The critical isotherm in this mean-field approximation is that for which $c\theta = 4$. The shaded areas are equal, by symmetry; ρ^g and ρ^1 are the densities of the coexistent gas and liquid phases.

transformable one into the other by reversing the occupancy of each cell: changing occupied to unoccupied and vice versa. The mean-field approximation is therefore faithful to the model's symmetry. The separate ρ^g and ρ^l, from (5.27), (5.29), and (5.30), are, however, the two solutions (other than the trivial solution $\rho = \frac{1}{2}$) of the transcendental equation

$$\ln[\rho/(1-\rho)] - (c\varepsilon/kT)(\rho - \tfrac{1}{2}) = 0; \tag{5.31}$$

and this equation for ρ^g and ρ^l, unlike the exact (5.30), holds only in mean-field approximation.

The symmetry of this model implies that each state is conjugate to another on the other side of the line of symmetry $\rho = \frac{1}{2}$. A state on this line is self-conjugate, and an orthobaric gas state is conjugate to the coexisting liquid state, and vice versa. The properties of conjugate states are related by the following equations, in which the properties of one state are primed, and of the other, unprimed.

$$T' = T$$
$$(\rho - \tfrac{1}{2})' = -(\rho - \tfrac{1}{2})$$
$$(\ln \zeta + \tfrac{1}{2}c\varepsilon/kT)' = -(\ln \zeta + \tfrac{1}{2}c\varepsilon/kT) \tag{5.32}$$
$$(pv_0/kT - \tfrac{1}{2}\ln \zeta)' = (pv_0/kT - \tfrac{1}{2}\ln \zeta)$$
$$(Uv_0/\varepsilon V + \tfrac{1}{2}c\rho)' = (Uv_0/\varepsilon V + \tfrac{1}{2}c\rho).$$

To this point we have outlined the properties of the homogeneous phases in the lattice-gas model and in (5.25) or (5.27), and (5.31) have applied the mean-field approximation. We turn now to the inhomogeneous lattice gas, and treat the liquid–vapour interface in the same approximation.

In the inhomogeneous fluid the ρ in (5.2) and (5.23) is $\rho(z)$ (we are again imagining a planar liquid–gas interface perpendicular to the z-direction) with the z coordinate now taking the discrete values $z = \ldots,$ $-2, -1, 0, 1, 2, \ldots$. The u_{mf} in (5.3) is $u_{mf}(z)$ and is to be evaluated from the discrete analogue of (5.6). Of the c neighbours of any cell at height z, let c' be in the layer at $z + 1$, let c' be in the layer at $z - 1$, and let the remaining $c - 2c'$ be in the same layer, at z. Then from (5.6),

$$u_{mf}(z) = -c\varepsilon\rho(z) - c'\varepsilon\Delta^2\rho(z) \tag{5.33}$$

where Δ^2 is the second-difference operator,

$$\Delta^2\rho(z) = \rho(z-1) - 2\rho(z) + \rho(z+1). \tag{5.34}$$

Thus, from (5.2), (5.3), (5.23), and (5.33), the density $\rho(z)$ must satisfy the difference equation

$$c'\Delta^2\rho(z) = -c\rho(z) + (kT/\varepsilon)\ln\{\rho(z)/[1-\rho(z)]\zeta\}. \tag{5.35}$$

Here ζ is the uniform value of the activity in the two-phase system at the temperature T, so it is that given by (5.30). The functional equation (5.35), with ζ given by (5.30), was first obtained by Ono and Kondo.[8]

In terms of the chemical potential $\mu(\rho, T)[= M(\rho)]$ of the homogeneous fluid of density ρ and temperature T, related to $\zeta(\rho, T)$ by the first of (5.17), and of the uniform chemical potential μ of the two-phase system, related to ζ by (5.16), the functional equation (5.35), with (5.25), is

$$\varepsilon c' \Delta^2 \rho(z) = \mu[\rho(z), T] - \mu. \qquad (5.36)$$

This is analogous to (5.19) in the model of attracting hard spheres, and to (3.7) of the general van der Waals theory, now with the second-difference operator Δ^2, appropriate to discrete distances z, in place of the second derivative. Unlike in (5.19), however, there is no additional small-gradient approximation in (5.36), only the mean-field approximation. With respect to a *continuous* variable z, the operator Δ^2, defined by (5.34), is non-local; and (5.36) is in that sense even more closely analogous to (5.18) than to (5.19).

Typical density profiles[6,8] obtained by solving (5.35) are shown in Fig. 5.2. Only those $\rho(z)$ at the positions z of the lattice planes have meaning here; the continuous curves in the figure are interpolations. The origin $z = 0$ of the z-scale has been arbitrarily taken to be the point about which $\rho(z)$ is antisymmetric: $\rho(z) + \rho(-z) \equiv 1$. The lattice planes are then either at $z = \ldots, -2, -1, 0, 1, 2, \ldots$ or at $z = \ldots, -\frac{3}{2}, -\frac{1}{2}, \frac{1}{2}, \frac{3}{2}, \ldots$. Those are the two cases for which (5.35) may be solved subject to the boundary conditions $\rho(\pm\infty) = \rho^{g,1}$, and they lead to similar figures. To solve (5.35) in the first case one specifies $\rho(0) = \frac{1}{2}$ and then searches for the value of $\rho(1)$ that, taken together with $\rho(0)$ as initial conditions, leads to a solution that

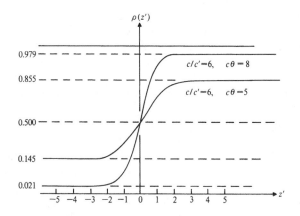

FIG. 5.2. Density profiles from (5.35).

satisfies $\rho(\infty) = \rho^l$. In the second case one searches for the value of $\rho(\frac{1}{2})$ that, taken together with $\rho(-\frac{1}{2}) = 1 - \rho(\frac{1}{2})$ as initial conditions, leads to such a solution. These two solutions for $\rho(z)$, one at integer z and the other at half-odd-integer z, are physically different: in the first there is a lattice plane in which $\rho = \frac{1}{2}$ (viz., that at $z = 0$), in the second there is not.

In the limit of small gradients the second difference may be approximated by a second derivative:

$$\Delta^2 \approx b^2 \, d^2/dz^2 \qquad (5.37)$$

where b is the distance between neighbouring cells in the z-direction, and where z is now continuous and has the dimensions of length. The parameter m in this lattice-gas model, as we see upon comparing (5.36) with (5.19) or (3.7), is then

$$m = \varepsilon v_0 c' b^2. \qquad (5.38)$$

In this limit the surface tension is again given by (5.22), where $W(\rho)$, which is such that $-dW(\rho)/d\rho = \mu(\rho, T) - \mu = kT \ln[\zeta(\rho, T)/\zeta]$, is explicitly calculable from (5.25) and (5.30) for this model. We shall recall these relations in Chapter 9 when we treat interfaces near critical points.

5.4 Lattice-gas model: two components

The two-component lattice-gas model we treat here provides our first example of an interface of composition defined by two independently varying densities, as in a two-component version of the general, many-component van der Waals theory of Chapter 3.

The volume V is divided into cells of volume v_0, as before, and each cell again has c neighbours. There are two species of molecules, a and b. No more than one molecule may occupy any one cell, and unlike molecules may not occupy neighbouring cells. There are no other interactions.

The readily derivable two-component generalization of (4.69) (with $v(\mathbf{r}) = 0$) is

$$\rho_a/\zeta_a = \langle \exp(-u_a/kT) \rangle, \qquad \rho_b/\zeta_b = \langle \exp(-u_b/kT) \rangle \qquad (5.39)$$

where $\zeta_a = v_0 \lambda_a/\Lambda_a^3$ and $\zeta_b = v_0 \lambda_b/\Lambda_b^3$; where u_a or u_b is the interaction energy of a test molecule a or b with the molecules of the fluid when these are in a typical equilibrium configuration unaffected by the test particle; and where the averages are taken uniformly over the whole volume. In the present model, in which the only interactions are infinite repulsions between any two molecules in the same cell and between unlike molecules in neighbouring cells, (5.39) becomes

$$\rho_a/\zeta_a = \omega e_b \qquad \rho_b/\zeta_b = \omega e_a \qquad (5.40)$$

where ω is the probability that a cell be empty, e_b is the probability that none of the cells neighbouring such an empty cell be occupied by a b particle, and e_a that none of them be occupied by an a particle. For this model

$$\omega = 1 - \rho_a - \rho_b \qquad (5.41)$$

is exact, just as (5.23) was exact for the earlier one-component lattice gas. We seek to approximate e_a and e_b.

There are no molecular attractions in this model, so no direct analogue of (5.3) or (5.6). The mean-field type of approximation we shall apply here is instead the Bethe–Guggenheim (quasi-chemical) approximation.[9,10] It shares with the earlier mean-field approximation the assumption that the cells neighbouring any empty cell in the lattice gas are occupied independently of each other. Thus, if q_a is the probability that one such neighbouring cell be occupied by an a particle, and q_b is the probability that it be occupied by a b particle, we take

$$e_a = (1 - q_a)^c, \qquad e_b = (1 - q_b)^c \qquad (5.42)$$

in the homogeneous fluid. Hence, from (5.40)–(5.42), for the homogeneous fluid,

$$\zeta_a(\rho_a, \rho_b) = \rho_a/(1 - \rho_a - \rho_b)(1 - q_b)^c$$
$$\zeta_b(\rho_a, \rho_b) = \rho_b/(1 - \rho_a - \rho_b)(1 - q_a)^c. \qquad (5.43)$$

The probabilities q_a and q_b must still be determined as functions of ρ_a and ρ_b. We cannot simply identify q_a with ρ_a and q_b with ρ_b, even though the analogous identification was made successfully in the mean-field approximation to the one-component lattice gas in § 5.3, where it led to $u_{mf} = -c\varepsilon\rho$ in the homogeneous fluid ((5.7) and (5.24), or (5.33) taken in the limit in which the fluid is homogeneous). We verified the thermodynamic consistency of the earlier mean-field theory in the discussion following (5.11). In the present model, had we identified q_a with ρ_a and q_b with ρ_b, the thermodynamic identity (Appendix 1)

$$(\partial \ln \zeta_a/\partial \rho_b)_{T,\rho_a} = (\partial \ln \zeta_b/\partial \rho_a)_{T,\rho_b} \qquad (5.44)$$

would not have been satisfied, as is easily seen from (5.43). (In this model, where the only interactions are infinitely strong repulsions, ζ_a and ζ_b depend only on ρ_a and ρ_b, so the subscripts T in (5.44) may be omitted.) Even if the cells neighbouring any central cell may be occupied independently of each other, their occupancy is not independent of the state of occupancy of the central cell. That dependence could with full thermodynamic consistency be ignored in the earlier model, but not in the present one. We shall see that here $q_a \simeq \rho_a$ only when $\rho_b \ll 1$, and $q_b \simeq \rho_b$ only when $\rho_a \ll 1$.

We may obtain q_a and q_b as functions of ρ_a and ρ_b from an extended form[3,10] of the principle in (5.39): Let P_i or P_j be the probability that a selected cell in a lattice gas be in the state of occupancy i or j, and let P_{ij} be the probability that one selected cell be in state of occupancy i and that a second be simultaneously in state of occupancy j; then

$$P_{ij}/P_iP_j = \langle\exp(-u_{ij}/kT)\rangle/\langle\exp(-u_i/kT)\rangle\langle\exp(-u_j/kT)\rangle, \qquad (5.45)$$

where u_i, u_j, and u_{ij} are, respectively, the interaction energies felt by test particles of type i and j, and by a 'diatomic' test particle compounded of one of type i and one of type j held at a fixed separation equal to that of the two cells in question. In the present application the two selected cells are neighbours, and the states of occupancy i and j are 'empty', which we shall call state 0, and 'occupied by an a particle', which we shall call state a, respectively. Here $\langle\exp(-u_0/kT)\rangle$ is just the probability that the 'test particle' (in this case, an 'empty') find the cell empty,

$$\langle\exp(-u_0/kT)\rangle = 1 - \rho_a - \rho_b; \qquad (5.46)$$

and, as in (5.39) and (5.43),

$$\langle\exp(-u_a/kT)\rangle = (1 - \rho_a - \rho_b)(1 - q_b)^c; \qquad (5.47)$$

while

$$\langle\exp(-u_{0a}/kT)\rangle = (1 - \rho_a - \rho_b)(1 - q_a - q_b)(1 - q_b)^{c-1}, \qquad (5.48)$$

the first factor being the probability that the first cell be empty, the second factor being the probability that the neighbouring second cell be empty when the first cell is, and the third factor being the probability that the neighbouring cell be accessible to a test particle of type a when the first cell is empty. (Here, as in (5.47), we continue to make the Bethe–Guggenheim approximation, in which the neighbours of any cell are taken to be independent of each other.) But also

$$P_{0a} = P_0q_a \qquad P_0 = 1 - \rho_a - \rho_b \qquad P_a = \rho_a. \qquad (5.49)$$

Then from (5.45)–(5.49),

$$(1 - q_a - q_b)/(1 - q_b)q_a = (1 - \rho_a - \rho_b)/\rho_a. \qquad (5.50)$$

The same argument with state j taken to be 'occupied by a b particle' instead of 'occupied by an a particle' leads to the same relation with a and b interchanged, so we obtain finally

$$\rho_a = q_a(1 - q_b)/(1 - 2q_aq_b), \qquad \rho_b = q_b(1 - q_a)/(1 - 2q_aq_b) \qquad (5.51)$$

as the relations between the qs and ρs. (The analogous relations for the more general model in which the repulsion between an a particle and a neighbouring b is finite are derived in Reference 10.) The inverse relations, giving the qs in terms of the ρs, are readily obtained from (5.51) as solutions of quadratic equations, but we shall not need them here. We see from (5.51) that $q_a \simeq \rho_a$ when $\rho_b \ll 1$ and $q_b \simeq \rho_b$ when $\rho_a \ll 1$.

Equations (5.43) with (5.51) give the activities as functions of the densities for this model in the Bethe–Guggenheim approximation. If the occupancies of the cells neighbouring any one cell were indeed independent of each other, these relations would be exact. They are therefore exact for the class of lattices called 'trees', on which there are no closed paths: between any two neighbours of one site there is no path other than via that site, so that, once the state of occupancy of the cell at that site is specified, the cells at the neighbouring sites are decoupled from, and independent of, each other. Because the Bethe–Guggenheim approximation is thus exact for one class of models, it is necessarily thermodynamically consistent; and, indeed, it may be verified by explicit calculation that (5.43) with (5.51) satisfies (5.44).

The one-component lattice gas of § 5.3 may also be treated in the Bethe–Guggenheim approximation, which is a generalization of, and improvement upon, the simple mean-field theory. The latter follows from the former in the limit of large c and small ε. The resulting mean-field theory is then necessarily thermodynamically consistent, because the Bethe–Guggenheim approximation is consistent for all c and ε. In the present two-component model, in which the only interactions are infinitely strong repulsions, there is no simplification we can make beyond the Bethe–Guggenheim approximation and still retain thermodynamic consistency; there is no parameter ε, and, while the coordination number c is at our disposal, there is no limit to which we can usefully take it.

Because in this model unlike molecules cannot occupy neighbouring cells, we may expect that at high total densities $\rho_a + \rho_b$ the mixture will separate into two phases, one almost pure a and the other almost pure b, approaching the separate pure components in the limit $\rho_a + \rho_b \to 1$, where every cell is occupied. At very low total densities we may expect a single, homogeneous phase to be stable because of the favourable entropy of mixing. Thus, we expect there to be a critical point for the phase separation, and a coexistence curve in the ρ_a, ρ_b-plane like that shown schematically in Fig. 5.3. The lines are tielines connecting the pairs of phases that may coexist. Clearly, by symmetry, with single and double primes referring to coexisting phases,

$$\rho_a' = \rho_b'', \qquad \rho_a'' = \rho_b', \qquad \zeta_a' = \zeta_b'', \qquad \zeta_a'' = \zeta_b'; \qquad (5.52)$$

and since $\zeta_a' = \zeta_a''$ and $\zeta_b' = \zeta_b''$ are required for equilibrium, the latter two of

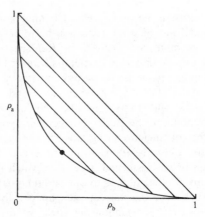

FIG. 5.3. Coexistence curve and critical point for the model two-component lattice gas in the ρ_a,ρ_b-plane. The lines are tielines, connecting coexisting phases, and the dot is the critical point.

the symmetry relations (5.52) imply that the equation of the coexistence curve in Fig. 5.3 is

$$\zeta_a(\rho_a, \rho_b) = \zeta_b(\rho_a, \rho_b) \quad \text{(coexistence curve)} \quad (5.53)$$

and that, more generally, on each tieline

$$\zeta_a = \zeta_b = \zeta \quad \text{(on the tieline)} \quad (5.54)$$

where ζ is the common value of ζ_a and ζ_b at coexistence. The relations (5.52)–(5.54) are exact for this model.

We have from (5.43), (5.51), and (5.53) that in the Bethe–Guggenheim approximation the coexistence curve is determined by the roots (other than the trivial root $q_a = q_b$) of

$$q_a(1 - q_a)^{c-1} = q_b(1 - q_b)^{c-1}. \quad (5.55)$$

A graphical solution of (5.55) is illustrated in Fig. 5.4, where $x(1-x)^{c-1}$ is plotted against x. The curve has a maximum at $x = 1/c$, which corresponds to the critical point. Every horizontal line below the maximum makes two intersections with the curve, at x_1 and x_2, say, with $x_1 < x_2$. These are then q_a and q_b at one point of the coexistence curve. The corresponding ρ_a, ρ_b follow from (5.51). The identification $x_1 = q_a$, $x_2 = q_b$ yields that branch of the coexistence curve on which $\rho_a < \rho_b$, the opposite identification yields the other branch.

The common value $x_1 = x_2 = 1/c$ at the maximum in Fig. 5.4 is the common value of q_a and q_b at the critical point. Therefore, from (5.51), the common value of ρ_a and ρ_b at the critical point in the Bethe–Guggenheim approximation is

$$\rho_a = \rho_b = (c-1)/(c^2-2) \quad \text{(critical point)}. \quad (5.56)$$

From (5.43) and (5.51) there follows also the corresponding value of the activities,

$$\zeta_a = \zeta_b = \frac{1}{c-2}\left(\frac{c}{c-1}\right)^{c-1} \quad \text{(critical point)}. \quad (5.57)$$

When the coordination number c is large, the right-hand side of (5.57) may be approximated by e/c.

In the interface (assumed flat, and perpendicular to the z-direction, as before) we still have (5.40) and (5.41), with uniform $\zeta_a = \zeta_b = \zeta$, but now with ρ_a, ρ_b, e_a, e_b, and ω all z-dependent. Here $e_a(z)$ and $e_b(z)$ are still the probabilities that the cells neighbouring an empty cell at z contain no a or b particles, respectively. We shall again make the Bethe-Guggenheim approximation, and so approximate $e_a(z)$ and $e_b(z)$ by what they would have been if the cells neighbouring the empty cell at z had been occupied by a and b particles independently of each other. If, as in § 5.3, each cell has c' neighbours in the layer above, c' in the layer below, and $c-2c'$ in the same layer with itself, then this approximation is

$$e_a(z) = [1-q_a(z-1;z)]^{c'}[1-q_a(z;z)]^{c-2c'}[1-q_a(z+1;z)]^{c'}, \quad (5.58)$$

where $q_a(z';z)$ is the probability that a cell at z' that is the neighbour of an empty cell at z will be occupied by an a particle; and similarly for $e_b(z)$.

We shall now make the further simplifying assumption that $q_a(z';z)$ and $q_b(z';z)$, which are the probabilities of occupancy of a cell at z', depend only on the densities at z', and are independent of the location, z, of the empty cell of which the cell at z' is a neighbour; so that $q_a(z';z)$ and $q_b(z';z)$ may be called simply $q_a(z')$ and $q_b(z')$; and we take their dependence on the local densities $\rho_a(z')$ and $\rho_b(z')$ to be the same as in the homogeneous fluid, hence as in (5.51). This simplification in the

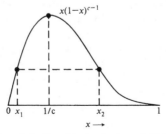

FIG. 5.4. Graphical solution of (5.55).

description of the inhomogeneous fluid may be expected to be quantitatively accurate only in the limit of small gradients, thus near the critical point of the phase equilibrium. With these assumptions, (5.58) becomes

$$e_a(z) = [1 - q_a(z)]^c \exp\{c'\Delta^2 \ln[1 - q_a(z)]\}, \qquad (5.59)$$

and similarly for $e_b(z)$, with $q_a(z)$ and $q_b(z)$ related to $\rho_a(z)$ and $\rho_b(z)$ by (5.51). Here Δ^2 is again the second-difference operator, defined as in (5.34) but now applied to $\ln[1 - q_a(z)]$ or $\ln[1 - q_b(z)]$. Equation (5.59) is the generalization of (5.42) and reduces to it when the fluid is homogeneous.

From (5.40), (5.41), and (5.59) there follow functional equations for the composition profile of the interface,

$$c'\Delta^2 \ln[1 - q_a(z)] = \ln\{\zeta_b[\rho_a(z), \rho_b(z)]/\zeta\}$$
$$c'\Delta^2 \ln[1 - q_b(z)] = \ln\{\zeta_a[\rho_a(z), \rho_b(z)]/\zeta\}, \qquad (5.60)$$

where $\zeta_a(\rho_a, \rho_b)$ and $\zeta_b(\rho_a, \rho_b)$ are the functions given in (5.43)—the activities as functions of the densities in the homogeneous fluid in the mean-field approximation—while ζ, as in (5.54), is the common, uniform value of ζ_a and ζ_b in the two-phase system. The equations (5.60), with q_a and q_b expressible in terms of ρ_a and ρ_b by (5.51), are a pair of coupled, second-order difference equations which may be solved for $\rho_a(z)$ and $\rho_b(z)$. The form of (5.60) reflects the repulsion between neighbouring unlike molecules in the model, which makes the activity of b depend largely on the density of a, and vice versa.

With (5.43) and (5.51), the equations (5.60) become

$$c'\Delta^2 \ln(1 - q_a) = \ln[q_b/(1 - q_a - q_b)(1 - q_a)^{c-1}\zeta]$$
$$c'\Delta^2 \ln(1 - q_b) = \ln[q_a/(1 - q_a - q_b)(1 - q_b)^{c-1}\zeta], \qquad (5.61)$$

which is the form in which they are most simply expressed and solved. Once $q_a(z)$ and $q_b(z)$ are determined from (5.61), the density profiles $\rho_a(z)$ and $\rho_b(z)$ follow from (5.51). The solution for the representative case $c' = 1$, $c = 7$, $\zeta = 1$ is shown in Fig. 5.5. To obtain it, we first calculate the probabilities $q_a(\pm\infty)[= q_b(\mp\infty)]$ in the bulk phases as the q_a and q_b for which the right-hand sides of (5.61), with the given ζ, vanish. We define $z = 0$ by $q_a(0) = q_b(0)$, their common value being so far unknown. With this choice of $z = 0$, we must have by symmetry $q_a(z) = q_b(-z)$. We then guess the values of $q_a(0)[= q_b(0)]$ and $q_a(-1)[= q_b(1)]$, and thence, from (5.61), we calculate $q_a(z)$ and $q_b(z)$ at all integral z. Only with the right $q_a(0)$ and $q_a(-1)$ do the resulting $q_a(z)$ and $q_b(z)$ approach the required limiting values for large $|z|$, so the former are found by trial-and-error. The result is the solution of (5.61) for all integral z, with $z = 0$ defined by $q_a(0) = q_b(0)$.

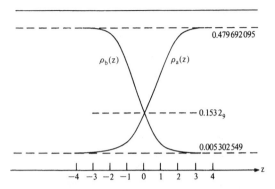

FIG. 5.5. $\rho_a(z)$ and $\rho_b(z)$ from (5.61) and (5.51), for $c' = 1$, $c = 7$, $\zeta = 1$.

In the illustration in Fig. 5.5 the probabilities q_a and q_b in the bulk phases are $q_a(\infty) = q_b(-\infty) = 0.479\,888\,882\ldots$ and $q_a(-\infty) = q_b(\infty) = 0.010\,096\,240\ldots$; the right initial values prove to be $q_a(0) = q_b(0) = 0.174\,3_7$ and $q_a(-1) = q_b(1) = 0.030\,7_8$; and then $q_a(1) = q_b(-1) = 0.405_5$, $q_a(-2) = q_b(2) = 0.012_7$, $q_a(2) = q_b(-2) = 0.47_1$. The corresponding $\rho_a(z)$ and $\rho_b(z)$ are then found from (5.51). The curves shown in Fig. 5.5 were interpolated and extrapolated for other values of z. As in § 5.3, the densities ρ_a and ρ_b in successive layers of this lattice gas could, alternatively, be calculated at $z = \ldots, -\frac{3}{2}, -\frac{1}{2}, \frac{1}{2}, \frac{3}{2}, \ldots$. That solution would be similar, but with the physical difference that there would then be no layer in which $\rho_a = \rho_b$.

Figure 5.6 shows the coexistence curve for $c = 7$; the tieline that

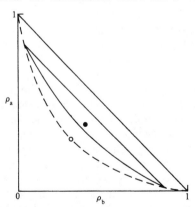

FIG. 5.6. Dashed curve, coexistence curve for $c = 7$; solid curve, variation of ρ_a and ρ_b through the interface (the 'trajectory') for $c' = 1$, $c = 7$, $\zeta = 1$; solid line, tieline for $c = 7$, $\zeta = 1$; open circle, critical point; filled circle, coordinates of the saddle point on the free-energy surface.

connects the compositions of the bulk phases for $c = 7$ and $\zeta = 1$; and the trajectory that gives the variation of ρ_a with ρ_b through the interface for $c' = 1$, $c = 7$, and $\zeta = 1$, as constructed from the curves in Fig. 5.5.

For this model the potential W of (3.21) or $\tilde{W}$ of (3.23)—which we shall here take in dimensionless form and call $W(\rho_a, \rho_b)$—is such that

$$-\partial W/\partial \rho_a = \ln[\zeta_a(\rho_a, \rho_b)/\zeta]$$
$$-\partial W/\partial \rho_b = \ln[\zeta_b(\rho_a, \rho_b)/\zeta] \tag{5.62}$$

with $\zeta_a(\rho_a, \rho_b)$ and $\zeta_b(\rho_a, \rho_b)$ given by (5.43) and (5.51). The saddle point of the potential surface is then at a point ρ_a, ρ_b that satisfies

$$\zeta_a(\rho_a, \rho_b) = \zeta_b(\rho_a, \rho_b) = \zeta \quad \text{(saddle point)}. \tag{5.63}$$

By symmetry we must have $\rho_a = \rho_b$ at the saddle point. This is what distinguishes (5.63) from the equation for the densities of coexisting phases, (5.53), of which one seeks the solutions *other* than $\rho_a = \rho_b$; those other solutions give the maxima in the W surface. If q^* is the common value of q_a and q_b at the saddle point, then from (5.43), (5.51), and (5.63),

$$q^*/(1 - 2q^*)(1 - q^*)^{c-1} = \zeta. \tag{5.64}$$

For $c = 7$ and $\zeta = 1$ this gives $q^* = 0 \cdot 185. \ldots$ Then from (5.51), with ρ^* the common value of ρ_a and ρ_b at the saddle point, we find $\rho^* = 0 \cdot 162$, shown by the filled circle in Fig. 5.6.

On leaving the origin in Fig. 5.6 and moving along the line $\rho_a = \rho_b$, therefore, we encounter first the coexistence curve at the critical point, which by (5.56) is at $\rho_a = \rho_b = 0 \cdot 128$ when $c = 7$; we encounter next the midpoint of the interface-composition trajectory, where $\rho_a = \rho_b = 0 \cdot 153$; next the saddle point of the potential surface at $\rho_a = \rho_b = 0 \cdot 162$; next the midpoint of the $\zeta = 1$ tieline at the mean of the densities $\rho_a(\infty) = \rho_b(-\infty) = 0 \cdot 480$ and $\rho_a(-\infty) = \rho_b(\infty) = 0 \cdot 005$ of the coexisting phases, hence at $\rho_a = \rho_b = 0 \cdot 2425$; and finally the upper limit $\rho_a = \rho_b = \frac{1}{2}$, where there are two phases, one pure a and the other pure b, with all cells occupied.

We note in Fig. 5.6 that the interface-composition trajectory is concave toward the saddle point of the W surface. That is the opposite of what we would expect if there were a dynamical analogy in which W was the potential energy and in which the trajectory was that of a particle of positive mass moving subject to that potential; for then, because of the centrifugal force, the trajectory would be convex toward the saddle point—as illustrated in Figs 3.5 and 3.7, for example. We shall see that in the present problem there is again a dynamical analogy, but now to a system described by generalized rather than rectangular coordinates, in which the relation of the kinetic energy to the velocities is of an uncommon and surprising form.

At low densities $q_a \sim \rho_a$ and $q_b \sim \rho_b$, so that $\ln(1-q_a) \sim -\rho_a$ and $\ln(1-q_b) \sim -\rho_b$; while near the critical point, where the gradients are small, the second difference Δ^2 is practically the same as the second derivative $b^2 \, d^2/dz^2$, as in (5.37). The densities at the critical point in the Bethe–Guggenheim approximation are given by (5.56) as $\rho_a = \rho_b = (c-1)/(c^2-2)$, which are indeed small when the coordination number c is large. Thus, near the critical point, the coupled equations (5.60), with (5.62), are roughly of the form

$$
\begin{aligned}
-c'b^2\ddot{x} &= -\partial W/\partial y \\
-c'b^2\ddot{y} &= -\partial W/\partial x
\end{aligned}
\tag{5.65}
$$

where a dot means derivative with respect to z, and where ρ_a and ρ_b are now called x and y. As in (5.37), z is now continuous and has the dimensions of a length, although in the dynamical analogy it again plays the role of the time, as in Chapter 3.

We may now interpret (5.65) as the equations-of-motion of a mechanical system of two degrees of freedom, with generalized coordinates x and y. Here the y-component of the generalized force, $-\partial W/\partial y$, is proportional to the x-component of acceleration, and vice versa. The constant of proportionality is the negative mass m related to the parameters of the original model by

$$
m = -c'b^2 < 0.
\tag{5.66}
$$

We saw following (5.60) that it was the repulsion between unlike molecules in the model that was responsible for the form of (5.60), so it is that which is now also responsible for the form of (5.65). In the two-component (primitive) version of the penetrable-sphere model, which we treat in §§ 5.5–5.7, we see the equivalent of (5.65) again in (5.133), where it has a similar physical origin. The negative mass m in that case is (in dimensionless form) $-1/2(s+2)$, with s the dimensionality of the potential; see (5.134) below.

The equations-of-motion (5.65) have a first integral

$$
W + m\dot{x}\dot{y} = E
\tag{5.67}
$$

with some constant energy E (which, in our application to the interface, is again, as in Chapter 3, equal to the value of W at its maxima—0, by convention).

The kinetic energy K, the Lagrangian L, the generalized momenta p_x and p_y conjugate, respectively, to x and y, and the Hamiltonian H, are

$$
\begin{aligned}
K &= m\dot{x}\dot{y}, \qquad L = m\dot{x}\dot{y} - W, \\
p_x &= m\dot{y}, \qquad p_y = m\dot{x}, \qquad H = W + p_x p_y/m.
\end{aligned}
\tag{5.68}
$$

The x-component of momentum is here the mass times the y-component of velocity, and vice versa. The equations-of-motion (5.65) are the Euler–Lagrange equations for the minimization of the action

$$\sigma v_0/kT = \int_{-\infty}^{\infty} (-W + K)\,dz. \tag{5.69}$$

In the interface problem $(-W + K)kT/v_0$ is the excess free-energy density due to the inhomogeneity; the minimized σ is therefore again the equilibrium interfacial tension, just as in the dynamical analogy to the original van der Waals theory. In the model fluid the repulsions between unlike molecules are such as to make ρ_a decrease with increasing ρ_b in the interface, and vice versa, as seen in Figs 5.5 and 5.6. In the mechanical system, analogously, $\dot{x}$ and $\dot{y}$ are of opposite sign; so with $m < 0$, as in (5.66), the kinetic energy $m\dot{x}\dot{y}$ is always positive: the density gradients in the interface still make positive contributions to the interfacial tension, as in the earlier theories.

The analogues of (3.11) and (3.12) are

$$\sigma v_0/kT = \int_{-\infty}^{\infty} -2W\,dz = \int_{-\infty}^{\infty} 2K\,dz \tag{5.70}$$

(because $E = 0$ in (5.67)). The second of these is now equivalent to

$$\sigma v_0/kT = 2\int_{y^\alpha}^{y^\beta} p_y\,dy = 2\int_{x^\alpha}^{x^\beta} p_x\,dx = \int_{x^\alpha}^{x^\beta} p_x\,dx + \int_{y^\alpha}^{y^\beta} p_y\,dy \tag{5.71}$$

where x^α, y^α and x^β, y^β (the compositions of the bulk α and β phases in the original model fluid) are the end-points of the trajectory in the x, y-plane. These express the action as integrals of the momenta over the conjugate coordinates in this system. If $y = y(x)$ is the trajectory in the x, y-plane, and if we write $dy(x)/dx = y'(x)$, then, since $K = -W$ on the trajectory, we have $-W = m\dot{x}\dot{y} = m\dot{y}^2/y'(x) = p_x^2/my'(x)$, and similarly $-W = p_y^2/mx'(y)$. Then from either of the first two of (5.71), we see that the action may be calculated from the potential $W(x, y)$ when the trajectory $y(x)$ (or $x(y)$) is known:

$$\sigma v_0/kT = 2(-m)^{\frac{1}{2}} \int_{y^\alpha}^{y^\beta} \{x'(y)\,W[x(y), y]\}^{\frac{1}{2}}\,dy \tag{5.72}$$

$$= 2(-m)^{\frac{1}{2}} \int_{x^\beta}^{x^\alpha} \{y'(x)\,W[x, y(x)]\}^{\frac{1}{2}}\,dx. \tag{5.73}$$

For (5.72) or (5.73) to hold, y or x, respectively, must be single-valued on the trajectory; and we have assumed $y^\alpha < y^\beta$ or $x^\beta < x^\alpha$, respectively. All of m, $x'(y)$, $y'(x)$, and W are negative, so the square-roots are real, and we take them to be positive.

Had x and y been ordinary rectangular coordinates and m an ordinary positive mass, we would have had $-W = p^2/2m$ on the trajectory, with p the ordinary momentum; and in place of (5.72) and (5.73) we would have had formulae for the action expressing it as an integral of the momentum over the distance

$$(2m)^{\frac{1}{2}} \int_{y^{\alpha}}^{y^{\beta}} \{-W[x(y), y][1 + x'(y)^2]\}^{\frac{1}{2}} \, dy \tag{5.74}$$

or

$$(2m)^{\frac{1}{2}} \int_{x^{\beta}}^{x^{\alpha}} \{-W[x, y(x)][1 + y'(x)^2]\}^{\frac{1}{2}} \, dx. \tag{5.75}$$

These formulae for the action in the more familiar system, or the corresponding formulae (5.72) and (5.73) in the system at hand, are analogous to the one-dimensional (3.13) or (5.22), but with one crucial difference due to the difference in dimensionality: whereas, from (3.13) or (5.22), the action in a one-dimensional system may be obtained from the potential alone, without first solving the dynamical problem, that is usually not possible in systems of two or more dimensions, where we must also know the trajectory—$y(x)$ or $x(y)$ in this instance.

Now we turn to the penetrable-sphere models of fluids and treat them as we have done the lattice-gas models and the model of attracting hard spheres; but there are some respects in which we shall be able to go beyond the mean-field approximation, and so find the limits of its applicability.

5.5   Penetrable-sphere model: theory

In a lattice gas the molecules move continuously but are subject to discrete potentials. The density profile $\rho(z)$ and other properties of the system can be found only at fixed points separated by the lattice spacing, as in Figs 5.2 and 5.5. The penetrable-sphere model[11] is a true continuum model which has much of the tractability of the lattice gas, shares with it a symmetry similar to the hole–particle symmetry of (5.32), but differs from it in that $\rho(z)$ etc. can now be calculated for all values of z. In this and the following sections we describe the model briefly and use it to illustrate the application of some of the results of the last chapter.

Consider a volume V containing N molecules each at the centre of a sphere of volume v_0. These spheres are freely penetrable and serve only to define the configurational energy. If the volume covered by these spheres is $W(\mathbf{r}^N)$ then the energy $\mathcal{U}(\mathbf{r}^N)$ is given by

$$v_0 \mathcal{U}(\mathbf{r}^N)/\varepsilon = W(\mathbf{r}^N) - Nv_0 \leqslant 0. \tag{5.76}$$

Clearly $\mathcal{U}$ is a multi-body potential, and is also short-ranged, since if the molecules are so widely separated that none of the spheres overlap each other, then $W = N v_0$ and $\mathcal{U}$ is zero.

This model is isomorphous with a binary system with the pair potentials

$$
\begin{array}{ll}
u_{aa}(r) = 0 & u_{bb}(r) = 0 \\
u_{ab}(r) = \infty \quad (r < l) & u_{ab}(r) = 0 \quad (r \geq l)
\end{array} \tag{5.77}
$$

where $v_0 = \frac{4}{3}\pi l^3$. The obvious a–b symmetry of this, the *primitive* system, is maintained on transcribing its grand partition function to that of the *penetrable-sphere* model. We return in the next section to the primitive version whose potential (5.77) has a clear analogy with that of the two-component lattice gas of the last section.

The equations analogous with (5.32) are

$$
\begin{aligned}
(\theta + \ln \theta \zeta)' &= (\theta + \ln \theta \zeta) \\
[\theta + \ln(\zeta/\theta)]' &= -[\theta + \ln(\zeta/\theta)] \\
(\theta + \pi)' &= (\theta + \pi) \\
[\theta(1 - \rho - \phi) - \rho]' &= -[\theta(1 - \phi - \rho) - \rho] \\
[\theta(1 - \rho - \phi) + \rho]' &= [\theta(1 - \rho - \phi) + \rho]
\end{aligned} \tag{5.78}
$$

where

$$
\rho = N v_0 / V \qquad \theta = \varepsilon / kT \qquad \pi = p v_0 / kT \qquad \phi = v_0 U / \varepsilon V. \tag{5.79}
$$

At high densities of both species the primitive version of the system is driven into two fluid phases by the a–b repulsion of (5.77), with a phase diagram similar to Fig. 5.3 but with the orthobaric lines asymptotic to the ρ_a and ρ_b axes. Similarly at high density and low temperature the attractive potential of (5.76) draws the penetrable-sphere model into two phases, as shown in Fig. 5.7. The line of symmetry of the analogue of Fig. 5.3 is the line $\rho_a = \rho_b$, and orthobaric states are mutually conjugate. The isomorphism of the transcription preserves the line of symmetry in Fig. 5.7; for the penetrable-sphere model it also passes through the critical point and is the diameter of the conjugate orthobaric states. It follows from (5.78) that

$$
\phi^l = 1 - \rho^l - \theta^{-1}\rho^g \qquad \phi^g = 1 - \rho^g - \theta^{-1}\rho^l \tag{5.80}
$$

and that throughout the two-phase region

$$
\zeta = \theta e^{-\theta}, \tag{5.81}
$$

an equation which is the analogue of the second part of (5.30) for the lattice gas.

Let us now introduce a mean-field description of the liquid–gas

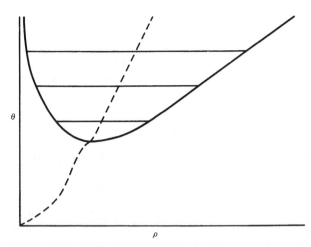

FIG. 5.7. Sketch of the phase diagram of the penetrable-sphere model. The full curve is the orthobaric line and the dashed curve is the line of symmetry.

surface, in which $\rho(z)$ is a continuous function of height z, similar to the discrete functions of Fig. 5.2. We measure z in units of l, the radius of the sphere of volume v_0. The essence of the mean-field approximation is the assumption that on each plane of height z the molecular centres are distributed at random with a density $\rho(z)$.

In a uniform random distribution of density ρ, the probability that any point is not covered by one or more spheres is $e^{-\rho}$; hence the probability that it is covered is $1 - e^{-\rho}$, and from (5.76), the mean-field approximation to the energy density is

$$\phi = 1 - \rho - e^{-\rho}. \tag{5.82}$$

In a non-uniform fluid in which the spheres are distributed at random with a local density $\rho(z)$, it is clear that the probability of a point at height z not being covered depends not only on $\rho(z)$ but on all values of this function between $z-1$ and $z+1$. Simple geometric arguments[12] show that the probability is $e^{-\sigma(z)}$, where $\sigma(z)$ is an effective (or non-local) density given by

$$\sigma(z) = \frac{3}{4} \int_{-1}^{1} (1 - u^2)\rho(z + u) \, du. \tag{5.83}$$

(We retain the notation $\sigma(z)$ for this density to accord with the original paper, and to preserve the progression $\rho(z)$, $\sigma(z)$, $\tau(z)$ below. It is always written with an argument and so should not be confused with the surface tension, σ.)

The activity $\zeta(z) = v_0 \lambda(z)/\Lambda^3$ can now be calculated by the potential distribution theorem from the mean change of energy on adding a $(N+1)$th molecule at random on the plane of height z. If the mean value of the volume covered by the new molecule but not covered by any of the existing N molecules is denoted $w(z)$ then, by further simple geometrical arguments,[12]

$$\frac{w(z)}{v_0} = e^{-\tau(z)} = \frac{3}{4} \int_{-1}^{1} (1-u^2)e^{-\sigma(z+u)} \, du \tag{5.84}$$

where $\tau(z)$ is a second effective density which depends on the value of $\sigma(z)$ between $z+1$ and $z-1$, and so on $\rho(z)$ over the range $z+2$ to $z-2$. Within the mean-field approximation the exponential of a mean is the mean of the exponential, and so from (4.71)

$$\rho(z) = \zeta(z) \exp[\theta(1 - e^{-\tau(z)})]. \tag{5.85}$$

The activity is constant throughout the two-phase system and given exactly by (5.81), so that the last equation becomes

$$\rho(z) = \theta \exp[-\theta e^{-\tau(z)}]. \tag{5.86}$$

This equation together with the definition of $\tau(z)$ and $\sigma(z)$, (5.83) and (5.84), constitute an integral equation for $\rho(z)$, which takes the place of the difference equation (5.35) obtained for the lattice gas.

These equations can be shown to have a hidden symmetry about the plane $z = 0$ similar to that of the equations for uniform states, (5.78) and (5.80). Define a function $f(z)$ by

$$\rho(z)f(z) = \theta e^{-\sigma(-z)}. \tag{5.87}$$

From (5.84)

$$\theta e^{-\tau(z)} = \frac{3}{4} \int_{-1}^{1} (1-u^2)\rho(-z+u)f(-z+u) \, du; \tag{5.88}$$

but from (5.86) the left-hand side of this equation is $\ln[\theta/\rho(z)]$, which from (5.87) is $[\sigma(-z) + \ln f(z)]$. Hence

$$\sigma(z) = -\ln f(-z) + \frac{3}{4} \int_{-1}^{1} (1-u^2)\rho(z+u)f(z+u) \, du. \tag{5.89}$$

If this function is to satisfy (5.83) then $f(z) = 1$; that is the three effective densities are linked by the chain of equations

$$\rho(z) = \theta \exp[-\sigma(-z)] = \theta \exp[-\theta \exp[-\tau(z)]]. \tag{5.90}$$

In the limit $z \to \infty$, $\rho(z)$, $\sigma(z)$, and $\tau(z)$ go to ρ^1, and $\rho(-z)$, $\sigma(-z)$, and

$\tau(-z)$ go to ρ^g. Equation (5.90) then reduces to[11]

$$\theta = \rho^g \exp(\rho^l) = \rho^l \exp(\rho^g) \tag{5.91}$$

which is the analogue of (5.31) for the lattice gas.

This equation has a parametric solution[13] in terms of $\Delta = \frac{1}{2}(\rho^l - \rho^g)$

$$\rho^l = \Delta \coth \Delta + \Delta, \qquad \rho^g = \Delta \coth \Delta - \Delta,$$
$$\theta = \Delta \operatorname{cosech} \Delta \, \exp(\Delta \coth \Delta). \tag{5.92}$$

A natural definition of $\phi(z)$, the local energy density, which satisfies the symmetry condition is

$$\phi(z) = 1 - \rho(z) - \theta^{-1}\rho(-z). \tag{5.93}$$

In the limits $z \to \pm\infty$ this reduces to the exact equations (5.80). From (5.90) it is equivalent to

$$\phi(z) = 1 - \rho(z) - e^{-\sigma(z)} \tag{5.94}$$

and this is just the definition of the local energy density which is consistent with the hypothesis that molecules are randomly distributed on each plane of height z, and that $\phi(z)$ is measured by the mean area on that plane covered by the spheres. This is a natural definition of a local energy density but such definitions are not unique, as we have seen in § 4.10. In particular it is not the value[14] of $\phi(z)$ given by the extension of the potential distribution theorem, (4.73). We thus have the paradox that the calculation of $\rho(z)$, from the condition of constancy of $\zeta(z)$ by the potential distribution theorem, leads to a definition of $\phi(z)$ which is inconsistent with an energy density calculated from a natural extension of that theorem.

The function $\phi^{(2)}(\rho)$, which is the mean value of ϕ in a two-phase system of overall density ρ, and of orthobaric densities ρ^l and ρ^g, is a linear interpolation in ρ between ϕ^l and ϕ^g, and so from (5.80)

$$\phi^{(2)}(\rho) = 1 - \theta^{-1}(\rho^l + \rho^g) + \rho(\theta^{-1} - 1). \tag{5.95}$$

A surface excess energy density can be defined as the difference of $\phi(z)$ and $\phi^{(2)}[\rho(z)]$, whence from (5.93),

$$\phi^s(z) = \phi^s(-z) = \theta^{-1}[\rho^l + \rho^g - \rho(z) - \rho(-z)]. \tag{5.96}$$

The dividing surface of zero excess number density, z_e, is defined by

$$\int_{z_e}^{\infty} [\rho(z) - \rho^l] \, dz + \int_{-\infty}^{z_e} [\rho(z) - \rho^g] \, dz = 0, \tag{5.97}$$

and the total excess surface energy by[15]

$$\phi^s = \int_{-\infty}^{\infty} \phi^s(z)\,dz. \tag{5.98}$$

Substitution of (5.96) into this last equation leads after some rearrangement to

$$\theta\phi^s = 2z_e(\rho^l - \rho^g). \tag{5.99}$$

The excess surface energy obtained by integrating the different $\phi(z)$ found from the potential distribution theorem, has the same value[14] of ϕ^s as (5.99). Since, moreover, the values of ϕ^s are the same it follows from (2.49) that both functions $\phi(z)$ have all the same dividing surface, z_ϕ. We have (with $z_\rho \equiv z_e$)

$$z_\phi/z_\rho = (\theta + 1)/(\theta - 1), \tag{5.100}$$

or z_ϕ lies on the liquid side of z_ρ, as for argon (§ 2.3).

The surface tension in units of $(\varepsilon l/v_0)$ is equal to the surface excess free energy (2.23), defined with respect to the dividing surface of (5.97). From the Gibbs–Helmholtz equation

$$\theta\psi^s = \theta\sigma = \int_{\theta^c}^{\theta} \phi^s\,d\theta' = 2\int_{\theta^c}^{\theta} \frac{z_e(\rho^l - \rho^g)}{\theta'}\,d\theta'. \tag{5.101}$$

Equations (5.95)–(5.101) are formally exact in that they do not invoke the mean-field approximation, and that the values of ϕ^s and σ are independent of the convention used to define $\phi(z)$. In the limit of zero temperature the liquid phase is a uniform infinitely dense phase for $z > z_e = \frac{1}{2}$ (the numerical value of z_e is determined by symmetry[12]) and the gas phase ($z < z_e$) has zero density. As this limit is approached ρ^l behaves as θ, and so it follows that ϕ^s and σ both go to unity at zero temperature. This result is also exact.

At non-zero temperatures the surface tension can be calculated only in the mean-field approximation, after solving the integral equation for $\rho(z)$. This has been done numerically[16] for the penetrable-sphere model, and analytically[12] for a penetrable-cube model, in which the spheres are replaced by oriented cubes of volume v_0. The effective density $\sigma(z)$ is then given by an equation simpler than (5.83), viz.

$$\sigma(z) = \frac{1}{2}\int_{-1}^{1} \rho(z + u)\,du \tag{5.102}$$

and with a parallel change in the definition of $\tau(z)$. For this potential the integral equation has the solution

$$2\rho(z) = (\rho^l + \rho^g) + (\rho^l - \rho^g)\tanh[\tfrac{1}{2}(\rho^l - \rho^g)(z - \tfrac{1}{2})]. \tag{5.103}$$

Thus z_e here retains at all temperatures its low-temperature limit of $\frac{1}{2}$. Integration of the integral equation for $\rho(z)$ shows that for the penetrable-sphere model, z_e falls from $\frac{1}{2}$ at zero temperature to $\sqrt{(3/20)} = 0 \cdot 387$ at the critical point.

In the last chapter it was shown that there are two quite different routes to the surface tension from a knowledge of the intermolecular forces and the direct and total correlation functions. The first is the virial route (4.104), and the second the route via the direct correlation function (4.173). The result above, (5.101), is obtained from a third route, which we may call the energy route, following the analogy with the three similar routes to the pressure of a homogeneous fluid.[17] This route is useful for this model because of the symmetry argument which leads directly to an energy density $\phi(z)$, and so to a surface excess ϕ^s. In general, however, this is not a useful route since we rarely know more about ϕ^s than we do about σ itself.

The two other routes to σ can also be followed for this model, although not without heavy algebra, most of which we omit. The derivation of the virial equation of § 4.4 can be extended to the multibody potential of this model, and leads[16] to an expression for σ which, in the mean-field approximation, is a double integral over $\rho(z)$:

$$\sigma = \frac{3}{4\theta} \int_{-\infty}^{\infty} dz\, \rho(z) \int_{-1}^{1} du(3u^2 - 1)\rho(-z + u). \qquad (5.104)$$

The third route requires the direct correlation function, for which the mean-field approximation is[16,18]

$$c(\mathbf{r}_1, \mathbf{r}_2) = \int_{\bar{v}} \rho(-z_3)\, d\mathbf{r}_3, \qquad (5.105)$$

where r, like z, is now measured in units of the radius of the sphere, and $\bar{v}$ is the volume common to two spheres of unit radius centred at $\mathbf{r}_1$ and $\mathbf{r}_2$. This expression for $c(\mathbf{r}_1, \mathbf{r}_2)$ can be used with (4.52) to give an integral equation for $\rho(z)$. If the external potential is zero, and if the gradient of density is only in the z-direction, then (4.52) becomes in reduced units,

$$\rho'(z_1) = \rho(z_1)\frac{3}{4\pi} \int c(\mathbf{r}_1, \mathbf{r}_2)\rho'(z_2)\, d\mathbf{r}_2. \qquad (5.106)$$

Substitution of the correlation function (5.105) and changing the order of integration gives

$$\rho'(z_1) = \rho(z_1)\left(\frac{3}{4\pi}\right)^2 \int d\mathbf{r}_3 \rho(-z_3)f_{13} \int d\mathbf{r}_2 \rho'(z_2)f_{23} \qquad (5.107)$$

where f_{ij} is a unit step-function

$$f_{ij} = -1 \quad (r_{ij} < 1) \qquad f_{ij} = 0 \quad (r_{ij} \geq 1). \tag{5.108}$$

The integrations over the planes of fixed z_2 and z_3 can be made in cylindrical polar coordinates:

$$\rho'(z_1) = \rho(z_1) \frac{9}{16} \int_{-1}^{1} dv(1-v^2)\rho(-z_1+v) \int_{-1}^{1} du(1-u^2)\rho_u'(z_1-v+u) \tag{5.109}$$

where

$$u = z_2 - z_3 \qquad v = z_1 - z_3 \quad \text{and} \quad \rho_u' = (\partial\rho/\partial u). \tag{5.110}$$

From (5.83), the inner integral is

$$\tfrac{4}{3}\rho_v'(-z_1+v)/\rho(-z_1+v)$$

and so

$$\frac{d \ln \rho(z_1)}{dz_1} = \frac{3}{4} \int_{-1}^{1} (1-v^2)\rho_v'(-z_1+v)\, dv = -d\sigma(-z_1)/dz_1. \tag{5.111}$$

This equation is the derivative of the first part of (5.90), so we see that the condition of symmetry for this model leads, via the mean-field approximation for the direct correlation function, to the general integral equation for the profile (4.52).

The surface tension is given in terms of the direct correlation function by (4.173) which in reduced units is

$$4\theta\sigma = \frac{3}{4\pi} \int_{-\infty}^{\infty} dz_1 \int d\mathbf{r}_2 \rho'(z_1)\rho'(z_2)(x_{12}^2 + y_{12}^2)c(r_{12}, z_1, z_2). \tag{5.112}$$

A change to the variables u and v of (5.110), the substitution of the direct correlation function (5.105), and integration over the planes at z_2 and z_3 give

$$\sigma = \frac{9}{128\theta} \int_{-\infty}^{\infty} dz_1\rho'(z_1) \int_{-1}^{1} dv\, \rho(-z_1+v) \int_{-1}^{1} du\, \rho_u'(z_1-v+u)f(u,v), \tag{5.113}$$

where

$$f(u,v) = (1-u^2)^2(1-v^2) + (1-u^2)(1-v^2)^2. \tag{5.114}$$

Further partial integrations and the use of (5.111) allow this triple integral to be reduced to the double integral of (5.104).

This proof of the identity of the virial and correlation function routes to the surface tension has been made within the mean-field approximation; but at $T = 0$ (where, if any difference were to exist it would be expected to be greatest), the value of σ given by (5.104) is unity, which is

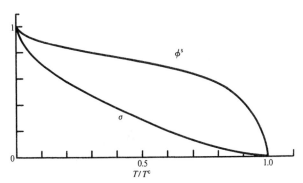

FIG. 5.8. The surface tension and surface excess energy for the penetrable-sphere model in the mean-field approximation.

also the exact value. The proof is, of course, not so general as that of § 4.7.

The integral equation for $\rho(z)$ has been solved numerically and the resulting $\rho(z)$ and z_e used to calculate ϕ^s from (5.99) and σ from (5.101) and from (5.104), that is, via the energy route and via the equation common to the virial and correlation function routes. Both equations lead to the same value of σ which, with ϕ^s, is shown in Fig. 5.8.

The mean-field approximation used throughout this section is exact for a system of infinite dimensionality, and over-emphasizes the propensity of the model to separate into two phases for systems of finite dimensionality. Thus the critical temperature for infinite dimensionality, and so also for the mean-field approximation for all dimensions, which is $kT^c/\varepsilon = 1/e$, is an upper bound for T^c for systems of finite dimensionality. For a one-dimensional system there is no phase separation, for T^c has become zero.[19] Similarly, the mean-field approximation underestimates the thickness of the surface layer between the fluid phases in a three-dimensional system. Nevertheless its high degree of internal consistency, and qualitative correctness make it a useful approximation for systems of three (or more) dimensions.

5.6   Penetrable-sphere model: applications

Tractable models may be used to test approximations whose effect on the calculation of the properties of realistic systems is difficult to assess. We illustrate this point by using the results of the last section to test several versions of the van der Waals or density-gradient theory of Chapters 3 and 4. This theory, even in its most general form, is to be thought of as a set of approximations (smallness of $\rho'(z)$, constancy of the

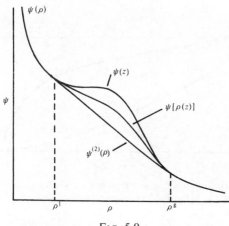

FIG. 5.9

coefficient m, etc.) which are additional to the mean-field approximation. Hence the results of the last section can be used as a standard by which the less accurate theories of this section can be judged.

Figure 5.9 repeats the definitions of § 3.1 in the form we need here; $\psi(\rho)$ is the reduced configurational Helmholtz free-energy density $Fv_0/\varepsilon V$ of the one-phase system; $\psi[\rho(z)]$ is its analytic continuation through the two-phase region, and may be abbreviated to $\psi(\rho)$ when no confusion can arise; $\psi^{(2)}(\rho)$ is the free-energy density of the two-phase system of mean density ρ; and $\psi(z)$ is the van der Waals approximation to the actual free-energy density at height z;

$$\psi(z) = \psi[\rho(z)] + \tfrac{1}{2}m[\rho'(z)]^2 \tag{5.115}$$

where the coefficient m is independent of density in the original theory of van der Waals, but is more properly taken to be the second moment of the direct correlation function of the homogeneous fluid (4.162). The functions Ψ and W of § 3.1 are the differences

$$\begin{aligned}\Psi(z) &= \psi(z) - \psi^{(2)}[\rho(z)], \\ -W(z) &= \psi[\rho(z)] - \psi^{(2)}[\rho(z)].\end{aligned} \tag{5.116}$$

The free-energy density of the homogeneous fluid can be found in the mean-field approximation by integrating the Gibbs–Helmholtz equation, since we know the energy density $\phi(\rho)$, (5.82).

$$\psi(\rho) = 1 - \rho - e^{-\rho} - \theta^{-1}(\rho - \rho \ln \rho), \tag{5.117}$$

whence

$$-\theta W(\rho) = \rho^l + \rho^g + \rho^l \rho^g - \theta e^{-\rho} - \rho \ln(e\theta/\rho). \tag{5.118}$$

The choice of a profile $\rho(z)$ that minimizes the free energy leads to the equalities,

$$\Psi(z) = -2W(z) = m[\rho'(z)]^2, \tag{5.119}$$

and the surface tension is then the integral of any one of these functions, (3.3), (3.11), and (3.12).

For simplicity we now restrict the discussion to a one-dimensional system of penetrable rods moving on a line. There is still a phase transition since we are using a mean-field approximation. The choice of a one-dimensional system is made in order to have explicit expressions for $\rho(z)$ and $c(z_1, z_2)$; a three-dimensional system would require much numerical integration which would obscure the simplicity of the equations without changing the conclusions qualitatively.

Near the critical point $W(z)$ is of order $[\rho'(z)]^2$, as can be seen by expanding (5.118) and using the profile (5.103). So, from (5.119), the limiting value of m at the critical point is

$$m = \tfrac{1}{3}\theta^{-1}. \tag{vdW} \tag{5.120}$$

This value is independent of ρ and is used with (3.13) to give the van der Waals approximation to the surface tension,

$$\sigma_{\text{vdW}} = \left(\frac{2}{3\theta}\right)^{\frac{1}{2}} \int_{\rho^{\text{g}}}^{\rho^{\text{l}}} [-W(\rho)]^{\frac{1}{2}} \, d\rho. \tag{5.121}$$

The coefficient m of (5.120), although independent of ρ, is proportional to T and so in calculating adsorptions we must use the functions denoted $\tilde{W}$ and $\tilde{K}$ in §§ 3.3 and 3.4.

The more accurate expression for m, (4.162), is

$$m(\rho) = \frac{1}{2\theta} \int z_{12}^2 c(z_{12}; \rho) \, dz_{12}, \tag{5.122}$$

where $c(z_{12}; \rho)$ is the one-dimensional version of (5.105)

$$c(z_{12}; \rho) = \frac{1}{2} \int_{z_2 - 1}^{z_1 + 1} \rho(-z_3) \, dz_3 \qquad (0 < (z_2 - z_1) < 2) \tag{5.123}$$

or

$$m(\rho) = \tfrac{1}{3} e^{-\rho}, \tag{YFG} \tag{5.124}$$

which reduces to (5.120) at the critical point $\rho^{\text{c}} = \ln \theta^{\text{c}} = 1$. This form of m is now independent of T and so is to be associated with W and K of § 3.3, not $\tilde{W}$ and $\tilde{K}$. From (3.13)

$$\sigma_{\text{YFG}} = \left(\frac{2}{3}\right)^{\frac{1}{2}} \int_{\rho^{\text{g}}}^{\rho^{\text{l}}} e^{-\frac{1}{2}\rho} [-W(\rho)]^{\frac{1}{2}} \, d\rho \tag{5.125}$$

where the initials are those of Yang, Fleming, and Gibbs[20] who suggested that an expression of the form of (5.125) might be a significant improvement on (5.121). An upper bound to the integral in (5.125) is obtained by omitting the term $\ln(e\theta/\rho)$ from $W(\rho)$ in (5.118). This upper bound becomes the limiting value of σ as the temperature becomes zero. The integration can then be made analytically to give

$$\lim_{\theta \to \infty} \sigma_{YFG} = \pi/\sqrt{6} = 1.283. \qquad (5.126)$$

Figure 5.10 shows the three results; σ from the mean-field approximation alone (whose zero temperature limit is exact), σ_{vdW}, and σ_{YFG}. At $T/T^c = 0.5$, a temperature near which many liquids solidify, the error introduced by the van der Waals theory is about 20 per cent, and that by the modification of Yang *et al.* is 8 per cent. At zero temperature σ_{vdW} is infinite, but σ_{YFG} exceeds the exact value by only 28 per cent, a remarkable result for an approximation which rests on the supposition that the gradient of the density is everywhere small, when used for a state for which $\rho(z)$ is an infinite step-function.

A second line of improvement of the original van der Waals approximation, which can also be tested on the penetrable-sphere model,[21] is the substitution of a 'two-density' for a 'one-density' theory, as set out in § 3.3 and as applied to the two-component lattice gas in § 5.4.

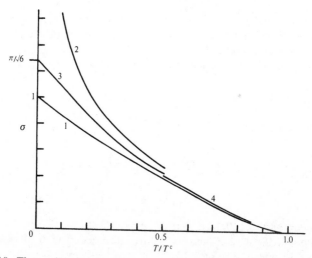

FIG. 5.10. The surface tension for a one-dimensional potential ($s = 1$) as calculated, (1) from the mean-field approximation, (2) from the original van der Waals approximation, (3) from the modification of Yang *et al.*, and (4) from the 'two-density' theory.

The equations above for $W(\rho)$ could be generalized to give $W(\rho, \eta)$ of (3.21) or $\tilde{W}(\rho, \phi)$ of (3.23), where η is the entropy density, but the symmetry of the problem is more apparent if we move to the entirely equivalent two-component or primitive version defined by (5.77), and consider $W(\rho_a, \rho_b)$. This choice preserves, moreover, the analogy with the two-component lattice gas. The isomorphism[11] of the primitive and transcribed (or penetrable-sphere) versions means that a problem solved in one is also solved in the other. Both have two independent densities, ρ and η or ρ and ϕ for the penetrable-sphere model, and ρ_a and ρ_b for the primitive version. Although the latter is a system of two components it has only two independent densities, not the expected three, since ϕ is always zero and temperature an irrelevant variable—consequences of the form of the Hamiltonian (5.77).

The primitive system separates into two phases at high densities of both components, with continuous profiles $\rho_a(z)$ and $\rho_b(z)$ similar to those of Fig. 5.5. In a mean-field approximation the molecules of species b (say) are distributed at random on each plane of height z with a density $\rho_b(z)$, and each is surrounded by a volume v_0 from which the molecules of species a are excluded. By the arguments of the last section

$$\rho_a(z)/\zeta = \exp(-\sigma_b(z)) \qquad \rho_b(z)/\zeta = \exp(-\sigma_a(z)) \qquad (5.127)$$

where $\zeta = \zeta_a = \zeta_b$ is the constant value of the reduced activity of each species throughout the system. In the homogeneous phases α and β, the effective density $\sigma(z)$ is equal to $\rho(z)$, and (5.127) reduces to

$$\zeta = \rho_a^\alpha \exp(\rho_a^\beta) = \rho_a^\beta \exp(\rho_a^\alpha), \qquad \rho_a^\alpha = \rho_b^\beta, \qquad \rho_a^\beta = \rho_b^\alpha. \qquad (5.128)$$

This is the equation which transcribes to (5.91) and which is equivalent to (5.54) for the lattice gas. The rules for the transcription are as follows,[11] with the transcribed variables on the left and the primitive or two-component variables on the right;

$$\rho = \rho_a, \qquad \theta = \zeta_b, \qquad \zeta = \zeta_a \exp(-\zeta_b),$$
$$\pi(\zeta, \theta) + \theta = \pi(\zeta_a, \zeta_b). \qquad (5.129)$$

The potential $W(\rho_a, \rho_b)$ which determines the free-energy density of the hypothetical homogeneous fluid in the interface has derivatives which satisfy (5.62). This potential transcribes into θW of (5.116). The activities $\zeta_a(\rho_a, \rho_b)$ and $\zeta_b(\rho_a, \rho_b)$ of (5.62) are those of the homogeneous fluid and so, from (5.128),

$$\zeta_a(\rho_a, \rho_b) = \rho_a \exp(\rho_b), \qquad \zeta_b(\rho_a, \rho_b) = \rho_b \exp(\rho_a). \qquad (5.130)$$

Integration of (5.62), with constants chosen to make $W(\rho_a^\alpha, \rho_b^\alpha)$ and

$W(\rho_a^{\beta}, \rho_b^{\beta})$ both zero, gives

$$-W(\rho_a, \rho_b) = \rho_a^{\alpha} + \rho_a^{\beta} + \rho_a^{\alpha}\rho_a^{\beta} + \rho_a\rho_b + \rho_a \ln \rho_a + \rho_b \ln \rho_b + (\rho_a + \rho_b)(1 + \ln \zeta)$$
(5.131)

which is the transcription of (5.118).

The Euler–Lagrange equations for $\rho_a(z)$ and $\rho_b(z)$ that minimize the free energy are the equations of motion of the mechanical analogue. If the 'masses' m_{ij} are again assumed to be independent of densities, we have

$$m_{aa}\ddot{\rho}_a + m_{ab}\ddot{\rho}_b = -\partial W(\rho_a, \rho_b)/\partial\rho_a$$
$$m_{ab}\ddot{\rho}_a + m_{bb}\ddot{\rho}_b = -\partial W(\rho_a, \rho_b)/\partial\rho_b$$
(5.132)

where, as before, the dots denote differentiation with respect to the 'time' z.

From the first of (5.62) and the expression for ζ_a in (5.130) and ζ in (5.127) we have

$$-\partial W(\rho_a, \rho_b)/\partial\rho_a = \rho_b(z) - \sigma_b(z) = \frac{-1}{2(s+2)}\ddot{\rho}_b(z) + O(\dddot{\rho}_b(z)). \quad (5.133)$$

The second part follows by expanding the integrand of (5.83) ($s = 3$) or (5.102) ($s = 1$) etc. in a Taylor series about $\rho_b(z)$. Here s is the dimensionality of the potential; $s = 1$ for rods, oriented squares, and oriented cubes etc., $s = 2$ for discs and oriented cylinders etc. and $s = 3$ for spheres. There is a similar equation for $\partial W/\partial\rho_b$, and comparison with (5.132) gives the masses

$$m_{aa} = m_{bb} = 0, \qquad m_{ab} = -\frac{1}{2(s+2)}. \quad (5.134)$$

The first integral of (5.133) gives for the kinetic energy K,

$$K = m_{ab}\dot{\rho}_a\dot{\rho}_b. \qquad \text{(cf. (5.68))} \quad (5.135)$$

As for the two-component lattice gas, the mechanical analogy has become somewhat strained since the only mass in these equations is negative, but nevertheless K of (5.135) is always positive or zero, since $\dot{\rho}_a$ and $\dot{\rho}_b$ have opposite signs, see Fig. 5.5.

The profiles $\rho_a(z)$ and $\rho_b(z)$ can now be found by minimizing the free energy, or in the mechanical analogy, by solving the equations of motion. The kinetic energy can be diagonalized

$$K = \tfrac{1}{2}m\dot{x}^2 - \tfrac{1}{2}m\dot{y}^2, \quad (5.136)$$

where mass m is

$$m = -2m_{ab} = 1/(s+2), \quad (5.137)$$

and

$$x = \tfrac{1}{2}(\rho_a - \rho_b), \qquad y = \tfrac{1}{2}(\rho_a + \rho_b). \tag{5.138}$$

This mass m, like W, also differs from that of the transcribed version (5.120) by a factor of θ. The equations of motion are

$$-\partial W(x, y)/\partial x = m\ddot{x}, \qquad -\partial W(x, y)/\partial y = -m\ddot{y}, \tag{5.139}$$

which becomes on introduction of (5.131) in the new variables,

$$\ln\left(\frac{y + x}{y - x}\right) - 2x = m\ddot{x} \tag{5.140}$$

$$\ln(y^2 - x^2) + 2y - 2x_0 \coth x_0 - 2 \ln(x_0 \operatorname{cosech} x_0) = -m\ddot{y}$$

where $\pm x_0$ are the values of x in the coexisting phases and are related to ρ_a^α and ρ_b^β etc. by

$$\rho_a^\alpha = x_0 \coth x_0 + x_0, \qquad \rho_b^\beta = x_0 \coth x_0 - x_0. \tag{5.141}$$

This pair of differential equations can be reduced to one equation by elimination of the variable z. If y' denotes (dy/dx) along the trajectory through the interface, then by differentiation of $y' = \dot{y}/\dot{x}$, we have $\dot{x}^2 y'' = \ddot{y} - y'\ddot{x}$ while from constancy of the total energy, potential plus kinetic,

$$W + \tfrac{1}{2}m\dot{x}^2 - \tfrac{1}{2}m\dot{y}^2 = 0. \tag{5.142}$$

Substitution in (5.140) gives

$$2Wy''(1 - y'^2)^{-1} = \ln(y^2 - x^2) + 2y - y'\left(\ln\frac{y + x}{y - x} + 2x\right)$$

$$- 2x_0 \coth x_0 - 2 \ln(x_0 \operatorname{cosech} x_0). \tag{5.143}$$

This equation can be solved to give the trajectory $y(x)$ for each value of the parameter x_0, that is, $\rho_a(\rho_b)$ for each value of the coexisting densities $\rho_a^\alpha = \rho_b^\beta$.

In transcribing these results into those for the penetrable-sphere model we identify ρ for that model with ρ_a, and obtain the surface tension for the penetrable-sphere model without knowing explicitly the function $\rho(z)$;

where

$$\theta\sigma = \int_{-x_0}^{x_0} (-2mW)^{\frac{1}{2}}(1 - y'^2)^{\frac{1}{2}}\, dx \quad \text{(cf. 5.22)} \tag{5.144}$$

$$\theta = x_0 \operatorname{cosech} x_0 \exp(x_0 \coth x_0). \tag{5.145}$$

The trajectory $y(x)$ or $\rho_a(\rho_b)$ has been found, with difficulty,[21] from (5.143) for values of x_0 up to $2\cdot75$, which is equivalent to values of T/T^c down to $0\cdot481$. Figure 5.11 shows the W-surface for $x_0 = 2\cdot5$. As for the lattice gas, the trajectory is on the outer slope of the two arêtes that join

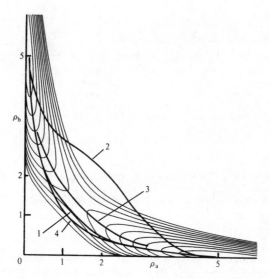

FIG. 5.11. Contours of the W-surface for $x_0 = 2 \cdot 5$. The four paths shown are, (1) the mean-field approximation, (2) the original van der Waals theory, (3) the 'one-density' theory of (5.146), and (4) the 'two-density' theory of (5.143).

the peaks $W^{\alpha} = W^{\beta} = 0$, and this route is a consequence of the negative sign of the mass m_{ab} of (5.134). A particle of positive mass moving on such a surface would roll along the inner slope, and pass below the col or saddle-point with its maximum speed, and experience there a centrifugal force which just balances $(\partial W/\partial y)_{x=0}$, the gradient of the potential along the line of symmetry.

The relation between ρ_a and ρ_b that follows from the mean-field approximation alone (that is, without the use of the gradient expansion common to all van der Waals-like theories) lies also on the outer slope, but is closer to the col than the trajectory from (5.143). This relation is drawn here for $s = 1$, while the solution of (5.143) is independent of s.

A third approximation is a path that passes down one arête, crosses the col, and climbs the other. On this path, we have, to a good approximation

$$\partial W(x, y)/\partial y = 0. \qquad \text{(cf. 3.19).} \quad (5.146)$$

It is only an approximation to $y(x)$ from (5.143) but may be an acceptable one since it can be obtained directly from the W-surface without solving a differential equation. It is discussed further in § 9.1.

Finally, the path of the original van der Waals theory can be found for the penetrable-sphere model by integrating[14]

$$z(\rho) = \left(\frac{1}{2(s+2)}\right)^{\frac{1}{2}} \int [-\theta W(\rho)]^{-\frac{1}{2}} \, d\rho \quad \text{(cf. 3.9).} \quad (5.147)$$

The constant of integration is found from knowing the separation of the dividing surface $z = z_e$ and the plane of symmetry $z = 0$. This curve, for $s = 1$, differs greatly from the other three. Near the critical point, $x_0 \to 0$, it coincides with the path defined by (5.146) since the original van der Waals theory is a one-density theory, but for large values of x_0 the two paths differ.

The surface tension from the two-density trajectory is calculated from (5.144) and is shown in Fig. 5.10 for $s = 1$. It is better than the original van der Waals result, but the improvement is obtained only after difficult computations. Finally we could, in principle, combine both methods of improvement suggested in this section; that is, use a two-density trajectory but with m a density-dependent parameter determined from the second moment of the local value of the direct correlation function. Such a theory would be a very sophisticated form of the van der Waals theory but it may be doubted whether its accuracy would justify its computational difficulties.

5.7 Penetrable-sphere model: spherical surfaces

We have seen in § 4.8 the difficulty of reconciling the arbitrariness of the surface of tension, z_s, defined mechanically (or quasi-thermodynamically) by (2.89) and the planar limit of the surface of tension of a drop, R_s, introduced through the thermodynamic arguments of § 2.4—arguments which fix its position with respect to the equimolar surface, R_e. One way of analysing problems of this kind is by exact calculations for a model system. The mean-field treatment of the penetrable-sphere model is not exact, but, as we have seen, it becomes so in the two limits of (1) infinite dimensionality at all temperatures, and (2) zero temperature for all dimensions. Here we examine the three-dimensional spherical drop (and bubble) in the mean-field approximation and show that the results resolve some of the difficulties of § 4.8.

It is simplest to use the primitive form of the model, that is the system described by the pair potentials of (5.77), and to consider a drop rich in component a surrounded by a fluid rich in b. The transcription to the one-component version of the model is then made by using (5.129).

Consider first, the two homogeneous phases α (inside the drop) and β (outside the drop) at points well removed from the interface. The pressures are given by[13]

$$\pi^\alpha = \rho_a^\alpha + \rho_b^\alpha + \rho_a^\alpha \rho_b^\alpha, \qquad \pi^\beta = \rho_a^\beta + \rho_b^\beta + \rho_a^\beta \rho_b^\beta, \qquad (5.148)$$

so that the pressure difference $\Delta\pi = \pi^\alpha - \pi^\beta$ is known in terms of the four densities ρ_a^α etc. The conditions of equilibrium between the two phases are, cf. (5.128)

$$\zeta_a = \rho_a^\alpha \exp(\rho_b^\alpha) = \rho_a^\beta \exp(\rho_b^\beta),$$
$$\zeta_b = \rho_b^\alpha \exp(\rho_a^\alpha) = \rho_b^\beta \exp(\rho_a^\beta). \tag{5.149}$$

The physical state of the system can be specified by two independent variables for which we choose $\Delta\pi$ and ζ_b. From these ζ_a, ρ_a^α, ρ_a^β, ρ_b^α, and ρ_b^β can be found by solving the five equations formed from (5.149) and a subtraction of the two equations in (5.148).

The density profiles, $\rho_a(r)$ and $\rho_b(r)$, where r is the distance from the centre of the drop, are determined by constraining the activities of each component to be constant for all values of r; that is, cf. (5.127),

$$\rho_a(r)/\zeta_a = \exp(-\sigma_b(r)), \qquad \rho_b(r)/\zeta_b = \exp(-\sigma_a(r)), \tag{5.150}$$

where

$$\sigma_a(r) = \frac{3}{4} \int_{-1}^{1} (1-u^2)\left(1+\frac{u}{r}\right)\rho_a(r+u)\,du \qquad (r>1), \tag{5.151}$$

and there is a similar equation for $\sigma_b(r)$. These equations differ from (5.83) by the term in $(1+u/r)$, which is a geometrical factor characteristic of spherical surfaces. Equations (5.150) and (5.151) form a pair of integral equations for $\rho_a(r)$ and $\rho_b(r)$ which can be solved for known values of ζ_a and ζ_b, and with the known limits ρ_a^α and ρ_b^α for $r \to 0$, and ρ_a^β and ρ_b^β for $r \to \infty$. Figure 5.12 shows a typical pair of profiles for a drop of radius of about 10 molecular diameters. This model has an unusually compressible liquid state and so the excess of π^α over π^β causes ρ_a^α to be about 10 per cent larger than ρ_b^β, or than ρ_a under a planar interface.

These results can be transcribed to yield the profile $\rho(r)$ of a spherical drop of the liquid of the one-component version of the model by using (5.129); that is, $\rho_a(r)$ becomes $\rho(r)$ and ζ_b becomes the reciprocal temperature θ; $\Delta\pi$ remains unchanged on transcription. The profile for a bubble could be obtained by inverting the transcription, so that $\rho_b(r)$ became $\rho(r)$, and ζ_a became θ.

The surface tension at the surface of tension, that is σ_s, can be calculated from (4.228) and (4.231), which are the extension to a spherical surface of the calculation of σ for a plane surface via the direct correlation function. For this model in the mean-field approximation these equations give[22]

$$\sigma_s = \frac{3}{4} \int dr\,\rho(r) \int_{-1}^{1} du\,(3u^2-1)\exp(-\sigma(r-u)). \tag{5.152}$$

This expression reduces to (5.104) when the curvature is zero since the

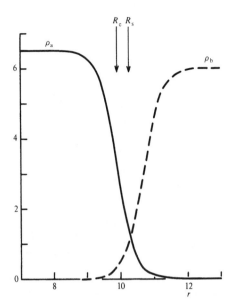

FIG. 5.12. The density profiles for a drop, rich in component a, of radius $r \sim 10$, at $T/T^c = 0.4452$.

exponential term in the integrand is, by the symmetry of the planar surface, then equal to $\theta^{-1}\rho(-z + u)$, cf. (5.90).

These equations yield values of $p^l - p^g$, of R_e and of σ_s for the curved surface, and, as described in § 5.5, it is straightforward to obtain σ_∞ for the planar surface at the same temperature. We recall that these quantities are linked by (2.57), (2.59), and (2.74), viz.,

$$p^l - p^g = \frac{2\sigma_s}{R_s}, \qquad \frac{\sigma_s}{\sigma_\infty} = 1 - \frac{2\delta}{R_e}, \qquad \delta = R_e - R_s. \tag{5.153}$$

We have therefore two independent equations for R_s, the radius of the surface of tension, or, what is equivalent, for δ, its separation from the equimolar surface. Figure 5.13 shows σ_s/σ_∞ as a function of R_e^{-1}. The points[22] are the values of σ_s calculated from (5.152) and the straight line has a slope -2δ, with δ calculated by eliminating σ_s between the two thermodynamic equations (5.153), viz.,

$$\frac{\delta}{R_e} = \frac{2\sigma_\infty - R_e(p^l - p^g)}{4\sigma_\infty - R_e(p^l - p^g)}. \tag{5.154}$$

The limiting slope of the curve through the points agrees well with the

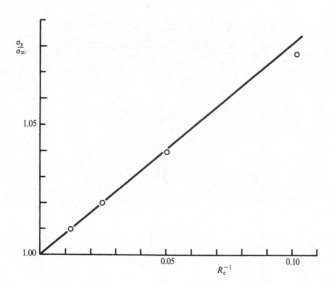

FIG. 5.13. The surface tension of a drop as a function of reciprocal radius at $T/T^c = 0\cdot4452$.

slope calculated from (5.154), thus showing the consistency of the thermodynamic equations and the statistical calculation of σ_s. This consistency is destroyed if σ_s is calculated by any equation other than (4.228) and (4.231), for example, one based on the bending of the interface about a dividing surface other than the equimolar surface used to obtain (4.228).

It is noteworthy that σ_s is greater than σ_∞, and that δ is therefore negative. For this model the surface of tension is on the gas-side of the equimolar surface. This can be confirmed directly at zero temperature since δ can then be shown[22] to be exactly $-\frac{1}{2}$. Calculations for different sizes of drop and for different temperatures show that δ is always negative and that it changes only slowly with these variables. Within the accuracy of the calculations, the apparent limit at $T = 0$ and $R_e^{-1} = 0$ agrees with the exact result of $-\frac{1}{2}$. At $T = 0$ the mean-field approximation becomes exact, and there is no uncertainty in the calculation of σ_s by (4.228) since the interface is a step-function. Hence the confirmation of the consistency of the statistical and thermodynamic calculations is strong evidence for the correctness of both. Towards the critical temperature δ tends to a finite non-zero length, but this result is a consequence of the mean-field approximation.

It has hitherto been accepted that δ is positive and hence that σ_s for a drop is less than σ_∞. There is scarcely any experimental evidence on this point (§ 2.4), and the theoretical evidence (§ 7.6) is based on the identification, which we believe to be faulty, of z_s of (2.89) with the planar limit of R_s of the thermodynamic treatment, and the calculation of the former by the arbitrary choice of the pressure tensor calculated by Irving and Kirkwood. Nevertheless we cannot conclude that δ is negative for real liquids, or for more realistic model liquids, since this result may be a peculiarity of the model. Such a peculiarity, if it be one, does not, of course, detract from the value of the consistency test.

Notes and references

1. Brout, R. *Phase Transitions*, pp. 10–11. Benjamin, New York (1965).
2. de Boer, J. *Physica Norvegica* **5,** 271 (1971); *Physica* **73,** 1 (1974).
3. Widom, B. *J. chem. Phys.* **39,** 2808 (1963).
4. That was emphasized also by Cotter, M. A. *Mol. Cryst. Liq. Cryst.* **39,** 173 (1977), for the mean-field theory of the phase equilibrium between an isotropic liquid and a nematic liquid crystal.
5. Longuet-Higgins, H. C. and Widom, B. *Mol. Phys.* **8,** 549 (1964); Guggenheim, E. A. *Mol. Phys.* **9,** 43, 199 (1965); Rigby, M. *Quart. Rev. chem. Soc.* **24,** 416 (1970).
6. Widom, B. *J. stat. Phys.* **19,** 563 (1978). It is possible to improve on or dispense with this further approximation, but only at the cost of added complexity; Robledo, A. *J. chem. Phys.* **72,** 1701 (1980).
7. Lee, T. D. and Yang, C. N. *Phys. Rev.* **87,** 410 (1952); Hill, T. L. *Statistical Mechanics*, Chapter 7, McGraw-Hill, New York (1956).
8. Ono, S. and Kondo, S. in *Encyclopedia of Physics* (ed. S. Flügge) Vol. 10, p. 134. Springer, Berlin (1960).
9. Guggenheim, E. A. *Mixtures*, Oxford University Press (1952); Domb, C. *Adv. Phys.* **9,** 149, 245 (1960); Rushbrooke, G. S. *Introduction to Statistical Mechanics* pp. 300–307, Oxford University Press (1949).
10. Wheeler, J. C. and Widom, B. *J. chem. Phys.* **52,** 5334 (1970).
11. Widom, B. and Rowlinson, J. S. *J. chem. Phys.* **52,** 1670 (1970); Rowlinson, J. S. *Adv. chem. Phys.* **41,** 1 (1980).
12. Leng, C. A., Rowlinson, J. S. and Thompson, S. M. *Proc. Roy. Soc. A* **352,** 1 (1976).
13. Guerrero, M. I., Rowlinson, J. S., and Morrison, G. *J. chem. Soc. Faraday Trans. II* **72,** 1970 (1976).
14. Harrington, J. M. and Rowlinson, J. S. *Proc. Roy. Soc. A* **367,** 15 (1979).
15. For a generalization of (5.96) and (5.98) to give $\Gamma_{i(j)}$ as an integral over z of a smooth difference of two values of the density ρ_i at height z, see Widom, B. *Physica* **95A,** 1 (1979).
16. Leng, C. A., Rowlinson, J. S. and Thompson, S. M. *Proc. Roy. Soc. A* **358,** 267 (1978).
17. Barker, J. A. and Henderson, D. *Rev. mod. Phys.* **48,** 587 (1976).
18. Guerrero, M. I., Rowlinson, J. S. and Sawford, B. L. *Mol. Phys.* **28,** 1603 (1974).

19. Widom, B. *Mol. Phys.* **25,** 657 (1973).
20. Yang, A. J. M., Fleming, P. D. and Gibbs, J. H. *J. chem. Phys.* **64,** 3732 (1976).
21. Hemingway, S. J., Rowlinson, J. S. and Severin, E. S. *J. chem. Soc. Faraday Trans. II* **76,** 936 (1980).
22. Hemingway, S. J., Henderson, J. R. and Rowlinson, J. S. *Faraday Sympos. chem. Soc.* **16** (1981) in press.

COMPUTER SIMULATION OF THE
LIQUID–GAS SURFACE

6.1 The experimental background

The aim of a molecular theory is the interpretation of the behaviour of part of the physical world in terms of the properties of its constituent molecules. The property we need to know in order to use the statistical theory of Chapter 4 is $\mathcal{U}(\mathbf{r}^N)$, the configurational energy of an assembly of N molecules as a function of their positions $\mathbf{r}^N$ (and of their orientations ω^N if the molecules are not spherical). In Chapter 5 we evaded the problem of determining $\mathcal{U}$ for a real system by using somewhat unrealistic model systems chosen so that the form of $\mathcal{U}$ leads to tractable expressions when used in the equations of Chapter 4. This is useful because of the enlightenment that comes from exact, or mathematically well-defined approximate solutions of the equations. The history of the statistical theory of phase transitions and related problems teaches us that progress has come more often from 'good' solutions for idealized models than from solutions for realistic systems that require ill-defined approximations. The Ising model, the lattice gas, and the penetrable-sphere model are examples of such idealized but tractable systems. Nevertheless our ultimate aim must always be the interpretation of the real world.

We can approach the problem from the other end and ask what we know experimentally, first about $\mathcal{U}$, and secondly about the structure and properties of the liquid–gas surface. A short answer to the first question is that it is only for the inert gases and their mixtures that we know $\mathcal{U}$ with reasonable accuracy, say to 1 per cent in the dilute gas, and to 3 per cent in the liquid.[1] For non-spherical molecules we know much about the functional form[2] of $\mathcal{U}(\mathbf{r}^N, \omega^N)$ but even for such simple cases as N_2, O_2, CO, and CH_4 we cannot determine this function to better than about 10 per cent for typical liquid configurations.

The answer to the second question is no more encouraging: the gross thermodynamic properties of surface tension σ, surface energy ϕ^s, and adsorptions $\Gamma_{i(j)}$ can be measured satisfactorily, but there is scarcely any information on the structure of the interface. Little is known even of $\rho(z)$, except in some special cases, and nothing at all of any higher distribution function. The usual way of measuring distribution functions in uniform fluids is by the scattering of radiation, but there are great difficulties in applying these methods to the interface. The radiation scattered from a two-phase system comes, unless special precautions are taken, almost entirely from the molecules in the bulk phases. One precaution is to use a geometry in which the radiation scattered at a certain angle comes almost exclusively from the surface region. Thus if a beam of visible light meets a liquid surface from below at an angle that exceeds that for total internal reflection then the angular and spectral distribution of light in the reflected beam contains information on the structure and dynamics of the surface. If the light falls on the gas side, only a part is reflected but that part contains similar information. The reflection of light is therefore a useful tool for studying the properties of a surface, both of pure liquids[3] and of those with adsorbed layers,[4] on a scale of length down to that of the wavelength of the light, or about 100 nm. In principle the technique of ellipsometry[5] can extend the range down to about 10 nm but it has so far been used mainly for the study of adsorbates on solids, and moreover the length is still too large by a factor of 10–100 to tell us anything in detail about $\rho(z)$ for liquids near their triple points. Near the critical temperature the thickness of the interface is comparable with the wavelength of light and optical techniques can then be used to determine $\rho(z)$; Webb and his colleagues have used this method but we defer their results to § 9.4.

Some assumption is needed to relate the 'profile of refractive index' to that of density, when the latter is neither a step-function, on the one hand, nor, on the other, a function which varies only over lengths comparable with the wavelength of the light.[6] Thus only one imperfectly defined parameter, a thickness, can be measured, and this is typically[7] $0 \cdot 79 \pm 0 \cdot 05$ nm for argon at 90 K. In a mixture the adsorptions can be estimated by this technique.[8]

X-rays and neutrons have wavelengths of the order of $0 \cdot 1$ nm but to avoid the scattering from the bulk phases it is necessary to use an angle of incidence that exceeds that for total reflection. If this is to be practicable we need a liquid whose refractive index, with respect to these waves, is less than unity,† so that the angle of total reflection lies on the gas side of the surface. Examples are, mercury with X-rays, and isotope ^{36}Ar with neutrons. Lu and Rice[9] have carried out the first experiment, the reflection of X-rays from the surface of mercury at almost grazing incidence, and find a profile that has, within the accuracy of their measurements, the form of a hyperbolic tangent, and a thickness of $0 \cdot 56 \pm 0 \cdot 05$ nm. The reflection of neutrons from argon is more difficult and has not yet been attempted.[10] It is clearly going to be some time before there is a substantial body of reliable information, even on $\rho(z)$, for simple liquids of the kind to which the theory of Chapter 4 is applicable.

† The phase velocity but not the group velocity of the X-rays then exceeds the speed of light in vacuo.

This gap, between theories that can be applied only to idealized systems, and experiments that are restricted to liquids or to circumstances that are not readily treated theoretically, can be filled by the method of computer simulation. Here $\mathcal{U}(\mathbf{r}^N)$ is specified and, within reason, can have as realistic a form as we wish. Computer simulation is then a technique which tells us the macroscopic physical properties such a realistic model system would possess. It has two aspects; first, with respect to statistical theories, it is an 'experiment' which tells us the properties of a system of given configurational energy. This has been its primary use and the one we exploit in this Chapter. A second aspect, with which we are less concerned, is its use in the determination of intermolecular forces. By comparing the properties of the simulated system with those of a real system we can judge how closely the chosen form and strength of $\mathcal{U}$ match those of the real molecules.

6.2 The methods of computer simulation

The two methods of computer simulation are known by the labels of Molecular Dynamic (MD) and Monte Carlo (MC) simulation. In the first the evolution of an assembly of N molecules is followed by numerical solution of Newton's equations of motion. The system is one of fixed N, V, and U and so is the simulation of a micro-canonical ensemble, but since the sequence of states is that of 'real time' both equilibrium and dynamic information can be obtained. In the second method a sequence of states is generated such that each state occurs with a probability proportional to its Boltzmann factor, $\exp(-\mathcal{U}(\mathbf{r}^N)/kT)$. The sequence is (usually) specified by fixed values of N, V, and T, and so the ensemble represented is canonical. The ordering of the steps of the sequence is arbitrary (that is, it contains no information) and so only thermodynamic properties can be calculated. The principles and practice of these techniques are described elsewhere;[11] both have been used to study the liquid–gas surface and here we describe only the special problems which these studies involve.

Unless otherwise stated, all simulations discussed in this chapter have used the same form of $\mathcal{U}(\mathbf{r}^N)$, namely a Lennard–Jones (12, 6) potential function between each pair of molecules,

$$\mathcal{U}(\mathbf{r}^N) = \sum_{i<j} \sum u(r_{ij}), \tag{6.1}$$

$$u(r_{ij}) = 4\varepsilon\left[\left(\frac{d}{r_{ij}}\right)^{12} - \left(\frac{d}{r_{ij}}\right)^6\right]. \tag{6.2}$$

This potential has its zero value at $r = d$ and its minimum of $u = -\varepsilon$ at

$r^6 = 2d^6$. It is a widely used and reasonably realistic potential for simple liquids, but results obtained for this potential cannot be equated with those for (say) liquid argon without incurring errors that cannot be ignored.[1] With this potential we use the reduced units of $\tau = kT/\varepsilon$, and $\rho = Nd^3/V$.

Even the largest computer cannot handle a system of $N \sim 10^{23}$ molecules, but is restricted to $N \sim 10^3$. It is this restriction that causes most of the difficulties of simulating realistically a system with a planar liquid–gas surface. Some of the first studies of surfaces were done with drops containing $N \sim 10^1$–10^3 molecules,[12] which are of interest in their own right, relevant to problems of the nucleation, growth, and evaporation of drops, but which tell us little about planar surfaces. Simulations of uniform fluid systems overcome the problem of the small size of the system by confining $N \sim 10^2$–10^3 molecules to a cubic cell of side L and using a convention of repetition of the Cartesian coordinates; that is, if a molecule leaves the cell by passing through the wall at $z = L$, it is deemed to re-enter it simultaneously through the wall at $z = 0$. The same convention is applied to all six faces so that the sample behaves as if it were immersed in an infinitely repeating set of replicas of itself; for most purposes it can be treated as a part of an infinite system. The potential energy is calculated by allowing each molecule to interact with those in its own cell and with their images in the surrounding boxes. To avoid spurious correlations this interaction is subject to a *minimum image* convention whereby molecule i interacts either with j, or with one of its images, j', j'', etc., whichever is the closest. This convention is obviously most appropriate if the range of the intermolecular potential is less than $\frac{1}{2}L$. In practice this is not a serious restriction for non-ionic and non-polar systems, since although intermolecular potentials go only asymptotically to zero at large r_{ij}, in practice they are negligible beyond about 1 nm.

If the system is not uniform but separated into two fluid phases then the convention of repeating coordinates is adequate for the calculation of bulk thermodynamic properties but not for the study of the interface which would be of irregular, ever-changing, and unidentifiable shape. A simple change of the boundary conditions leads to a much more useful configuration; repetition of the coordinates is retained in the x and y directions, but the cell is bounded by reflecting walls at $z = 0$ and $z = L$. If now the simulation is started with a flat liquid surface in the x, y (or horizontal) plane, then the repetition of these coordinates tends to maintain its position. Other constraints can be added to enhance the stability.

The commonest additional constraint has been a short-ranged attractive field near the bottom of the cell ($z \sim 0$) which serves to anchor the liquid in the lower half. This field is chosen to mimic simply the attractive forces that would emanate from a uniform liquid phase below the bottom boundary. However the restriction that no molecule in the cell crosses this boundary leads to a distortion of $\rho(z)$ from its uniform value of ρ^l for values of z less than a few molecular diameters. Croxton and Ferrier[13] made a MD simulation with 200 molecules of a two-dimensional system (with a one-dimensional 'surface') and tried to avoid this distortion by allowing random fluctuations in the shape of the bottom of the cell. Their results show, however, strong oscillations in $\rho(z)$ from $z = 0$ through the whole of the liquid state out to the liquid–gas surface, and it is now apparent that they were unsuccessful in preventing the propagation of the distortion through the fluid. The constraint of a steady field and a flat bottom has been used more successfully in three-dimensional systems by Liu[14] (MC simulation, 129 molecules), by Opitz[15] (MD simulation, 300 molecules), and by Chapela et al.[16,17] (MC and MD simulations, 255–4080 molecules). In each case the strong density distortions near $z = 0$ have become negligible, or at least much less serious, at the height of the liquid–gas surface, which is at values of z/d of 5 to 15 where d is the molecular diameter, (6.2). Liu claims that his results show oscillations, but they are hard to see in his published graphs.

Barker and his colleagues have used thick films of liquid with two planar surfaces near (say) $z = \pm l$, where $l \sim 6d$, so that $2l$ is sufficiently large for the liquid near $z = 0$ to be in a state close to that of the bulk liquid. At first,[18] they confined the film to a specified range of z by applying an artificially strong quasi-gravitational attraction towards the film to any molecule which strayed into the regions of large $|z|$. Although this field did not act directly on the liquid nor its surface, it seems to have been a constraint which, together with MC sequences that were later found to be insufficiently long, induced oscillations in $\rho(z)$ even at depth of $9d$ below the surface. In later simulations[19,20] they dispensed with the quasi-gravitational field and averaged over longer MC sequences. These results were freer from oscillations, that is $\rho(z)$ became essentially uniform throughout the bulk liquid and fell monotonically to the (negligible) density of the bulk gas on passing through either surface of the film. The theoretical disadvantage that the centre-of-mass of the film was free to move in the z-direction, thereby inducing an artificial thickening of the surface when $\rho(z)$ is calculated from a long MC sequence, was avoided by having a sufficiently large number of molecules (256). A similar configuration has been used by Rao and Levesque[21] for a MD simulation; here the film does not move if the initial configuration is chosen so that there is no net momentum in the z-direction. Clearly if the equimolar dividing surface is free to move in 'laboratory' coordinates then any oscillations, if present, would be eliminated in the average profile in these coordinates.[22] Osborn and Croxton[23] have argued that this may have happened in some simulations, although in the results of Chapela et al. the net movement was less than $0 \cdot 1d$, and it is unlikely from the geometrical constraints that the variance of this shift was much larger.

Typical dimensions of the cells in which the anchored liquid layer or the thick film is contained are 4 to $7d$ in the x- and y-directions and $20d$ or more in the z-direction, of which a height of 6 to $15d$ is filled with liquid. Such a cell contains typically about 250 molecules. A few simulations[17,21,24] have been made with significantly larger numbers of molecules (1000–4000), principally in order to study the change of $\rho(z)$ with the cellular area of the surface.

Two tests that have been used to monitor the correctness of the simulation are the constancy of temperature[21,25] and chemical potential[25] through the interface. The former (see § 4.10) can be measured by the local density of kinetic energy and is an appropriate test of a MD simulation, that is, of a micro-canonical ensemble. The precision of the test diminishes as one passes from liquid to gas, since the fall in $\rho(z)$ means that the average is taken over a falling number of molecules. The chemical potential can be measured by using the potential distribution theorem (4.71), and here the precision is greater when $\rho(z)$ is low. Within these limitations, both $T(z)$ and $\mu(z) = kT \ln \lambda(z)$ have been found to be constant.

6.3 The density profile

The first aim of most simulations is the determination of the density profile, that is, of the singlet density $\rho(z, T)$. The limiting value of this function when z is well below the surface is $\rho^l(T)$, the orthobaric liquid density. The limit is hard to estimate since most simulations show $\rho(z)$ rising perceptibly as z falls, even when z is several units of d below the surface. The apparent values of ρ^l, estimated from $\rho(z)$ with z about $4d$ below the surface, are below the best estimates of ρ^l from simulations of homogeneous systems[26] (Fig. 6.1). This discrepancy arises from the use of truncated potentials. It can be avoided, without using potentials of infinite range in the simulation, by treating an average of the missing or truncated part of the Lennard–Jones potential as a density-dependent one-body potential.[27] But if this is not done the results for $\rho(z)$ in the surface layer will not be as accurate as their apparent precision suggests.

Three typical profiles are shown in Fig. 6.2. The first[21] is for a film with two surfaces (MD, 1728 molecules) at $\tau = 0 \cdot 701$, and the second and third[17] are for systems in which the liquid is anchored to the bottom of the cell, and which therefore have oscillations in $\rho(z)$ near the wall at $z = 0$. The second is at $\tau = 0 \cdot 785$ (MD, 1020 molecules) and the third at $\tau = 1 \cdot 127$ (MC, 255 molecules). As the temperature rises, ρ^l falls, ρ^g rises and the surface becomes thicker. In every case the shape of $\rho(z)$ is that of

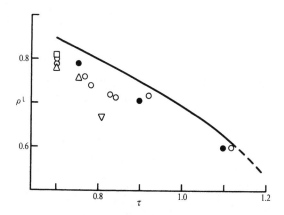

FIG. 6.1. The orthobaric liquid density of a Lennard–Jones fluid as found from computer simulations of the liquid–gas surface. Open circles, Chapela *et al.*;[17] closed circles, Liu;[14] triangles, Rao and Levesque;[21] inverted triangle, Opitz;[15] squares, Lee *et al.*[18] The curve is the best estimate of ρ^l from simulations of homogeneous systems.[26]

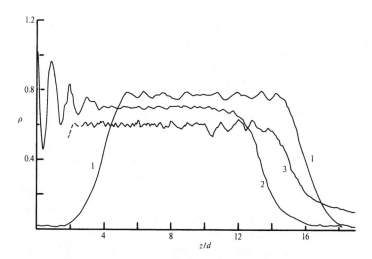

FIG. 6.2. Three profiles, $\rho(z)$, for a Lennard–Jones fluid. Curve 1 is a two-sided film (MD, 1728 molecules) at $\tau = 0{\cdot}701$; curve 2 is for a portion of liquid against a wall (MD, 1020 molecules) at $\tau = 0{\cdot}785$; curve 3 is for a similar configuration but is a MC simulation (255 molecules) at $\tau = 1{\cdot}127$.

a monotonic, apparently antisymmetric function which can, for example, be fitted by an expression of the form

$$\rho(z) = \tfrac{1}{2}(\rho^1 + \rho^g) - \tfrac{1}{2}(\rho^1 - \rho^g)\tanh(2(z - z_e)/D). \qquad (6.3)$$

Here z_e is the height of the equimolar dividing surface, and D is a measure of the thickness. Although a hyperbolic tangent arises naturally for the penetrable-sphere model (§ 5.5) and in the van der Waals theory of a system near its gas–liquid critical point (§ 9.1) its use here is purely empirical; indeed, we shall see in § 7.5 that an exponential decay of $\rho(z)$ at large values of $|z - z_e|$ is not correct for a Lennard–Jones potential that runs to $r = \infty$. This equation is, however, a convenient one since it can be fitted to 'experimental' points by inverting it to give

$$\ln\left(\frac{\rho^1 - \rho(z)}{\rho(z) - \rho^g}\right) = \frac{4(z - z_e)}{D}, \qquad (6.4)$$

so that a graph of the expression on the left, as a function of z, has an intercept of z_e and a slope of $(4/D)$. The parameter D, the surface thickness, can be defined generally by

$$D = -(\rho^1 - \rho^g)[d\rho(z)/dz]_{z=z_e}^{-1}, \qquad (6.5)$$

a definition of which the parameter D in (6.3) is a special case. Other measures of surface thickness have been advocated[28] but D in (6.5) is the parameter most easily obtained from simulation results. The '10–90' thickness (that is, the distance from $[\rho^g + (0\cdot1)(\rho^1 - \rho^g)]$ to $[\rho^g + (0\cdot9)(\rho^1 - \rho^g)]$) is less sensitive to the exact shape of the profile than D of (6.5), but it cannot be estimated with any precision from the 'noisy' results of computer simulation.

Figure 6.3 shows the thickness D as a function of the reduced temperature τ for cells of different widths. The thickness increases as the temperature rises and is infinite at the critical temperature. Computer simulation is not a useful technique for studying critical behaviour since the properties of a fluid near that point are determined by correlations of a range long compared with the range of the intermolecular forces, whilst the cells used in simulation cannot be larger than about 2 to 5 times this range. Hence Fig. 6.3, although showing the approach of D to infinity at $\tau = \tau^c$, gives no useful information about the functional dependence of D on $(\tau^c - \tau)$.

More interesting is that D increases also with the width of the cell. The repetition of the Cartesian coordinates of molecules in the x- and y-directions means that a simulated surface in a cell of width l is effectively tied to a horizontal square grid of mesh $l \times l$. Capillary fluctuations of wavelength longer than l are suppressed. As the width of the cell

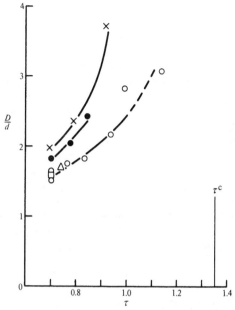

Fig. 6.3. Surface thickness for cells of different widths, l. Square, $l/d = 4\cdot4$; open circles, $l/d = 5\cdot0$; triangle, $l/d = 6\cdot78$; closed circles, $l/d = 10$; crosses, $l/d = 20$.

increases, therefore, we see the progressive excitation of these fluctuations of ever-increasing wavelength. The arguments set out in § 4.9 show that such waves lead to a thickening of the surface (in the absence of a gravitational field) that is proportional to $(\ln A)^{\frac{1}{2}}$, where $A = l^2$ is the cellular area. This thickness is, it is presumed, to be added to the intrinsic width of $\rho(z)$ unaffected by capillary waves. The small number of results in Fig. 6.3 is consistent with this interpretation, but cannot be used to test it quantitatively.

Kalos *et al.*[24] have analysed the results of Rao and Levesque[21] in order to study explicitly the correlations in the x,y-plane. They find the strongest correlations at low values of the wave-vector q, and confirm qualitatively the form of $H(q; z_1, z_2)$ given by Wertheim's expression (4.261).

6.4   The surface tension

There are two principal routes from the molecular distributions to the most important thermodynamic property of the interface—the surface tension. The first (§ 4.4) is via the virial theorem and requires a knowledge of the two-body distribution function $g(r_{12}, z_1, z_2)$; the second

(§ 4.6) requires a knowledge of the gradient of the density $\rho'(z)$ and of the direct two-body correlation function $c(r_{12}, z_1, z_2)$. The first needs elaboration if the system is one with multi-body potentials but we restrict our attention for the moment to simulations based on the Lennard–Jones pair potential (6.1)–(6.2). The equivalence of these two routes was shown in § 4.7, so the choice between them is a matter of convenience, and here the first has everything in its favour. The virial of the pair potential, $r(\mathrm{d}u/\mathrm{d}r)$, is readily calculated at each step of a MC or MD simulation, and although it is difficult to obtain $g(r_{12}, z_1, z_2)$ explicitly, the distribution of the molecules which it describes is again just what is generated naturally in a simulation. The second route requires the direct correlation function, c, which could only be found from the distribution function, g, by solving the Ornstein–Zernike equation by Fourier transforms. Even the determination of the gradient $\rho'(z)$ can be done only at the end of a simulation, and then not with high accuracy, as Fig. 6.2 shows. The second route has, therefore, not been used and most of the results below are obtained from the first route. Equation (4.81) can be expressed

$$\sigma A = \left\langle \sum_{i<j} (r_{ij} - 3z_{ij}^2/r_{ij})u'(r_{ij}) \right\rangle \qquad (6.6)$$

where the summation is taken over all pairs of molecules in a sample of area A, and where the angle brackets denote an average over a MD (micro-canonical) or MC (canonical) simulation. In the bulk phases $3z_{ij}^2/r_{ij}$ has an average of r_{ij} so that the whole of the effective contribution to σ comes from pairs of molecules near the interface. The boundary conditions on the sample and the use of truncated potentials introduce minor complications[17] into the calculation of the average which need not be discussed here.

The accuracy of the surface tension calculated from (6.6) is low, perhaps ± 10 per cent for a MC simulation of 250 molecules and 10^6 steps. A more accurate result can be obtained from a MC calculation of the free energy needed to cut a block of liquid into two parts, separate them, and allow the profiles to relax to their equilibrium shapes. Bennett[29] devised a MC scheme for the calculation of the change in free energy in the first step, and the whole operation was carried out by Miyazaki et al.[20] The accuracy in σ for a Lennard–Jones liquid near its triple point is improved to about ± 2 per cent.

Figure 6.4 shows the values of σ calculated in the main by the virial route, but with the one point of Miyazaki et al. The greater accuracy of this point has influenced the chosen course of the continuous curve drawn through all the points. (The results of Rao and Levesque[21] and of Rao and Berne[30] are for a potential truncated at $r = 2\cdot5d$, and have been increased[17] by $0\cdot50$ and $0\cdot40$ respectively to make them comparable with

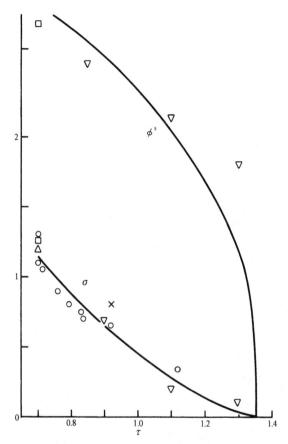

FIG. 6.4. The values of surface tension, σ, and surface excess energy, ϕ^s, from the computer simulations of Chapela *et al.*,[17] circles; Miyazaki *et al.*,[20] squares; Rao and Levesque,[21] triangles; Liu,[14] inverted triangles; and Rao and Berne,[30] crosses.

those for the complete Lennard–Jones potential.) The temperature range shown is that of the whole of the liquid state, from the triple point at $\tau^t = 0.70$ to the critical point at $\tau^c = 1.35$. The results in the upper half of the figure are determinations of the surface excess energy from an expression similar to (6.6) but with $u(r)$ in place of $ru'(r)$. The upper line is the value of this energy that is thermodynamically consistent with the full line drawn through the points for the surface tension; that is, the two lines are related by (2.41).

The results in Fig. 6.4 are compared with theoretical calculations in the next chapter. They can be compared also with experimental results for the inert gases, but the comparison is inevitably a crude one since the Lennard–Jones potential is not a good representation of the potential between a pair of argon atoms.[1] The best that can be done in this direction is to choose $d = 0.340$ nm and $\varepsilon/k = 119.4$ K, when the second virial coefficient and other macroscopic properties of gaseous and solid argon are reproduced as well as is possible with a Lennard–Jones potential. The surface tension in Fig. 6.4 then becomes the curve numbered 2 in Fig. 6.5, in which the customary units of kelvin (for T) and mN m^{-1} (for σ) replace the reduced scales of kT/ε and $\sigma d^2/\varepsilon$. The curve numbered 1 represents the experimental results of Fig. 1.5; it lies below the Lennard–Jones curve. Better agreement with experiment is obtained by using a more accurate pair potential for argon, u_{ij}, and introducing the correction for the three-body dipole potential, u_{ijk}. There is now general agreement[1] that the appropriate values of d and ε for the pair potential are $d = 0.336$ nm and $\varepsilon/k = 142$ K, and one potential that has these parameters is that of Barker, Fisher, and Watts.[31] They have also esti-

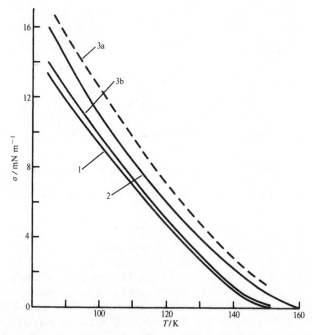

FIG. 6.5. The surface tension of argon (1), and its representation by a Lennard–Jones potential (2), by the pair potential of Barker, Fisher, and Watts (3a), and by this potential with a three-body correction (3b).

mated the strength of the three-body term, which cannot be ignored for calculations on condensed phases. Barker and his colleagues[18,20,32] have used a combination of MC simulations of a Lennard–Jones liquid with perturbation theory (§ 7.5) to estimate the differences in σ on changing from this potential to the more accurate one of Barker, Fisher, and Watts, and on adding a three-body potential. The results are shown in Fig. 6.5 and are seen to be much closer to the experimental curve than those calculated from the Lennard–Jones potential. The large size of the three-body correction is particularly noteworthy; its effect on the surface tension is much larger than that on the energy or free energy of the homogeneous liquid.[32]

6.5 Further work

The results discussed in the last two sections are the main part of the work so far done by the method of computer simulation, but other problems have been tackled and there are still further ones that are within the scope of modern computers. Here we note some of this work, and suggest how it might be extended.

A binary mixture of Lennard–Jones molecules, a and b has been studied by MC and MD simulation.[17] It has equal diameters ($d_{aa} = d_{ab} = d_{bb}$) and energies in the ratio of T^c for argon to T^c for krypton [$\varepsilon_{aa}/\varepsilon_{bb} = 0.763$ and $\varepsilon_{ab} = (\varepsilon_{aa}\varepsilon_{bb})^{1/2}$]. Component a, being that of lower surface tension at a given temperature, is preferentially adsorbed at the surface, in accord with the predictions of the Gibbs adsorption equation (2.45). The new information given by the computer simulation is where the adsorption occurs—a matter on which a purely thermodynamic treatment is silent. It appears (Fig. 6.6) that component a has a similar profile to that of b, but one which projects about a molecular diameter further into the gas phase. The amount adsorbed is related to the isothermal change of surface tension with composition by the Gibbs equation. Figure 6.6 shows that, within the large errors of estimation of σ and $\Gamma_{a(b)}$ the computer estimates of these quantities are mutually consistent. Such simulation could usefully be done at other mole fractions and temperatures.

Thompson and Gubbins[33] have studied diatomic molecules, using a site–site Lennard–Jones potential with two sites on each molecule. If these sites are close together, as in a potential chosen to represent N_2, then there is no discernible orientation in the surface layer. If they are well-separated, as in a potential with a separation of $0.608d$ chosen to represent Cl_2, then at low temperatures those molecules just below the equimolar surface tend to be aligned perpendicular to the surface, and

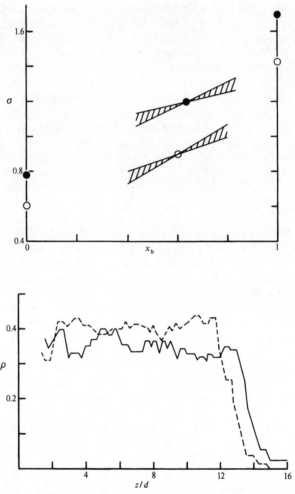

FIG. 6.6. The lower figure shows the profiles $\rho_a(z)$ (full line), and $\rho_b(z)$ (dashed line), for a mixture of two Lennard–Jones liquids with $\varepsilon_{aa}/\varepsilon_{bb} = 0.763$, at $\tau_a = 0.918$ and $\tau_b = 0.701$. The upper figure shows the surface tensions of the pure substances and of two mixtures at $\tau_a = 0.918$, $\tau_b = 0.701$ (open circles), and at $\tau_a = 0.933$, $\tau_b = 0.713$ (filled circles). The shaded zones are the slopes $(\partial\sigma/\partial x_b)$ estimated from the adsorptions calculated from the lower figure, and from a similar figure (not shown) for the higher temperature.

those above have a preference for a parallel orientation. Such orientations have been found also by Weber and Helfand[34] for simulated molecules of n-octane at the surface of a cavity. Again we may expect extensions of this work to other shapes and polarities.

The simulation[35] of the surface of a molten salt, chosen to represent KCl, shows a surface similar to but a little thinner than that of argon. There is no detectable adsorption of either ion at the surface.

Computer simulation has been used to obtain such local thermodynamic functions as the energy density,[21] $\phi(z)$, the transverse component of Irving and Kirkwood's definition of the pressure tensor,[30] $p_T(z)$, and hence the height of the related surface of tension in a planar interface.[30] The essential arbitrariness of these calculations has been discussed in §§ 4.3 and 4.10.

There have been a few attempts to obtain the pair distribution function $g(r_{12}, z_1, z_2)$ in the surface layer for monatomic fluids,[15,36,37] but the published results are too fragmentary to be of much value. They are certainly not good enough, for example, to use directly in the virial expression for the surface tension, or to Fourier transform to obtain the direct correlation function. Clearly this work and that on Gibbs adsorption and preferential orientation is only at an early stage; more results are needed and will surely be forthcoming in the next few years.

Finally we note, but do not discuss, some of the papers in which the form of $\rho(z)$ has been obtained by simulation for liquids packed against their own crystals,[38] or against other, usually more artificial walls.[39]

Notes and references

1. Maitland, G. C., Rigby, M., Smith, E. B. and Wakeham, W. A. *Intermolecular Forces: their origin and determination.* Oxford University Press (1981).
2. Gray, C. G. and Gubbins, K. E. *Theory of Molecular Fluids,* Vol. 1, Oxford University Press (in press).
3. Bouchiat, M.-A., Meunier, J. and Brossel, J. *C.R. Acad. Sci. Paris* **266,** B255 (1968); Cruchon, D., Meunier, J. and Bouchiat, M.-A. *C.R. Acad. Sci. Paris* **267,** B32 (1968); Meunier, J., Cruchon, D. and Bouchiat, M.-A. *C.R. Acad. Sci. Paris* **268,** B92 (1969); Katyl, R. H. and Ingard, K. U. *Light Scattering from Thermal Fluctuations of a Liquid Surface,* Chapter 6 of *In Honor of Philip M. Morse* (ed. H. Feshbach and K. U. Ingard) MIT Press, Cambridge, Mass. (1969); Langevin, D. *J. chem. Soc. Faraday Trans. I* **70,** 95 (1974); Byrne, D. and Earnshaw, J. C. *J. Phys. D* **12,** 1133 (1979); see also the Supplement to *J. Physique* **33,** p. C1 *et seq.* (1972). The use of light scattering for measuring surface tension is ·discussed briefly at the end of § 1.2.
4. McQueen, D. and Lundström, I. *J. chem. Soc. Faraday Trans. I* **69,** 694 (1973); Hammarlund, N., Ilver, L., Lundström, I. and McQueen, D. *J. chem. Soc. Faraday Trans. I* **69,** 1023 (1973); Lundström, I. and McQueen, D. *J.*

chem. Soc. Faraday Trans. I **70,** 2351 (1974); Byrne, D. and Earnshaw, J. C. *J. Phys. D* **12,** 1145 (1979).

5. *Optical Studies of Adsorbed Layers at Interfaces,* Symp. Faraday Soc. **4** (1970).
6. Castle, P. J. and Lekner, J. *Physica* **101**A, 99 (1980).
7. Beaglehole, D. *Phys. Rev. Lett.* **43,** 2016 (1979); *Physica* **100**B, 163 (1980). For earlier work on ellipsometry at the liquid–gas surface, see Bruce, H. D. *Proc. Roy. Soc. A* **171** 411 (1939); McBain, J. W., Bacon, R. C. and Bruce, H. D. *J. chem. Phys.* **7,** 818 (1939); Buff, F. P. and Lovett, R. A. in *Simple Dense Fluids* (ed. H. L. Frisch and Z. W. Salsburg) Chapter 2, Academic Press, New York (1968).
8. Beaglehole, D. *J. chem. Phys.* **73,** 3366 (1980).
9. Lu, B. C. and Rice, S. A. *J. chem. Phys.* **68,** 5558 (1978).
10. This experiment and some related studies on the use of neutron reflection to study interfacial properties, are discussed by Hayter, J. B., Highfield, R. R., Pullman, B. J., Thomas, R. K., McMullen, A. I. and Penfold, J. *J. chem. Soc. Faraday Trans I* **77,** 1437 (1981).
11. Hansen, J. P. and McDonald, I. R. *Theory of Simple Liquids,* Chapter 3, Academic Press, London (1976); Valleau, J. P. and Whittington, S. G. (Part A; Chapter 4), Valleau, J. P. and Torrie, G. M. (Part A; Chapter 5), Erpenbeck, J. J. and Wood, W. W. (Part B; Chapter 1), and Kushick, J. and Berne, B. J. (Part B; Chapter 2) in B. J. Berne (ed.), *Statistical Mechanics,* Parts A and B, Plenum, New York (1977).
12. Lee, J. K., Barker, J. A. and Abraham, F. F. *J. chem. Phys.* **58,** 3166 (1973); McGinty, D. J. *J. chem. Phys.* **58,** 4733 (1973); Abraham, F. F., Lee, J. K. and Barker, J. A. *J. chem. Phys.* **60,** 246 (1974); Abraham, F. F. *J. chem. Phys.* **61,** 1221 (1974); Binder, K. *J. chem. Phys.* **63,** 2265 (1975); Abraham, F. F. and Barker, J. A. *J. chem. Phys.* **63,** 2266 (1975); Miyazaki, J., Pound, G. M., Abraham, F. F. and Barker, J. A. *J. chem. Phys.* **67,** 3851 (1977); Rao, M., Berne, B. J. and Kalos, M. H. *J. chem. Phys.* **68,** 1325 (1978); Binder, K. and Stauffer, D. *J. stat. Phys.* **6,** 49 (1972); Briant, C. L. and Burton, J. J. *J. chem. Phys.* **63,** 2045, 3327 (1975); Farges, J., De Feraudy, M. F., Raoult, D. and Torchet, G. *J. Physique* **38,** C2-47 (1977); Etters, R. D. and Kaelberer, J. B. *Phys. Rev.* **11**A, 1068 (1975); Kaelberer, J. B. and Etters, R. D. *J. chem. Phys.* **66,** 3233 (1977); Nauchitel, V. V. and Pertsin, A. J. *Mol. Phys.* **40,** 1341 (1980).
13. Croxton, C. A. and Ferrier, R. P. *J. Phys. C* **4,** 2447 (1971).
14. Liu, K. S. *J. chem. Phys.* **60,** 4226 (1974).
15. Opitz, A. C. L. *Phys. Lett.* **47**A, 439 (1974).
16. Chapela, G. A., Saville, G. and Rowlinson, J. S. *Faraday Disc. chem. Soc.* **59,** 22 (1975).
17. Chapela, G. A. Saville, G., Thompson, S. M. and Rowlinson, J. S. *J. chem. Soc. Faraday Trans. II* **73,** 1133 (1977).
18. Lee, J. K., Barker, J. A. and Pound, G. M. *J. chem. Phys.* **60,** 1976 (1974).
19. Abraham, F. F., Schreiber, D. E. and Barker, J. A. *J. chem. Phys.* **62,** 1958 (1975).
20. Miyazaki, J., Barker, J. A. and Pound, G. M. *J. chem. Phys.* **64,** 3364 (1976).
21. Rao, M. and Levesque, D. *J. chem. Phys.* **65,** 3233 (1976).
22. Parsonage, N. G. *Faraday Disc. chem. Soc.* **59,** 51 (1975); Rowlinson, J. S. *Faraday Disc. chem. Soc.* **59,** 52 (1975).
23. Osborn, T. R. and Croxton, C. A. *Mol. Phys.* **34,** 841 (1977).
24. Kalos, M. H., Percus, J. K. and Rao, M. *J. stat. Phys.* **17,** 111 (1977).

25. Thompson, S. M. Thesis, Oxford University (1977).
26. Adams, D. J. *Mol. Phys.* **32,** 647 (1976); Nicolas, J. J., Gubbins, K. E., Streett, W. B. and Tildesley, D. J. *Mol. Phys.* **37,** 1429 (1979).
27. Barker, J. A., personal communication (1981).
28. Henderson, J. R. and Lekner, L. *Physica* **94A,** 545 (1978).
29. Bennett, C. H. *J. comput. Phys.* **22,** 245 (1976).
30. Rao, M. and Berne, B. J. *Mol. Phys.* **37,** 455 (1979).
31. Barker, J. A., Fisher, R. A. and Watts, R. O. *Mol. Phys.* **21,** 657 (1971).
32. Barker, J. A. and Henderson, D. *Rev. mod. Phys.* **48,** 587 (1976), see § IX. See also Present, R. D. and Shih, C. C. *J. chem. Phys.* **64,** 2262 (1976).
33. Thompson, S. M. *Faraday Disc. chem. Soc.* **66,** 107 (1978). (The vertical scales in Figs 5 and 6 should be divided by 2π.) Thompson, S. M. and Gubbins, K. E. *J. chem. Phys.* **74,** 6467 (1981).
34. Weber, T. A. and Helfand, E. *J. chem. Phys.* **72,** 4014 (1980).
35. Heyes, D. M. and Clarke, J. H. R. *J. chem. Soc. Faraday Trans. II* **75,** 1240 (1979).
36. Borštnik B. and Ažman, A. *Zeit. Naturforsch.* **34a,** 1236 (1979).
37. Thompson, S. M., Rowlinson, J. S. and Saville, G. (1978), preliminary results published in Croxton, C. A. *Statistical Mechanics of the Liquid Surface,* § 10.8, Wiley, Chichester (1980).
38. Borštnik, B. and Ažman, A. *Chem. Phys. Lett.* **32,** 153 (1975) (a two-dimensional simulation); Toxvaerd, S. and Praestgaard, E. *J. chem. Phys.* **67,** 5291 (1977); **70,** 600 (1979); Hiwatari, Y., Stoll, E. and Schneider, T. *J. chem. Phys.* **68,** 3401 (1978); Ladd, A. J. C. and Woodcock, L. V. *Mol. Phys.* **36,** 611 (1978); Cape, J. N. and Woodcock, L. V. *J. chem. Phys.* **73,** 2420 (1980); Bonissent, A., Gauthier, E. and Finney, J. L. *Phil. Mag. B* **39,** 49 (1979); Broughton, J. Q., Bonissent, A. and Abraham, F. F. *J. chem. Phys.* **74,** 4029 (1981).
39. Liu, K. S., Kalos, M. H. and Chester, G. V. *Phys. Rev.* **A10,** 303 (1974); Saville, G. *J. chem. Soc. Faraday Trans II* **73,** 1122 (1977); Abraham, F. F. *J. chem. Phys.* **68,** 3713 (1978); Snook, I. K. and Henderson, D. *J. chem. Phys.* **68,** 2134 (1978); Snook, I. K. and van Megen, W. *J. chem. Phys.* **70,** 3099 (1979); **72,** 2907 (1980); Rao, M., Berne, B. J., Percus, J. K. and Kalos, M. H. *J. chem. Phys.* **71,** 3802 (1979); Lane, J. E. and Spurling, T. H. *Chem. Phys. Lett.* **67,** 107 (1979); *Aust. J. Chem.* **33,** 231 (1980); Grigera, J. R. *J. chem. Phys.* **72,** 3439 (1980); Toxvaerd, S. *J. chem. Phys.* **72,** 3640 (1980); Torrie, G. M. and Valleau, J. P. *J. chem. Phys.* **73,** 5807 (1980); Heyes, D. M. and Clarke, J. H. R. *J. chem. Soc. Faraday Trans. II* **77,** 1089 (1981); Sullivan, D. E., Barker, R., Gray, C. G., Streett, W. B. and Gubbins, K. E. *Mol. Phys.* **44,** 597 (1981). This field is reviewed by E. Dickinson and M. Lal, *Adv. mol. Relax. Inter. Proc.* **17,** 1 (1980).

Additions to References

Computer simulation continues to be widely used to study surfaces of all kinds. We note one important innovation, a purpose-built MD simulator that can handle up to 12 000 atoms; see Sikkenk, J. H., Indekeu, J. O., van Leeuwen, J. M. J., and Vossnack, E. O. *Phys. Rev. Lett.* **59,** 98 (1987).

For a general review, see Allen, M. P. and Tildesley, D. J. *Computer Simulation of Liquids,* Chap. 11, Oxford University Press (1987).

CALCULATION OF THE DENSITY PROFILE

7.1 Introduction

In Chapter 4 we obtained several equations which relate the density profile of a planar surface, $\rho(z)$, to the pair potential, $u(r)$, or to functionals of it such as the two-body distribution function $\rho^{(2)}(r_{12}, z_1, z_2)$ or the direct correlation function $c(r_{12}, z_1, z_2)$. None of the equations, however, yields an explicit solution for $\rho(z)$. In this chapter we describe some of the extra assumptions that have been made to enable them to be solved, and discuss the results for realistic forms of $u(r)$. We consider primarily the Lennard–Jones $(12, 6)$ potential, (6.2), since it has been the most widely used, since there are several computer simulations for it, and since it is a reasonably realistic potential for simple fluids.[1]

The exact equations at our disposal are summarized at the end of § 4.2. They are, (1) the first YBG equation (4.85) which relates the gradient of $\rho(z_1)$ to the derivative of $u(r_{12})$; (2) the integral equation (4.86) which relates the gradients of $\rho(z_1)$ and $\rho(z_2)$ via the direct correlation function $c(r_{12}, z_1, z_2)$; and (3) the potential distribution theorem (4.87). The first YBG equation is an expression of a condition of mechanical equilibrium; that is, that $p_N(z)$ is a constant (§ 4.3). The third equation is an expression of the condition of thermodynamic equilibrium; i.e. that $\mu(z)$ is a constant, independent of z. Once $\rho(z)$ and either $\rho^{(2)}(r_{12}, z_1, z_2)$ or $c(r_{12}, z_1, z_2)$ are known then the surface tension can be found as described in §§ 4.4 and 4.6.

7.2 Solution of the YBG equation

The two-body distribution function can be written

$$\rho^{(2)}(r_{12}, z_1, z_2) = \rho(z_1)\rho(z_2)g(r_{12}, z_1, z_2). \tag{7.1}$$

This equation defines the distribution function g, and it is an approximation to this that we need to solve the first YBG equation. Substitution of (7.1) in (4.85) and integration from a point deep in the liquid $(z_1 = z')$ to an arbitrary point $(z_1 = z)$ gives

$$\ln\left(\frac{\rho(z)}{\rho^1}\right) = \frac{2\pi}{kT}\int_{z'}^{z} dz_1 \int_{-\infty}^{\infty} dz_2 z_{12}\rho(z_2)\int_{|z_{12}|}^{\infty} dr_{12}u'(r_{12})g(r_{12}, z_1, z_2)$$

$$\text{(pp)} \quad (7.2)$$

where $z_{12} = z_2 - z_1$ but $r_{12} = |\mathbf{r}_2 - \mathbf{r}_1|$.

Since we know little about $g(r_{12}, z_1, z_2)$ there have been many different suggestions as to how it may be approximated. There are, however, some constraints on the forms that we can choose. If both points are in the liquid or both in the gas then g must reduce to the isotropic functions $g^l(r_{12})$ and $g^g(r_{12})$ respectively. At low vapour pressure the latter can be approximated by $\exp(-u(r_{12})/kT)$ without sensible error. The former is well-known for the Lennard–Jones liquid, both from computer simulation[2] and from perturbation theory.[3] A second constraint is that g be invariant to a renumbering of the molecules; that is,

$$g(r_{12}, z_1, z_2) = g(r_{12}, z_2, z_1). \tag{7.3}$$

The need for this condition was first pointed out by Pressing and Mayer[4] who expanded g in a Taylor series and used (7.3) to show that, to first order, it is a function only of $(\tfrac{1}{2}z_1 + \tfrac{1}{2}z_2)$ and of r_{12}. Their approximation was, however, developed further with the critical region in mind, and was not intended for lower temperatures. The condition was emphasized again by Borštnik and Ažman[5] who observed that it was violated by Toxvaerd's first suggestion[6] for an approximate form of g. His later proposal[7] is not open to this criticism.

If ρ^l and ρ^g are fixed a priori, as they should be for a given temperature, then the solution of (7.2) is essentially an eigenfunction problem, for there will only be a limited set of functions g for which the limit $z = z^g$, a point in the gas, gives $\ln(\rho^g/\rho^l)$ on the right-hand side. This condition has been expressed as a set of 'sum-rules' by Lekner and Henderson.[8] In practice the relatively unimportant value of ρ^g is often left to be determined only by the limiting value of the integral at large z, although it is then an undesirable consequence that the profile obtained by integrating from liquid to gas is not the same as that obtained by going from gas to liquid.[9]

Once a form of g is chosen then (7.2) can be solved by iteration. A trial form of $\rho(z)$, for example, a step-function from ρ^l to ρ^g at $z = 0$, is used on the right and a new profile obtained by integration. The successive iterations of $\rho(z)$ have to be mixed if stable solutions are to be obtained; often only 10 per cent of the new function is mixed with the old. Up to 100 iterations can be needed and it is hard to distinguish between true solutions, metastable solutions, and solutions that are diverging slowly because of the form of g or of imperfections in the numerical analysis.

Different choices can apparently lead to quite different types of solution, even if the above conditions on g are met. Thus Nazarian[10] claimed that two simple approximations led to oscillatory solutions for $\rho(z)$, but, from the more recent work of Toxvaerd,[11] it is almost certain that these are divergent, not convergent solutions. Oscillatory solutions were also found in the early work of Croxton and Ferrier,[12] but this has also been criticized by Toxvaerd.[11]

We consider three sets of solutions for the Lennard–Jones potential for which both the form of g and the method of solution seem, *a priori*, to be acceptable. These are Toxvaerd's second solution,[7] and the more recent work of Fischer and Methfessel[13] and of Osborn and Croxton.[14] Toxvaerd uses the isotropic approximation

$$g(r_{12}, z_1, z_2) = g(r_{12}; \bar{\rho}), \tag{7.4}$$

with

$$\bar{\rho} = \rho(\tfrac{1}{2}z_1 + \tfrac{1}{2}z_2). \tag{7.5}$$

He interprets $g(r_{12}; \bar{\rho})$ as the distribution function of a hypothetical uniform fluid of density $\bar{\rho}$, and approximates it by a perturbation expression of the kind used by Weeks, Chandler, and Andersen,[3,15]

$$g(r_{12}; \bar{\rho}) = g_{hs}(r_{12}; \eta)\exp(-u_0(r_{12})/kT) \tag{7.6}$$

where g_{hs} is the distribution function for a fluid of hard spheres of diameter d_{hs} and reduced density $\eta = (\pi \bar{\rho} d_{hs}^3/6)$, and where u_0 is the repulsive part of the Lennard–Jones potential;

$$u_0(r) = u(r) + \varepsilon \quad (r < r_m), \qquad u_0(r) = 0 \quad (r > r_m), \tag{7.7}$$

where $-\varepsilon$ is the minimum value of $u(r)$ at $r = r_m$. His results for $\tau \equiv kT/\varepsilon = 1$ are shown in Fig. 7.1.

Fischer and Methfessel start by dividing $u(r)$ into two parts, $u_0(r)$, as in (7.7), and the perturbation term $u_1(r) = u(r) - u_0(r)$. This separation gives two integrals on the right-hand side of (7.2), in the second of which they put $g = 1$; that is, they use a mean-field approximation for the attractive part of the potential, $u_1(r)$. In the first integral they replace g by g_0, the distribution function for the reference system whose potential is u_0. The function g_0 is, in turn, approximated by g_{hs}, where the diameter of the hard spheres, d_{hs}, is given by the prescription of Barker and Henderson,[3,16,17]

$$\int_0^\infty [\exp(-u_0(r)/kT) - \exp(-u_{hs}(r)/kT)]\, dr = 0. \tag{7.8}$$

Finally g_{hs} is taken to be that of a uniform hard-sphere fluid of density equal to the average density of the Lennard–Jones system in a sphere of diameter d centred at $(\tfrac{1}{2}\mathbf{r}_1 + \tfrac{1}{2}\mathbf{r}_2)$. Their results are compared with those of Toxvaerd in Fig. 7.1 and with those obtained by computer simulation[18] in Fig. 7.2. Their liquid densities, ρ^l, are lower than those of Toxvaerd or of the computer simulation, which, in turn, are a little lower than those found by simulation of a uniform liquid (Fig. 6.1). If these differences of ρ^l are discounted then there is tolerable agreement between the three sets of profiles. Co *et al.*[9] obtained similar results for a square-well fluid.

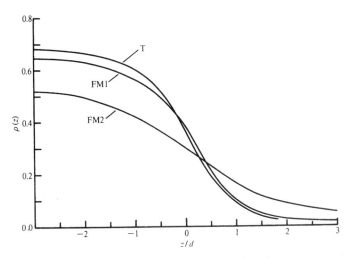

FIG. 7.1. The profiles calculated from the YBG equation by Toxvaerd, T, at $\tau = 1 \cdot 000$; and by Fischer and Methfessel, FM1, at $\tau = 0 \cdot 918$, and FM2, at $\tau = 1 \cdot 127$.

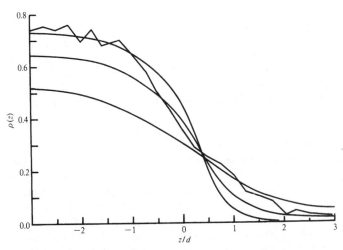

FIG. 7.2. The profiles from Fischer and Methfessel's solution of the YBG equation are the three smooth curves at temperatures, in decreasing size of ρ^l, of $\tau = 0 \cdot 759$, $0 \cdot 918$, and $1 \cdot 127$. The segmented curve is the result of computer simulation at $\tau = 0 \cdot 918$.

Osborn and Croxton use the isotropic approximation (7.4) but consider three definitions of the mean density $\bar{\rho}$, namely (7.5), and

and

$$\bar{\rho} = \tfrac{1}{2}\rho(z_1) + \tfrac{1}{2}\rho(z_2) \tag{7.9}$$

$$\bar{\rho} = \frac{1}{z_{12}} \int_{z_1}^{z_2} \rho(z) \, dz. \tag{7.10}$$

They also invoke the perturbation expansion of Weeks et al. to obtain $g(r_{12}; \bar{\rho})$. In the course of the iterative solution of (7.2) it has usually been necessary to redefine the zero of the z-axis between each step to stop the whole surface wandering meaninglessly. This can be done[6,7,9] by returning the equimolar dividing surface to $z = 0$ at each step. Osborn and Croxton use a different procedure; they choose a shift δz which minimizes the integral

$$\int [\rho_i(z) - \rho_{i+1}(z + \delta z)] \, dz$$

where $\rho_i(z)$ is the profile that results from the ith iteration. They claim that this procedure avoids destroying inadvertently any possible oscillatory structure in the profile. With $\bar{\rho}$ given by (7.5) and (7.10) they obtain smooth profiles, similar to those of Figs 7.1 and 7.2, for all reduced temperatures between $\tau = 0.6$ and $\tau = 1.2$. The use of (7.9) at temperatures below $\tau = 0.8$ leads to wildly oscillatory profiles in which the oscillations extend to about a depth of 12 molecular diameters into the liquid phase. It is hard to believe that these represent realistic solutions for the liquid–gas surface. They occur only for one form of $\bar{\rho}$, and then only for temperatures below or just above the triple point at $\tau = 0.70$. They are in conflict with the results of computer simulation and, we believe, it is unlikely that they would satisfy the requirement that $\mu(z) = \mu^l = \mu^g$ at every point, if $\mu(z)$ were to be calculated by the potential distribution theorem. We conclude that $\rho(z)$ for a Lennard–Jones fluid is a monotonic function of z.

We saw in §§ 5.5–5.7 that the potential distribution theorem was a valuable route to $\rho(z)$ for a simple model for which ρ^l and ρ^g were known. It has not yet been used for the liquid–gas surface of a Lennard–Jones fluid.[19] An older and less useful definition[20] of $\mu(z)$ is in terms of the function $g(r_{12}, z_1, z_2; \alpha)$ where α is a 'coupling parameter' which 'turns on' the interaction between molecules 1 and 2. By integrating from $\alpha = 0$ to $\alpha = 1$ it is possible to obtain a formally exact expression for $\mu(z)$ but one which has the great disadvantage that g is not known as a function of α. Plesner and Platz[21] used this route, with some simple approximation for g and for the equation of state of the liquid, to obtain profiles that are again similar to those of Figs 7.1 and 7.2.

The only attempt to calculate the position of the surface of tension from the YBG equation was made by Toxvaerd[6] with his earlier, and less satisfactory, approximation for g. He chose Irving and Kirkwood's defini-tion of $p_T(z)$ and obtained a value of $(z_e - z_s)$ of about 3 molecular diameters, which is about 3 times larger than the result found by computer simulation (§6.5). The shape of the curve for $[p_N(z) - p_T(z)]$ was similar—a positive lobe on the liquid side of z_e and a much smaller negative lobe on the gas side.

7.3 Approximations for the direct correlation function

Less is known of the direct correlation function in the interface $c(r_{12}, z_1, z_2)$ than of the total correlation function $h = g - 1$, except that it is almost certainly still a relatively short-ranged function, as it is in the homogeneous fluid. It is therefore natural to write, as in (7.4),

$$c(r_{12}, z_1, z_2) = c(r_{12}; \bar{\rho}) \qquad (7.11)$$

where $\bar{\rho}$ is given by (7.5) or (7.9). The only direct check on (7.11) is for the penetrable-sphere model[22] where, with $\bar{\rho}$ given by (7.5), the approxi-mation was found to be good to about ± 10 per cent at $r_{12} = d$, the diameter, but to be in error by up to 40 per cent at $r_{12} = 0$. The obvious step of substituting (7.11) into the exact equation (4.63) seems not to have been made except[23] as a way of estimating $\rho(z)$ near the limits ρ^l and ρ^g. It might be expected to be as useful as the substitution of (7.4) into the YBG equation. A more drastic form of this approximation, but not an unreasonable one since c depends but little on density in a homogeneous fluid, is to take $\bar{\rho}$ in (7.11) to be some constant density, as yet unspecified. With this assumption (4.86) becomes

$$\frac{d \ln \rho(z_1)}{dz_1} = 2\pi \int_{-\infty}^{\infty} dz_2 \rho'(z_2) \int_{|z_{12}|}^{\infty} dr_{12} r_{12} c(r_{12}; \bar{\rho}). \qquad (7.12)$$

An explicit form of c would now be needed to obtain $\rho(z)$ by numerical integration. However an explicit relation between the gas and liquid densities (analogous to the sum rule for the YBG equation[8]) follows on integration over z_1

$$\ln\left(\frac{\rho^g}{\rho^l}\right) = \lim_{L \to \infty} \int_{-L}^{L} dz_1 \int_{-L}^{L} dz_2 \rho'(z_2) \int_{|z_{12}|}^{\infty} dr_{12} 2\pi r_{12} c(r_{12}; \bar{\rho}) \qquad (7.13)$$

$$= \lim_{L \to \infty} \int_{-L}^{L} dz_1 \int_{0}^{\infty} dr_{12} 2\pi r_{12} c(r_{12}; \bar{\rho}) \int_{z_1 - r_{12}}^{z_1 + r_{12}} dz_2 \rho'(z_2)$$

$$= \lim_{L \to \infty} \int_{0}^{\infty} dr_{12} 2\pi r_{12} c(r_{12}; \bar{\rho}) \int_{-L}^{L} dz_1 [\rho(z_1 + r_{12}) - \rho(z_1 - r_{12})]$$

$$= -\int_{0}^{\infty} dr_{12} 2\pi r_{12} c(r_{12}; \bar{\rho}) \cdot 2r_{12}(\rho^l - \rho^g)$$

$$= -(\rho^l - \rho^g)v_0, \qquad (7.14)$$

where v_0 is the microscopic volume

$$v_0(T) = \int c(r; \bar{\rho}) \, d\mathbf{r}. \tag{7.15}$$

We see that (7.14) is formally the same equation for the orthobaric densities as was obtained for the penetrable-sphere model in the mean-field approximation, namely (5.91), and is the analogue of (5.31) for the lattice-gas. Like (5.91), it has the solution

$$\rho^c v_0(T^c) = 1 \tag{7.16}$$

at the critical point. But we know already from the compressibility equation (4.27), that v_0 defined by (7.15) is exactly $(1/\rho^c)$ at the critical point, since the critical compressibility κ^c is infinite. More realistic approximations to $c(r_{12}, z_1, z_2)$ should be rewarding.

Two other methods have used different approximations to the direct correlation function. One is an adaptation of the Percus–Yevick equations for a homogeneous binary mixture and the other is based on the functional expansions of § 4.5.

The adaptation of the Percus–Yevick approximation starts with the three Ornstein–Zernike equations which relate h_{aa}, h_{ab}, and h_{bb} to the set of direct functions c_{aa}, c_{ab}, and c_{bb}, in a homogeneous binary mixture of molecules a and b, which have hard cores but otherwise unspecified pair potentials. The limit is now taken in which the radius of the hard core of b becomes infinite and its concentration goes almost to zero, so that the system comprises a fluid of a molecules in contact with the flat wall of the one remaining b molecule. Only two Ornstein–Zernike equations remain, one for h_{aa} and one for the molecule–wall correlation, h_{ab}. These are solved by using the Percus–Yevick approximation,

$$c_{ij} = (h_{ij} + 1)(1 + e^{u_{ij}/kT}) \qquad (i, j = a, b). \qquad \text{(pp)} \quad (7.17)$$

This route is a useful way of determining the structure of a fluid near a solid wall with which it can interact, and has been so used.[24,25] Sullivan and Stell[26] have adapted this method to a wall which is penetrable, and has fluid on both sides of it. The component b has now become entirely hypothetical, and the a molecules interact freely across the wall, which remains only as a device to aid the calculation. They solved the two coupled integral equations for a system of hard spheres with a strong attractive force of negligible range (Baxter's 'sticky spheres'[27]). The profiles $\rho(z)$ are similar to those of Figs 7.1 and 7.2 but because of the difference of potentials no quantitative comparison can be made.

The other method of approximating the direct correlation function is based on the functional expansions of § 4.5 and is due principally to Ebner and Saam.[25,28–30] Closely related to (4.153) is an expansion of $F[\hat{\rho}]$

about the uniform local free energy in powers of the difference $\hat{\rho}(\mathbf{r}_1) - \hat{\rho}(\mathbf{r}_2)$. That is, to second order

$$F[\hat{\rho}] = \int \psi^0[\hat{\rho}(\mathbf{r}_1)] \, d\mathbf{r}_1 + \tfrac{1}{2}kT \int K(\mathbf{r}_1, \mathbf{r}_2)[\hat{\rho}(\mathbf{r}_1) - \hat{\rho}(\mathbf{r}_2)]^2 \, d\mathbf{r}_1 \, d\mathbf{r}_2, \quad (7.18)$$

where ψ^0 is again the free-energy density of a homogeneous fluid. There is, by symmetry, no term linear in $[\hat{\rho}(\mathbf{r}_1) - \hat{\rho}(\mathbf{r}_2)]$. If the kernel can be approximated by

$$K(\mathbf{r}_1, \mathbf{r}_2) = K(r_{12}; \bar{\rho}), \quad (7.19)$$

then consistency with (4.152)–(4.153) requires that[31]

$$K(\mathbf{r}_1, \mathbf{r}_2) = \frac{1}{2kT} \left(\frac{\delta^2 F[\hat{\rho}]}{\delta \hat{\rho}(\mathbf{r}_1) \delta \hat{\rho}(\mathbf{r}_2)} \right), \quad (7.20)$$

and so, from (4.134),

$$F[\hat{\rho}] = \int \psi^0[\rho(\mathbf{r}_1)] \, d\mathbf{r}_1 + \tfrac{1}{4}kT \int c(r_{12}; \bar{\rho})[\hat{\rho}(\mathbf{r}_1) - \hat{\rho}(\mathbf{r}_2)]^2 \, d\mathbf{r}_1 \, d\mathbf{r}_2. \quad (7.21)$$

From (4.131), with $v(\mathbf{r}) = 0$, we have

$$\mu = (\delta F[\hat{\rho}]/\delta \hat{\rho}(\mathbf{r}))_\rho. \quad (7.22)$$

Saam et al.[28,29] derived (7.21) and (7.22), and used them to calculate the equilibrium free energy of the two-phase system in terms of the presumed known values of T and μ. They first solved the Percus–Yevick integral equation for a homogeneous Lennard–Jones (12, 6) liquid to obtain $c(r_{12}; \rho)$ for $\rho \geq \rho^1$ and $\rho \leq \rho^g$. Between these limits they used a simple interpolation to obtain $c(r_{12}; \bar{\rho})$ for the hypothetical homogeneous fluid. They then substituted (7.21) in (7.22) and solved the equation, i.e. minimized $F[\hat{\rho}]$, for a two-phase system with a planar surface at $z = 0$, and subject to the limits $\rho(-\infty) = \rho^1$ and $\rho(+\infty) = \rho^g$. They did this by choosing trial solutions $\hat{\rho}(z)$ with up to six disposable parameters. The surface tension followed at once; it is the surface excess free energy with respect to the equimolar Gibbs surface. They observed that their procedure is similar to, but with fewer approximations than, a minimization of the free energy made earlier by Bongiorno and Davis.[32]

The approximation of Saam et al. can also be expressed in the form[25] (cf 4.140)

$$\int_0^1 d\alpha \int_0^\alpha d\beta \, c(r_1, r_2; \beta\rho) = \tfrac{1}{2}c(r_{12}; \rho). \quad (7.23)$$

This seems reasonable since the direct correlation function of a homogeneous fluid does not change rapidly with density. However it has been criticized by Lane et al.[33] on the grounds that it leads to results for Lennard–Jones molecules adsorbed on a solid surface which disagree with

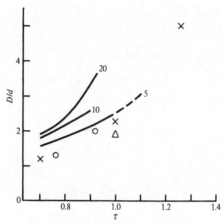

FIG. 7.3. The surface thickness as calculated from the YBG equation, circles[13] and triangle;[7] and from the direct correlation function, crosses.[29] The curves are the results of computer simulation (§ 6.3) for surfaces with cell widths of $5d$, $10d$, and $20d$.

those obtained by computer simulation. Such disagreement is not found with the Percus–Yevick theory of (7.11) and (7.17). Nevertheless the liquid–gas profiles of Ebner *et al.* seem to be satisfactory; that is, they are similar to those obtained by solving the YBG equation and by other methods discussed below.

Figure 7.3 compares the thickness of the surface of a Lennard–Jones liquid, as determined by the approximations discussed so far in this chapter, with those of the computer simulations discussed in § 6.3, using in each case[34] the thickness defined by (6.5). The theories agree tolerably well but yield profiles that correspond to thinner surfaces than those found by computer simulation. It seems as if the theories yield a 'bare' profile, and that this is thickened in practice by the capillary waves whose amplitude increases with the increasing cell-size of the computer simulations. This is a point of view that has been argued by Davis[35] and Abraham[36] (amongst others) and which we have discussed in more detail in § 4.8.

7.4 Modified van der Waals theories

We showed in §§ 4.4–4.5 that the functional expansion of the free energy in terms of the direct correlation function led naturally to a justification of the van der Waals approximation of Chapter 3, but with the coefficient of the square of the density gradient, $m(\rho)$, now a function of both density and temperature (4.162). In the original treatment of Rayleigh and van der Waals it was a constant, m, as in (1.38) or (4.164). If the coefficient is a function of density, and so of the height z, then it must be retained within all integrals over z, and not taken outside them. We saw in § 5.6 that this simple step greatly improves the original van der Waals theory when it is applied to the penetrable-sphere model. Telo da Gama and Evans[37] carried out parallel calculations for the Lennard–Jones fluid, using a parametrized trial function for $\rho(z)$, and the Percus–Yevick calculations of $c(r; \bar{\rho})$ of Ebner et al.[29] for obtaining $m(\rho)$. The profiles are thicker than those shown in Fig. 7.3, for example, $D/d \approx 3 \cdot 5$ at $\tau = 1 \cdot 0$. For mixtures of Lennard–Jones molecules they find profiles that are at least in qualitative agreement with the results of computer simulation[18] (Fig. 6.6).

An alternative expression for the free energy and surface tension in terms of the square of the density gradient can be made by starting with the virial expression for the tension, (4.104), writing $\rho(z_2)$ of (7.1) as $\rho(z_1 + z_{12})$ and expanding it about $\rho(z_1)$. This route, together with the usual replacement of $g(r_{12}, z_1, z_2)$ by $g(r_{12}; \bar{\rho})$ leads to the following expression for the coefficient,[38]

$$m(\rho) = \frac{1}{30} \int u'(r) g(r; \bar{\rho}) r^3 \, d\mathbf{r}. \qquad \text{(pp)} \quad (7.24)$$

The integrand of (7.24), like that of van der Waals's constant, m, is always short-ranged, and so $m(\rho)$ does not have the weak divergence at the critical point that is correctly given by (4.162). We discuss the significance of this divergence in § 9.2.

Bongiorno et al.[39] and Abraham[40,41] have also proposed gradient expansions that invoke explicitly the original suggestion of van der Waals that the entropy density of an inhomogeneous fluid is a function only of the local density (cf. § 3.4). This leads them (if we ignore any dependence of g on ρ, on which they disagree) to a further approximation for the coefficient, namely

$$m(\rho) = -\frac{1}{6} \int u(r) g(r) r^2 \, d\mathbf{r}. \qquad \text{(pp)} \quad (7.25)$$

This can be integrated by parts to give

$$m(\rho) = \frac{1}{30} \int [u'(r) g(r) + u(r) g'(r)] r^3 \, d\mathbf{r}, \qquad \text{(pp)} \quad (7.26)$$

and so it differs from (7.24). The numerical results obtained with both these approximations are quite reasonable, but have no advantage over the expression of $m(\rho)$ as an integral over $c(r; \rho)$, or of the use of the methods of the last two sections.

7.5 Perturbation theories

For the last fifteen years perturbation theory has been the most widely used tool for quantitative work on the structure and thermodynamic properties of homogeneous liquids. It is founded on the fact that the structure of a dense fluid of simple, reasonably spherical, molecules is determined primarily by the repulsive forces between the molecular cores. The attractive forces bind the system together and are responsible for the existence of states of high density at low external pressure, but they are not the primary determinants of the structure, as expressed in terms of the distribution functions of low order, $\rho^{(n)}(\mathbf{r}^n)$. The thermodynamic effects of the attractive forces can be found by averaging the attractive energy, $u_1(r)$, over the structure determined by the repulsive forces.

This simple idea lies behind van der Waals's equation of state, which was extended to the treatment of the surface of a liquid by Hill.[42] He recognized the value of dividing $u(r)$ into a repulsive part $u_0(r)$ and an attractive part $u_1(r)$, but used once again the simplification of van der Waals of assuming that the entropy density of the reference (or repulsive) system at $\mathbf{r}$ is determined solely by the local density $\rho(\mathbf{r})$.

Modern use of perturbation theory stems from the quantitative success of Barker and Henderson[16,17] in 1967 in calculating the structure and thermodynamic properties of a homogeneous Lennard–Jones fluid. A few years later Toxvaerd[11,43] extended this work to the liquid–gas surface. It is based, not on (7.7), but on the division $u(r) = u_0(r) + u_1(r)$ where

$$u_0(r) = u(r) \quad (r < d), \qquad u_0(r) = 0 \quad (r > d), \qquad (7.27)$$

and d is the collision diameter, $u(d) = 0$. Toxvaerd's expression for the free energy of an inhomogeneous system, to first order in $u_1(r)$, can, as Abraham[44] has shown, be based on an obvious first-order approximation to (4.146),

$$F = \int \psi(\mathbf{r})\, d\mathbf{r},$$

$$\psi(\mathbf{r}_1) = \psi_0[\rho(\mathbf{r}_1)] + \tfrac{1}{2}\rho(\mathbf{r}_1) \int \rho(\mathbf{r}_2) u_1(r_{12}) g_0(r_{12}; \bar{\rho})\, d\mathbf{r}_2, \quad \text{(pp)} \quad (7.28)$$

where $\psi_0[\rho(\mathbf{r}_1)]$ and $g_0(r_{12}; \bar{\rho})$ are the free energy density and pair distribution function of a hypothetical homogeneous fluid of intermolecular potential $u_0(r)$, but constrained to have a uniform density $\rho(\mathbf{r}_1)$. The integral is the average of $u_1(r_{12})$ over the distribution function $g_0(r_{12})$. Toxvaerd obtained also the second-order term in $[u_1(r)]^2$ which should involve distribution functions of higher order, but which he approximated in terms of a local compressibility $(\partial\rho(\mathbf{r})/\partial p)$, following the lead of Barker and Henderson. This term contributes little to the surface tension (about 6 per cent at the triple point according to Toxvaerd's original estimate, but only about $1\frac{1}{2}$ per cent according to Abraham's later calculation[40]) and so we do not record it. An alternative division of $u(r)$ into $u_0(r)$ and $u_1(r)$, namely (7.7), leads to the perturbation scheme of Weeks et al.,[15] which has been adapted to the study of the gas–liquid surface by Singh and Abraham.[45]

Toxvaerd minimizes the free energy of a system with a planar surface, with respect to the parameters of a trial function for $\rho(z)$, and subject to the constraints of fixed gas and liquid densities, $\rho(\pm\infty)$, and of fixed number of molecules, which is achieved by a suitable condition on the position of the equimolar dividing surface. His trial function for $\rho(z)$ comprises a constant term, $\frac{1}{2}(\rho^1 + \rho^g)$, and two tanh functions with arguments $[\frac{1}{2}A_{1,g}(z-b)]$, where A_1 is used for $z < b$ and A_g for $z > b$, and where $b = (A_1^{-1} - A_g^{-1})\ln 2$. Clearly there is a discontinuity in the derivative $\rho'(z)$ at $z = b$, but this is imperceptible in his published graphs. From the optimum profile and the minimized free energy he calculates the surface tension as the surface excess free energy with respect to the equimolar dividing surface.

This work was repeated and extended by Lee, Barker, and Pound[46] who used his method not only to calculate the surface profile and tension of a Lennard–Jones liquid, but also to estimate the difference in surface tension between that liquid and one of a liquid with a pair potential that accurately represents the interaction of argon molecules, and the further difference on adding the three-body attractive forces. In this way they were able to show that the experimental surface tension of argon could be matched to the best estimate from an accurate pair potential, if proper allowance is made for the three-body forces (§ 6.4 and Fig. 6.5).

Toxvaerd's calculations yielded a profile with a thicker surface than those shown by the theories displayed in Fig. 7.3, and roughly comparable with those obtained from the modified van der Waals theories. His curves indicate $D/d \sim 2$ at $\tau = 0.75$ and 3–3.5 at $\tau = 1.0$ but the form of his trial function makes these hard to determine. Singh and Abraham,[45] using the other form of perturbation theory,[15] obtain thinner interfaces, e.g. $D/d \sim 1.6$ at $\tau = 0.70$.

Perturbation theory is equally useful for non-spherical potentials. Gubbins and his colleagues[47] have used it to confirm the result obtained by computer simulation (§ 6.5), that diatomic molecules on the liquid side of the equimolar Gibbs surface tend to align perpendicularly to the surface.

An interesting link between perturbation theory and the results of the last section is what Abraham[40,41] calls a 'generalized' van der Waals theory. He rewrites (7.28) so as to incorporate the perturbation term for the homogeneous fluid into the first term; that is

$$\psi(\mathbf{r}_1) = \psi[\rho(\mathbf{r}_1)] + \tfrac{1}{2}\rho(\mathbf{r}_1) \int u_1(r_{12})[\rho(\mathbf{r}_2) - \rho(\mathbf{r}_1)]g_0(r_{12}; \bar{\rho}) \, d\mathbf{r}_2. \quad \text{(pp)}$$

$$(7.29)$$

This equation is a general case of the more specialized equations (5.14)–(5.18) which apply to a fluid of attracting hard spheres in the mean-field approximation. Such a generalization was foreseen, but not used, by van der Waals himself.[48]

It was the Taylor expansion of $\rho(\mathbf{r}_2)$ about $\rho(\mathbf{r}_1)$ that led Abraham to the square-gradient expansion, and so to (7.25), but if we do not make this step then (7.29) is a useful equation in its own right, which can, if necessary, be taken to second order in the perturbation expansion.[40]

The generalized van der Waals theory shows explicitly the non-local nature of the term that arises from the inhomogeneity of the fluid, which, as before, becomes a local term on making the gradient expansion. This distinction is conceptually important if we are concerned with the shape of the 'wings' of the density profile, that is, $\rho(z)$ for large values of $|z - z_e|$. If $u(r)$ varies as $-r^{-6}$ at large r, then a semi-infinite horizontal slab of liquid with a flat surface generates a potential field that falls off as z^{-3} where z is the height above the surface. This, in turn, induces[11,21] a limiting form of the density profile in which $\rho(z) - \rho^g$ also falls off as z^{-3}. Similarly, on the liquid side $\rho^l - \rho(z)$ also varies as $|z|^{-3}$. These results are implied by, and derivable from,[49] a non-local approximation to the free energy, such as (7.29), but they are lost if this is reduced to a local form by a gradient expansion, as has recently been emphasized by de Gennes.[50] The local approximation leads to profiles that decay exponentially.

The difference arises from the lack of convergence of the gradient expansion for $u(r) \sim -r^{-6}$ that was discussed at the beginning of § 4.6. The coefficients of the expansion are the moments of the direct correlation function, the higher ones of which diverge unless $u(r)$ is of finite range or decays exponentially. For such short-ranged potentials the exponential tails in $\rho(z)$ are correct, as is seen, for example, in (5.103) for the penetrable-cube model.

7.6 Surface tension

We conclude this chapter by comparing the surface tension of a Lennard–Jones liquid, as calculated from the theories above, with that found by computer simulation. Figure 7.4 shows the computer results of Fig. 6.4 as a 'best' single line. The four theoretical curves are one solution of the YBG equation,[13] one approximation based on the direct correlation function,[29] one modified van der Waals approximation,[37] and one

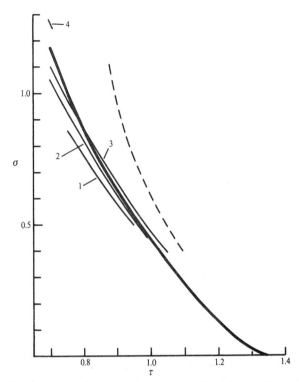

Fig. 7.4. The surface tension of a Lennard–Jones liquid from: line 1, a solution of the YBG equation;[13] line 2, an approximation for the direct correlation function;[29] line 3, perturbation theory;[43] line 4, perturbation theory.[45] The heavy line is the result of computer simulation (§ 6.4), and the dashed line from a modified van der Waals equation.[37]

perturbation theory[43] based on the treatment of Barker and Henderson. The single point is from a perturbation theory[45] based on the treatment of Weeks *et al.*

The modified van der Waals theory can itself be regarded as an approximation to the theory based on the direct correlation function. We saw in § 5.6 (Fig. 5.10) that this theory and its derivatives leads to too high a surface tension for the penetrable-sphere model. Here also its surface tension is larger than the results found by computer simulation, and it is clearly less accurate than the other theories, between which it is hard to discriminate on this basis. The surface tension is not sensitive to the details of how g or c has been approximated, provided that these functions satisfy the constraints set out above.

The surface of tension, as defined by (2.89), and with Irving and Kirkwood's definition chosen for $p_T(z)$, lies on the liquid side of the equimolar dividing surface. All treatments agree on this statement for a Lennard–Jones liquid, but they disagree on the size of the separation $\delta = z_e - z_s$. Computer simulation gives $(0.96 \pm 0.12)d$ at $\tau = 0.92$ (§ 6.5). The estimate based on a crude mean-field approximation (4.117) is $(r_m/3)$, or $0.374d$, at zero temperature. Toxvaerd's first solution of the YBG equation (§ 7.2) yields about $3d$ and his perturbation theory about $1.5d$, both at $\tau = 0.85$. Earlier theories, summarized by Uang,[51] and his own calculations based on linear and cubic profiles,[52] all lead to separations of around $(0.8-1.0)d$. We conclude that this separation is about a molecular diameter for a Lennard–Jones liquid at its triple point, but repeat the comment at the end of § 5.7 that the separation defined in this way is not the same as that defined as the planar limit of $R_e - R_s$, when R_s is determined thermodynamically. The relation between the two definitions is not yet known.

Notes and references

1. The subject of this chapter is similar to that of Chapter 2 of Croxton, C. A. *Statistical Mechanics of the Liquid Surface*, Wiley, Chichester (1980). His later chapters describe numerical calculations of $\rho(z)$ etc. for molecular fluids, mixtures, liquid metals, quantal fluids, water, polymers, and liquid crystals, many of which are outside the scope of this book. For a discussion of the surface of water, see also Croxton, C. A. *Physica* **106A,** 239 (1981).
2. The most extensive set of results are those deposited in the British Library by Nicolas, J. J., Gubbins, K. E., Streett, W. B. and Tildesley, D. J. *Mol. Phys.* **37,** 1429 (1979) (see their ref. 25). The most extensive published set are those of Verlet, L. *Phys. Rev.* **165,** 201 (1968).
3. See, for example, Hansen, J. P. and McDonald, I. R. *Theory of Simple Liquids*, Chapter 6. Academic Press, London (1976).
4. Pressing, J. and Mayer, J. E. *J. chem. Phys.* **59,** 2711 (1973).
5. Borštnik, B. and Ažman, A. *Mol. Phys.* **29,** 1165 (1975).
6. Toxvaerd, S. *Mol. Phys.* **26,** 91 (1973).
7. Toxvaerd, S. *J. chem. Phys.* **64,** 2863 (1976).
8. Lekner, J. and Henderson, J. R. *Mol. Phys.* **39,** 1437 (1980). These results have been extended to molecular fluids by Castle, P. J. *Mol. Phys.* **42,** 1157 (1981).
9. Co, K. U., Kozak, J. J. and Luks, K. D. *J. chem. Phys.* **66,** 1002 (1977).
10. Nazarian, G. M. *J. chem. Phys.* **56,** 1408 (1972).
11. Toxvaerd, S. *Statistical Mechanics* (ed. K. Singer) Vol. 2, Chap. 4. Specialist Periodical Report, Chemical Society, London (1975).
12. Croxton, C. A. and Ferrier, R. P. *J. Phys.* **C4,** 1909 (1971).
13. Fischer, J. and Methfessel, M. *Phys. Rev.* **A22,** 2836 (1980). See also Nieminen, R. M. and Ashcroft, N. W. *Phys. Rev. A* **24,** 560 (1981).
14. Osborn, T. R. and Croxton, C. A. *Mol. Phys.* **40,** 1489 (1980).
15. Weeks, J. D., Chandler, D. and Andersen, H. C. *J. chem. Phys.* **54,** 5237

(1971); Andersen, H. C., Weeks, J. D. and Chandler, D. *Phys. Rev. A* **4,** 1597 (1971).

16. Barker, J. A. and Henderson, D. *J. chem. Phys.* **47,** 2856, 4714 (1967).
17. Barker, J. A. and Henderson, D. *Rev. mod. Phys.* **48,** 587 (1976).
18. Chapela, G. A., Saville, G., Thompson, S. M. and Rowlinson, J. S. *J. chem. Soc. Faraday Trans. II.* **73,** 1133 (1977).
19. Robledo, A. *J. chem. Phys.* **72,** 1701 (1980) has used it for a hard-sphere fluid in contact with its solid, and Varea, C., Valderrama, A. and Robledo, A. *J. chem. Phys.* **73,** 6265 (1980), have used it to calculate the surface tension between two liquid phases composed of molecules with weak long-ranged attractive forces.
20. Hill, T. L. *J. chem. Phys.* **30,** 1521 (1959).
21. Plesner, I. W. and Platz, O. *J. chem. Phys.* **48,** 5361 (1968).
22. Leng, C. A., Rowlinson, J. S. and Thompson, S. M. *Proc. Roy. Soc. A* **358,** 267 (1978). The approximation (7.11) was apparently first suggested by Stillinger, F. H. and Buff, F. P. *J. chem. Phys.* **37,** 1 (1962).
23. Henderson, J. R. *Mol. Phys.* **39,** 709 (1980).
24. Perram, J. W. and White, L. R. *Chem. Soc. Faraday Discuss.* **59,** 29 (1975); Henderson, D., Abraham, F. F. and Barker, J. A. *Mol. Phys.* **31,** 1291 (1976); Waisman, E., Henderson, D. and Lebowitz, J. L. *Mol. Phys.* **32,** 1373 (1976); Percus, J. K. *J. stat. Phys.* **15,** 423 (1976); Blum, L. and Stell, G. *J. stat. Phys.* **15,** 439 (1976): Smith, E. R. and Perram, J. W. *J. stat. Phys.* **17,** 47 (1977); Snook, I. K. and Henderson, D. *J. chem. Phys.* **68,** 2134 (1978); Sullivan, D. E. and Stell, G. *J. chem. Phys.* **69,** 5450 (1978).
25. Saam, W. F. and Ebner, C. *Phys. Rev.* A**17,** 1768 (1978).
26. Sullivan, D. E. and Stell, G. *J. chem. Phys.* **67,** 2567 (1977).
27. Baxter, R. J. *J. chem. Phys.* **49,** 2770 (1968).
28. Ebner, C. and Saam, W. F. *Phys. Rev.* B**12,** 923 (1975).
29. Ebner, C., Saam, W. F. and Stroud, D. *Phys. Rev.* A**14,** 2264 (1976).
30. Saam, W. F. and Ebner, C. *Phys. Rev.* A**15,** 2566 (1977).
31. Evans, R. *Adv. Phys.* **28,** 143 (1979).
32. Bongiorno, V. and Davis, H. T. *Phys. Rev.* A**12,** 2213 (1975).
33. Lane, J. E., Spurling, T. H., Freasier, B. C., Perram, J. W. and Smith, E. R. *Phys. Rev.* A**20,** 2147 (1979); see also Ebner, C., Lee, M. A. and Saam, W. F. *Phys. Rev.* A**21,** 959 (1980); Ebner, C. *Phys. Rev.* A**22,** 2776 (1980).
34. Other measures of the thickness, and results for a wider range of theories, are compared by Lekner, J. and Henderson, J. R. *Physica* **94A,** 545 (1978).
35. Davis, H. T. *J. chem. Phys.* **67,** 3636 (1977); **70,** 600 (1979).
36. Abraham, F. F. *Chem. Phys. Lett.* **58,** 259 (1978).
37. Telo da Gama, M. M. and Evans, R. *Mol. Phys.* **38,** 367 (1979); **41,** 1091 (1980); *Faraday Sympos. chem. Soc.* **16** (1981), in press. The second and third papers are on mixtures.
38. Bongiorno, V., Scriven, L. E. and Davis, H. T. *J. Coll. Interf. Sci.* **57,** 462 (1976), eqn (3.7) and (3.8); Lekner, J. and Henderson, J. R. *Mol. Phys.* **34,** 333 (1977).
39. Bongiorno *et al.* Ref. 38, eqn (2.18) and (2.22); McCoy, B. F. and Davis, H. T. *Phys. Rev.* A**20,** 1201 (1979).
40. Abraham, F. F. *J. chem. Phys.* **63,** 157, 1316 (1975). The second paper is on mixtures.
41. Abraham, F. F. *Physics Rep.* **53,** 93 (1979), eqn (3.16) and (3.18).
42. Hill, T. L. *J. chem. Phys.* **20,** 141 (1952).

43. Toxvaerd, S. *J. chem. Phys.* **55**, 3116 (1971).
44. Abraham, ref. 41, eqn (3.26).
45. Singh, Y. and Abraham, F. F. *J. chem. Phys.* **67**, 537 (1977); Abraham, ref. 41, § 9.
46. Lee, J. K., Barker, J. A. and Pound, G. M. *J. chem. Phys.* **60**, 1976 (1974).
47. Haile, J. M., Gubbins, K. E. and Gray, C. G. *J. chem. Phys.* **64**, 1852 (1976); Thompson, S. M. and Gubbins, K. E. *J. chem. Phys.* **70**, 4947 (1979); Thompson, S. M., Gubbins, K. E. and Haile, J. M. *J. chem. Phys.* **75**, 1325 (1981).
48. It appears to be implicit in his paper of 1893, see § 15 of the English translation, *J. stat. Phys.* **20**, 197 (1979).
49. Barker, J. A. and Henderson, J. R. *J. chem. Phys.* in press.
50. de Gennes, P. G. *J. Physique Lett.* **42**, L-377 (1981).
51. Uang, Y.-H. *Phys. Rev.* A**22**, 758 (1980).
52. Fitts, D. D. *Physica* **42**, 205 (1969).

THREE-PHASE EQUILIBRIUM

8.1 Introduction

We have so far been concerned mainly with the structure and thermodynamics of the interface between two phases, and we have seen in outline, and sometimes in detail, the elements of the molecular theories that account for or predict that structure and thermodynamics. Macroscopically, the interface between two bulk phases is two-dimensional and locally planar—although at the molecular level it has a discernible three-dimensional structure, which we have studied and related to the thermodynamics.

In this chapter we consider the simultaneous equilibrium of three phases. We shall see that there are circumstances in which the phases meet in a line of three-phase contact. Macroscopically, this locus of points in which they meet is one-dimensional and locally linear; it is analogous to the macroscopically two-dimensional and locally planar interface between two phases. There is an excess free energy per unit length, or line tension, associated with the three-phase line, and we should in principle be able to calculate that tension from a microscopic theory. Such a line of three-phase contact should have a predictable, and in principle discernible, three-dimensional structure at the molecular level, and its structure and tension should be related, just as are the structure and tension of the interface between two phases.

In the next section we deduce the macroscopic condition for the existence of such a line of three-phase contact. The condition is that each of the three interfacial tensions be less than the sum of the other two; that is, that the three tensions be related to each other as are the sides of a triangle (the Neumann triangle). We shall see how the interfacial tensions are related to the contact angles, which are the dihedral angles occupied by the three phases at their line of mutual contact. The discussion in § 8.2 is wholly macroscopic and therefore not subject to the uncertainties that arose in §§ 2.5 and 4.8 when mechanical arguments were used at the microscopic level.

When the Neumann triangle collapses to a line, that is, when the largest of the three two-phase tensions is equal to the sum of the two smaller, there is no longer a line of three-phase contact. Instead, the equilibrium configuration of the three phases is that in which one of them—the one whose interfaces with the other two are those of lowest tension—spreads at and completely covers the high-tension interface. The relation of the three tensions in this case of spreading (complete wetting), viz. that the largest is the sum of the two smaller, is called Antonow's rule. There has historically been much contention over the status and generality of this relation. Clearly it does not hold when there is a line of three-phase contact, for then the three tensions are related as in the Neumann triangle and satisfy the triangle inequalities; but it does hold when one of the phases completely wets the interface between the two others. This is discussed in § 8.3, as is the related matter of the non-positivity of equilibrium spreading coefficients. The latter question has, if anything, an even more confused literature than Antonow's rule, but we shall risk an assessment. The matter is illuminated by an extension to three-phase equilibrium of the van der Waals theory, particularly by the three-phase versions of Figs 3.5 and 3.7.

In the spreading (Antonow's-rule) regime of three-phase equilibrium, the interface of highest tension, which we call the $\alpha\gamma$ interface, is unlike, in its structure and properties, either of the two interfaces of lower tension. The former is found in its equilibrium structure to contain—indeed, essentially to consist of—a layer of the wetting phase. The layer is of macroscopic thickness when the wetting phase is stable as a bulk phase; but, most interestingly, that layer is found to persist even into the two-phase region, where the wetting phase is no longer stable in bulk. The layer is then of microscopic rather than macroscopic thickness, but is still recognizable as the incipient third phase. That is the subject of § 8.4, where we again use the van der Waals theory as the theoretical framework of the discussion. Although we are concerned there with states in which only two thermodynamic phases are present, with the wetting phase not stable in bulk, the role of the latter as an incipient phase is central to the phenomenon, so we consider this a topic appropriate to this chapter on three-phase equilibrium.

With changing thermodynamic state the system may undergo a transition from the condition in which one phase wets the interface between the other two and in which the three tensions satisfy Antonow's rule, to the condition in which the three phases meet at a line of three-phase contact with measurable contact angles and in which the three tensions satisfy the relations of the Neumann triangle. Cahn has predicted that such transitions will occur, even in the two-phase region where the wetting phase is not stable in bulk. These Cahn transitions are first-order phase transitions occurring entirely within the $\alpha\gamma$ interface, unaccompanied by any singularities in bulk thermodynamic properties. Thus this, too, has to do with the properties of a two-phase interface, often (although not always) in the absence of a stable third phase; but, again, our understanding of the phenomenon is so dependent on our knowledge of the principles of three-phase equilibrium that it is appropriate to consider it in this chapter. Accordingly, the Cahn transition is the subject of § 8.5.

Finally, in § 8.6, we come to consider the nature of the three-phase contact line. We sketch briefly the thermodynamics of that line and the associated line tension, in parallel with our earlier discussion of the thermodynamics of two-phase interfaces and the interfacial tension. The statistical mechanics of the three-phase line, even at the phenomenological level of the van der Waals theory, is not nearly so extensively developed as that of the two-phase interface, but we outline what has been done and we mention some work in progress. Experimentally, also, the three-phase line is not nearly so well studied as is the two-phase interface. There are many fewer results on line tension than on interfacial tension; measurements of the former are intrinsically more difficult because the tensions are so small: 10^{-11} to 10^{-9} N, that is, excess free energies of 10^{-11} to 10^{-9} J m^{-1}. Unlike surface tension, line tension can be of either sign, as both theory and experiment show. Indeed, we shall refer to recent experiments that show that it can change sign with continuous change in the thermodynamic state.

Our first task is to derive the macroscopic conditions of equilibrium at a line of three-phase contact.

8.2 Contact angles and Neumann's triangle

We remarked in the previous section that macroscopically the locus of points in which three phases meet is one-dimensional and locally linear, even though at the molecular level it has a three-dimensional structure. In Fig. 8.1(a) are shown schematically three phases, α, β, and γ, occupying the dihedral angles between locally planar interfaces, which in turn meet in the three-phase line. We shall represent the dihedral angles by α, β, and γ, thus naming them after the phases they contain. Those dihedral angles are the *contact angles* in which the $\alpha\beta$, $\beta\gamma$, and $\alpha\gamma$ interfaces meet. We have

$$\alpha + \beta + \gamma = 2\pi. \tag{8.1}$$

Figure 8.1(b) is the same as 8.1(a), but looking down the three-phase line.

In Fig. 8.1(c) we see in cross-section a drop of β phase resting on the otherwise planar $\alpha\gamma$ interface. The three-phase line is then a circle, and is seen as such when the drop is viewed from above. On the circle, any arc much shorter than its radius may be treated as linear, just as on any of the two-phase interfaces any area with linear dimensions much smaller than

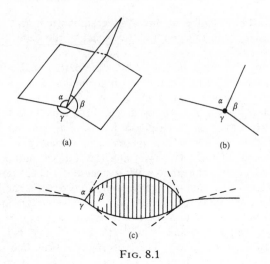

Fig. 8.1

its radii of curvature may be treated as planar. The interfaces in Fig. 8.1(c), and the circle in which they meet, are locally the two-phase planes and the three-phase line of Fig. 8.1(a). The dashed lines in Fig. 8.1(c) are the lines (planes seen edge-on) of Fig. 8.1(b).

At equilibrium, the net force on any element of the three-phase line vanishes. Resolving this force in directions that lie, respectively, in the $\alpha\beta$, $\beta\gamma$, and $\alpha\gamma$ interfaces and are perpendicular to the three-phase line (hence, in the directions of the lines in Fig. 8.1(b)), we have

$$\sigma^{\alpha\beta} + \sigma^{\beta\gamma} \cos\beta + \sigma^{\alpha\gamma} \cos\alpha = 0$$
$$\sigma^{\alpha\beta} \cos\beta + \sigma^{\beta\gamma} + \sigma^{\alpha\gamma} \cos\gamma = 0 \qquad (8.2)$$
$$\sigma^{\alpha\beta} \cos\alpha + \sigma^{\beta\gamma} \cos\gamma + \sigma^{\alpha\gamma} = 0$$

where $\sigma^{\alpha\beta}$ is the tension of the $\alpha\beta$ interface, etc.

The three equations (8.2) are a dependent set of homogeneous equations for the tensions—dependent because, by (8.1), the determinant of coefficients vanishes,

$$\begin{vmatrix} 1 & \cos\beta & \cos\alpha \\ \cos\beta & 1 & \cos\gamma \\ \cos\alpha & \cos\gamma & 1 \end{vmatrix} = 0; \qquad (8.3)$$

that is, any one of (8.2) follows from the other two. Because of their homogeneity, (8.2) determine only ratios of tensions, rather than the tensions themselves, in terms of the contact angles. The physical reason for this is that the forces on any element of the three-phase line would still balance, with the same contact angles, if all the tensions were multiplied by a common factor. Because of their non-independence, that is, because of (8.3), the equations (8.2) determine the ratios of tensions uniquely and consistently. The generic relation, from (8.2) and (8.1), is

$$\sigma^{\alpha\beta}/\sigma^{\beta\gamma} = \sin \gamma/\sin \alpha; \tag{8.4}$$

the others follow by permutation of α, β, γ. Conversely, the cosines of the contact angles are uniquely determined by the ratios of the interfacial tensions. The generic relation, from (8.2), is

$$
\begin{aligned}
\cos \beta &= \frac{(\sigma^{\alpha\gamma})^2 - (\sigma^{\alpha\beta})^2 - (\sigma^{\beta\gamma})^2}{2\sigma^{\alpha\beta}\sigma^{\beta\gamma}} \\
&= 1 - \frac{(\sigma^{\alpha\beta} + \sigma^{\beta\gamma} - \sigma^{\alpha\gamma})(\sigma^{\alpha\beta} + \sigma^{\beta\gamma} + \sigma^{\alpha\gamma})}{2\sigma^{\alpha\beta}\sigma^{\beta\gamma}} \\
&= \frac{1}{2} \left(\frac{\sigma^{\alpha\gamma}}{\sigma^{\alpha\beta}} \frac{\sigma^{\alpha\gamma}}{\sigma^{\beta\gamma}} - \frac{\sigma^{\alpha\beta}}{\sigma^{\beta\gamma}} - \frac{\sigma^{\beta\gamma}}{\sigma^{\alpha\beta}} \right),
\end{aligned}
\tag{8.5}
$$

in three alternative but equivalent forms.

From (8.4) and the first form of (8.5), the interfacial tensions and the supplements of the contact angles are related to each other as the sides and angles of the triangle in Fig. 8.2. This is Neumann's triangle.[1-4] The first equality in (8.5) and the two others obtained from it by permutation of α, β, and γ, are the law of cosines applied to this triangle; (8.4) and its permutations are the law of sines.

When the γ phase in Fig. 8.1(c) is non-deformable (as is a rigid solid), the angle γ is π. From (8.1) there is then only one independent contact angle, $\alpha = \pi - \beta$, and the second and third of (8.2) become

$$\sigma^{\beta\gamma} = \sigma^{\alpha\gamma} + \sigma^{\alpha\beta} \cos \alpha, \tag{8.6}$$

which is Young's equation, (1.21), with $\alpha = \theta$. (Compare Figs 8.1 and 1.3). The first of equations (8.2), which relates those components of the three forces that are in the direction lying in the $\alpha\beta$ interface and perpendicular to the three-phase contact line, does not take account of the added constraint of non-deformability of the γ phase and was thus inapplicable here; but the constraint does not contribute to the forces parallel to the surface of the γ phase, so the second and third of (8.2) could hold unchanged. For $\gamma = \pi$ to hold without constraint, and so to be consistent with (8.2) as they stand, would require either that $\sigma^{\alpha\gamma} = \sigma^{\beta\gamma} = \infty$ with any $\sigma^{\alpha\beta}$, or that $\sigma^{\alpha\gamma}$ and $\sigma^{\beta\gamma}$ have any finite value in common while $\sigma^{\alpha\beta} = 0$.

Because the three interfacial tensions form the sides of a triangle

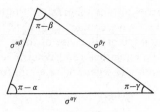

F<small>IG</small>. 8.2

(Fig. 8.2), they satisfy the triangle inequalities

$$\sigma^{\alpha\beta} < \sigma^{\alpha\gamma} + \sigma^{\beta\gamma}, \qquad \sigma^{\beta\gamma} < \sigma^{\alpha\gamma} + \sigma^{\alpha\beta}, \qquad \sigma^{\alpha\gamma} < \sigma^{\alpha\beta} + \sigma^{\beta\gamma}. \qquad (8.7)$$

These, finally, are the conditions for the equilibrium of three phases at a line of mutual contact. The three inequalities in (8.7) may be replaced by the single condition that the largest of the three tensions be less than the sum of the two smaller. When this condition is met, the equilibrium configuration of the three phases is as in Fig. 8.1; otherwise, not. We shall now see what that alternative is.

8.3 Spreading and Antonow's rule

When one of the inequalities (8.7) becomes an equality, that is, when the largest of the three tensions equals the sum of the two smaller, the Neumann triangle degenerates to a line: the vertex that was previously opposite the longest side comes to lie on that side, as the altitude of the triangle, measured from that vertex to that side, vanishes. Suppose $\sigma^{\alpha\gamma}$ is the largest tension. Then in the limit we are now contemplating

$$\sigma^{\alpha\gamma} = \sigma^{\alpha\beta} + \sigma^{\beta\gamma}; \qquad (8.8)$$

and the angles α, β, and γ are such that

$$\alpha = \gamma = \pi, \qquad \beta = 0. \qquad (8.9)$$

From Fig. 8.1(c) we see that the β phase is then spread as a film over the $\alpha\gamma$ interface, as in Fig. 8.3.

$$\underset{\gamma}{\overset{\alpha}{\rule{6cm}{0.4pt}}} \quad \beta$$

F<small>IG</small>. 8.3

It was suggested by Antonow,[5] as long ago as 1907, that (8.8) is the general rule. When β and γ are liquid phases and α is the vapour with which they are in equilibrium, (8.8) is the assertion that the interfacial tension $\sigma^{\beta\gamma}$ of the two liquids is the difference between the larger and smaller of the two liquid–vapour surface tensions: $\sigma^{\beta\gamma} = \sigma^{\alpha\gamma} - \sigma^{\alpha\beta}$. It was in that context, particularly, that Antonow proposed his rule.

How general is that rule? There is no doubt that three fluids do sometimes meet in the configuration of Fig. 8.1; so that, for example, one phase, β, may take the form of a lens ($\frac{1}{2}\pi < \alpha < \pi$, $\gamma \lesssim \pi$) or bead ($0 < \alpha < \frac{1}{2}\pi$, $\gamma \lesssim \pi$) between two other phases, α and γ, as in Fig. 8.1(c). In particular, that may happen when β and γ are liquids and α is their common vapour. This equilibrium configuration is described by the Neumann triangle of Fig. 8.2 with $\alpha < \pi$, $\gamma < \pi$, and $\beta > 0$, so the three tensions satisfy the triangle inequality $\sigma^{\alpha\gamma} < \sigma^{\alpha\beta} + \sigma^{\beta\gamma}$. Antonow's rule, (8.8), is then false.

But there is also no doubt that the equilibrium configuration of three phases is sometimes—indeed, often—that of Fig. 8.3, where one of the phases, β, spreads at (wets completely) the interface between the two others. In that case (8.9) holds, and with it Antonow's rule. That is not merely a limiting but unattainable case, and (8.8) is not merely an asymptotic law or an approximation.[6] The case is attainable, and (8.8) is exact throughout the range of three-phase-equilibrium states over which one of the phases spreads at the interface between the other two.

In the cases first studied by Antonow,[5] the two liquids β and γ were formed from two incompletely miscible, pure components not far from their consolute point (critical solution point). From our present knowledge[7-9] we are able to say that in just those circumstances we would expect one of the liquid phases to wet completely the interface between the other liquid phase and the vapour. (We shall see why, in § 8.5.) The picture in those cases would then be that of Fig. 8.3, so (8.8) would hold: Antonow was right. But we can imagine a hypothetical case, even of such a two-component liquid–liquid–vapour system near the consolute point of the liquids, in which Antonow's rule must fail.[10] Suppose the two components are an incompletely miscible enantiomeric pair, so that the two liquid phases β and γ, each slightly richer in one of the two components, are mirror images; and suppose α is their common vapour. Then $\sigma^{\alpha\gamma} = \sigma^{\alpha\beta}$ by symmetry, while $\sigma^{\beta\gamma} > 0$ by the supposed immiscibility, so $\sigma^{\alpha\gamma} < \sigma^{\alpha\beta} + \sigma^{\beta\gamma}$. As another hypothetical example, Varea et al.[11] have studied what they call a van der Waals fluid mixture, and treated a special case of it in which, by a symmetry inherent in the model, a vapour α is symmetrically related to two liquids β and γ; whereupon again $\sigma^{\alpha\gamma} = \sigma^{\alpha\beta} < \sigma^{\alpha\beta} + \sigma^{\beta\gamma}$, even if the liquids are close to a consolute point.

We may summarize[12] by saying that Antonow's rule, (8.8), is often right—exactly, not merely as an approximation; that when it holds, the three phases meet as in Fig. 8.3, and conversely; but that the rule is equally often wrong; and that when it fails, the inequality in (8.7) holds instead of the equality (8.8), and the three phases meet as in Fig. 8.1.

To say that (8.8) may hold not merely as an asymptotic limit but exactly, over a range of thermodynamic states, and yet not hold in all

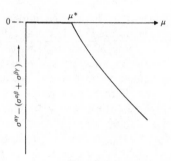

F<small>IG</small>. 8.4

states, is to say that $\sigma^{\alpha\gamma} - (\sigma^{\alpha\beta} + \sigma^{\beta\gamma})$ may change with the thermodynamic state, and be negative over some range and zero over another. We show this schematically in Fig. 8.4. There μ stands for some thermodynamic field, the temperature or a chemical potential, for example; indeed, it may stand for all the coordinates in a two- or higher-dimensional thermodynamic field space, in which case what is shown as the curve of $\sigma^{\alpha\gamma} - (\sigma^{\alpha\beta} + \sigma^{\beta\gamma})$ vs μ in Fig. 8.4 is really a surface having the dimensionality of that field space. We see in Fig. 8.4 a range of states (range of values of μ) over which the equality (8.8) holds, and a range over which the inequalities (8.7) hold. The transition between these two regimes, shown in Fig. 8.4 as occurring when $\mu = \mu^*$, is as predicted by Cahn[8] and studied experimentally by Moldover and Cahn.[9] We shall discuss this Cahn transition further in § 8.5.

We have so far excluded the possibility of the largest of the three tensions exceeding the sum of the two smaller: $\sigma^{\alpha\gamma} > \sigma^{\alpha\beta} + \sigma^{\beta\gamma}$; that is, we have admitted only the other sense of inequality, in (8.7), or the equality, in (8.8). Correspondingly, we have made no provision in Fig. 8.4 for a positive value of the difference $\sigma^{\alpha\gamma} - (\sigma^{\alpha\beta} + \sigma^{\beta\gamma})$. We shall presently argue that when the three tensions are the true equilibrium interfacial tensions, measured at full mutual saturation of the phases, that difference,

$$S^{\beta,\alpha\gamma} = \sigma^{\alpha\gamma} - (\sigma^{\alpha\beta} + \sigma^{\beta\gamma}), \tag{8.10}$$

called the *coefficient of spreading* of the β phase at the $\alpha\gamma$ interface, can never be positive—reports to the contrary notwithstanding.

The spreading coefficient defined by (8.10) clearly is closely related to (indeed, is only trivially different from) what we called the wetting coefficient, k, in (1.22). We can see why $S^{\beta,\alpha\gamma}$ is called a spreading coefficient; the larger it is, the greater is the excess of the free energy per unit area of an uncovered $\alpha\gamma$ interface, $\sigma^{\alpha\gamma}$, over the free energy per unit area of an $\alpha\gamma$ interface that is covered by a layer of β phase, $\sigma^{\alpha\beta} + \sigma^{\beta\gamma}$; so the greater is the tendency of the β phase to be spread at the $\alpha\gamma$

interface in the equilibrium configuration of the three phases. That is in fact correct, for the *non*-spreading regime, provided we understand that when we say 'the larger $S^{\beta,\alpha\gamma}$ is', or 'the greater is the excess free energy of an uncovered $\alpha\gamma$ interface', we mean the larger they are *algebraically*; that is, the less negative they are. In the second form of (8.5) we see that the negative of the spreading coefficient, that is, the positive quantity $-S^{\beta,\alpha\gamma}$, is the leading factor in the difference $1 - \cos\beta$. Then, indeed, the larger $S^{\beta,\alpha\gamma}$ is, that is, the closer this negative quantity is to 0, the closer the angle β is to 0, and so the more nearly is the β phase spread at the $\alpha\gamma$ interface.

The reason the equilibrium $S^{\beta,\alpha\gamma}$ can never be positive is that if, momentarily, it were, the $\alpha\gamma$ interface would immediately coat itself with a layer of the β phase, replacing the supposedly higher free energy per unit area of direct $\alpha\gamma$ contact, $\sigma^{\alpha\gamma}$, by the supposedly lower sum of the free energies per unit area of $\alpha\beta$ and $\beta\gamma$ contacts, $\sigma^{\alpha\beta} + \sigma^{\beta\gamma}$, thereby lowering the free energy of the system. The very structure of the $\alpha\gamma$ interface, at equilibrium, would then be that of the bulk β phase[13]; Fig. 8.3 would be not only a representation of a spread bulk β phase, but would be equally well a picture of the equilibrium $\alpha\gamma$ interface; and the free energy per unit area of that equilibrium interface would now be exactly $\sigma^{\alpha\beta} + \sigma^{\beta\gamma}$, which is (8.8). In the non-spreading regime, where $\sigma^{\alpha\gamma} < \sigma^{\alpha\beta} + \sigma^{\beta\gamma}$, the equilibrium structure of the $\alpha\gamma$ interface is not that of bulk β; the $\alpha\gamma$ interface has found an alternative structure of even lower free energy. But it always has available to it the resource of being coated with β phase; so it need never, and thus at equilibrium will never, have a higher free energy per unit area than $\sigma^{\alpha\beta} + \sigma^{\beta\gamma}$.

We shall presently give an alternative form of this argument in the framework of the two-density van der Waals theory, where it is most transparent, and where $\sigma^{\alpha\gamma} \leqslant \sigma^{\alpha\beta} + \sigma^{\beta\gamma}$ is virtually a truism.

Reported cases of $\sigma^{\alpha\gamma} > \sigma^{\alpha\beta} + \sigma^{\beta\gamma}$ can have either of two explanations, of which only the first and more obvious (albeit probably the more frequently applicable) is experimental error. The other is that at least one of the reported tensions, usually $\sigma^{\alpha\gamma}$, is not the equilibrium tension that would obtain with full mutual saturation of the phases, so that the corresponding $\sigma^{\alpha\gamma} - (\sigma^{\alpha\beta} + \sigma^{\beta\gamma})$ is not the equilibrium spreading coefficient. It has long been recognized that the *initial* spreading coefficient of a phase β at an $\alpha\gamma$ interface that has not previously been equilibrated with β, could well be positive. When it is, there is an initial rapid spreading of the β phase, followed (if the equilibrium spreading coefficient is negative rather than 0) by the retraction of the β film into a lens.[14] The initial spreading coefficient, in which the tension $\sigma^{\alpha\gamma}$ is that of an $\alpha\gamma$ interface that has not been equilibrated with, and may not even contain the chemical constituents of, the β phase, is often of interest and importance, but must be recognized as different from the equilibrium spreading coefficient with which we are concerned here.

The equilibration of the phases α and γ with the phase β ensures that each of the chemical substances present in β comes to be present in α and γ at the same thermodynamic activity as in β, even though, in cases of extreme insolubility, that may mean in very small amounts. But even those small amounts—which, spread through the bulk α and γ phases, might be in undetectably low concentration—could, if sufficiently concentrated in the $\alpha\gamma$ interface, affect profoundly the properties of that interface, and make it wholly different from one that was not equilibrated with β.

All this was known to Gibbs, and stated by him with characteristic succinctness.[15] (His A, B, and C are our α, γ, and β, respectively.)

Let A, B, and C be three different fluid phases of matter, which satisfy all the conditions necessary for equilibrium when they meet at plane surfaces. The components of A and B may be the same or different, but C must have no components except such as belong to A or B. Let us suppose masses of the phases A and B to be separated by a very thin sheet of the phase C. . . . The value of the superficial tension for such a film will be $\sigma_{AC} + \sigma_{BC}$, if we denote by these symbols the tensions of the surfaces of contact of the phases A and C, and B and C, respectively. . . . Now if $\sigma_{AC} + \sigma_{BC}$ is greater than σ_{AB}, the tension of the ordinary surface between A and B, such a film will be at least practically unstable. [It will retract into a lens.]. . . We cannot suppose that $\sigma_{AB} > \sigma_{AC} + \sigma_{BC}$, for this would make the ordinary surface between A and B unstable and difficult to realize. If $\sigma_{AB} = \sigma_{AC} + \sigma_{BC}$, we may assume, in general, that this relation is not accidental, and that the ordinary surface of contact for A and B is of the kind which we have described [that is, with A and B separated by a sheet of C].

With this weight of authority we may confidently dismiss all reports of positive equilibrium spreading coefficients as due to experimental error: incomplete equilibration or imprecise measurement.

An interesting side issue is the question of how the phases dispose themselves in the earth's gravitational field when $\sigma^{\alpha\gamma} = \sigma^{\alpha\beta} + \sigma^{\beta\gamma}$ and when β, which should then spread at the $\alpha\gamma$ interface, is the phase of greatest or least density. The answer is that the surface forces at first overwhelm the gravitational forces; the required spreading of β at the $\alpha\gamma$ interface occurs to a sufficiently great extent to produce a film of some thickness, in defiance of gravity; and only then, if there is present more than enough β phase to form the film, does the excess float to the top or sink to the bottom. That had been observed by Heady and Cahn,[7] and was recently beautifully demonstrated by Moldover and Cahn.[9] In a mixture of methanol and cyclohexane at temperatures below that of their consolute point the methanol-rich phase is the denser, and so the lower, of the two liquid phases; but it is also the liquid phase of lower surface tension, and it completely wets the interface between the upper, cyclohexane-rich phase and the vapour. Moldover and Cahn showed this by adding a dye that dissolves preferentially in methanol and so makes a brightly coloured lower phase, as well as a clearly visible, coloured border where the film at the interface between the upper liquid phase and the vapour meets the glass container wall.

De Gennes[16] has discussed the thickness of this film, taking explicit account of the long-range London forces (cf. § 7.5). He finds a contribution of the form H/h^2 to the free energy per unit area of such a film of thickness h. The coefficient H is proportional to the product of the difference in polarizabilities of the α and β phases and the difference in polarizabilities of the β and γ phases, and is thus the direct manifestation of the London forces. It is positive when the polarizability of β is intermediate between that of α and γ, negative otherwise. If the order of the mass densities is $\rho^\alpha < \rho^\gamma < \rho^\beta$, and if L is the height of the γ phase, so that the β film at the $\alpha\gamma$ interface is at a height L above the upper surface of bulk β, then in terms of H, L, the gravitational acceleration g, and the density difference $\Delta\rho = \rho^\beta - \rho^\gamma$, de Gennes finds the film thickness to be $h = (2H/Lg\Delta\rho)^{\frac{1}{3}}$, when $H > 0$. Experiment† suggests that h may be a few hundred ångstroms. Although the comparison is not yet conclusive, de Gennes' formula, which takes explicit account of the long-range London forces, seems in reasonable accord with the experiments (except, perhaps, for very large L or close to the consolute point). We have already remarked in § 7.5 that there might be circumstances in which we could not assume exponential variation of interfacial profiles.

After that digression we return to the inequality $\sigma^{\alpha\gamma} \leq \sigma^{\alpha\beta} + \sigma^{\beta\gamma}$ and see how it is to be interpreted in the phenomenology of the van der Waals theory.

In the language of the one-density van der Waals theory of Chapter 3, we have in a c-component system a density of excess free energy, $-W$, as a function of some density or composition variable, x, at fixed values of the $c + 1$ thermodynamic fields (of which only $c - p + 2$ are independent if p phases are in equilibrium). The function $W(x)$ here is like the $W(\rho)$ in Fig. 3.2, except that now, to describe three-phase equilibrium, it must have three equal maxima, as in Fig. 8.5. In this figure the variable x in the bulk α, β, and γ phases is shown to take the respective values x^α, x^β, and x^γ, as determined by the prescribed values of the $c - 1$ (because $p = 3$) independent fields. Those are the points x at which W has its three maxima, and where $W = 0$. The remaining c densities—those other than x—are imagined merely to follow the variations in x through the various interfaces just as they would vary with x in a bulk phase. That is the essence of this one-density version of the theory, as explained in § 3.3.

---

† O'D. Kwon, D. Beaglehole, W. W. Webb, B. Widom, J. W. Schmidt, J. W. Cahn, M. R. Moldover, and B. Stephenson, *Phys. Rev. Lett.* **48,** 185 (1982).

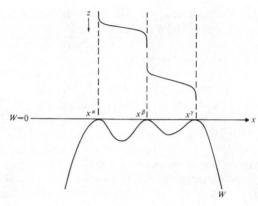

FIG. 8.5

The composition profile $x(z)$ as a function of the height z is then obtained in the usual way by considering x the coordinate and z the time for a particle of total energy 0 moving in the potential $W(x)$. The profile is shown in the upper part of Fig. 8.5. The particle, in the dynamical analogy, has vanishing velocity, and so spends infinite time, at $x = x^\alpha$, x^β, and x^γ, corresponding in the phase equilibrium to phases of macroscopic thickness in the z-direction. In the intervals $x^\alpha < x < x^\beta$ and $x^\beta < x < x^\gamma$ the function $x(z)$ of Fig. 8.5 is the composition profile of the $\alpha\beta$ and of the $\beta\gamma$ interface, respectively. They are profiles of the normal and expected kind, like (c) in Fig. 3.3. The composition profile of the $\alpha\gamma$ interface is given by the whole of $x(z)$, from x^α to x^γ, and is seen to have in it a step of indeterminately great height at the intermediate composition $x = x^\beta$. That says that the $\alpha\gamma$ interface incorporates in its equilibrium structure a macroscopically thick layer of a composition identical with that of the bulk phase β: a layer of β is adsorbed at the $\alpha\gamma$ interface. That is the picture in Fig. 8.3.

The interfacial tensions are the corresponding actions in the dynamical analogy. The action is the integral of the momentum over the coordinate, as in (3.13). But the integral from x^β to x^γ is the sum of the integrals from x^α to x^β and from x^β to x^γ. (We have $x^\alpha < x^\beta < x^\gamma$, by the convention in Fig. 8.5.) Thus, the three tensions satisfy Antonow's rule, (8.8). We see, then, that the one-density van der Waals theory requires that there be a layer of the third phase at the interface of highest tension, and that the three tensions be related in just the way that we had already concluded from macroscopic arguments they must be, when one phase spreads at the interface between the other two.

The theory also specifies that the high-tension interface is the one between the two phases that are least alike in composition, and that the spreading phase is the one that is of composition intermediate between the other two. In this one-density theory, by Fig. 8.5, as the composition in the $\alpha\gamma$ interface varies from that of bulk α to that of bulk γ it passes through that of bulk β. At that point all the thermodynamic conditions for the existence of β in bulk are met, so there is inevitably a macroscopic layer of β at the $\alpha\gamma$ interface.

To escape from this conclusion, and thus to allow non-spreading, with the three tensions related by the inequalities of the Neumann triangle, some way must be found of going from α to γ other than via β. Cahn[17] made the important remark that that can be done by allowing two or more composition variables to vary independently through the interface. In two or more dimensions there is no necessary betweenness: any point may be bypassed. That we might use a two- (or more-) density van der Waals theory to describe the case of non-spreading in three-phase equilibrium was anticipated in § 3.3, where it was stated as one of the motives for generalizing the one-density version of the theory.

Suppose, then, we consider the independent variation through the interface of two density or composition variables, say x and y. As in § 3.3, we now have a density of excess free energy, $-W(x, y)$, still defined at prescribed values of the $c - p + 2$ independent fields. Contours of constant W in the x, y-plane will be as in Figs 3.5 or 3.7, but now, with $p = 3$, they will show three peaks of equal height, as depicted in Fig. 8.6. The peaks

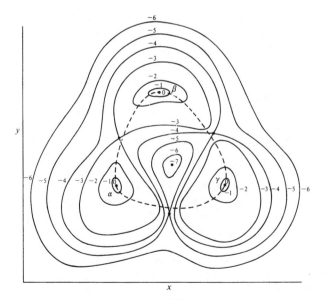

FIG. 8.6

in W, at which $W = 0$, occur at the points x, y that give the compositions of the bulk phases; those are the points labelled α, β, and γ.

We see in Fig. 8.6 two alternative trajectories that a particle might follow from α to γ, one an indirect path, via β, the other a more direct path, not via β. The indirect path always exists as a possible dynamical trajectory; the direct one only if the points α, β, and γ are not too nearly collinear (with β between α and γ).

Any trajectory is locally of minimum action, but when there are two, the action on one may be lower than that on the other. The one of lower action is the one that yields the stable, equilibrium structure of the $\alpha\gamma$ interface: x and y vary with each other through the interface as on the trajectory of minimum action in the x, y-plane. The interface may for some time, and with some degree of stability, assume a structure corresponding to a local but not absolute minimum in the interfacial tension (action), but it would then be only metastable, and would ultimately, and spontaneously, undergo transformation to the structure of absolutely lowest tension.

On the indirect path from α to γ via β, the particle in the dynamical analogy has vanishing velocity at the β peak, and so spends infinite time there. Thus, if that is the path of lower action, the equilibrium $\alpha\gamma$ interface consists of a macroscopically thick layer of β phase. Also, since the action on the indirect path is the sum of the actions on the $\alpha\beta$ and $\beta\gamma$ paths, we would then have $\sigma^{\alpha\gamma} = \sigma^{\alpha\beta} + \sigma^{\beta\gamma}$. This is exactly as in the one-density theory, though the description is now two-dimensional rather than one-dimensional. But if the action on the direct $\alpha\gamma$ path is the lower, the equilibrium structure of the $\alpha\gamma$ interface is as on the direct path; nowhere in that interface is there a point at which the composition is the same as in the bulk phase β; that interface is now of microscopic, not macroscopic, thickness; and, by present supposition, $\sigma^{\alpha\gamma} < \sigma^{\alpha\beta} + \sigma^{\beta\gamma}$. This is the alternative we sought, arising naturally in the two-density van der Waals theory.

We note that it is virtually a truism in this picture that the equilibrium $\sigma^{\alpha\gamma}$ cannot exceed $\sigma^{\alpha\beta} + \sigma^{\beta\gamma}$, since the indirect path, on which the action is only $\sigma^{\alpha\beta} + \sigma^{\beta\gamma}$, is always available. This is just a pictorial re-expression of the argument given earlier. The $\alpha\gamma$ interface could assume, and for some time retain, the structure given by the direct path even when the tension associated with that structure exceeds $\sigma^{\alpha\beta} + \sigma^{\beta\gamma}$; but that, as remarked earlier, would be a metastable, not an equilibrium, condition. It is when the actions on the two paths in Fig. 8.6 are equal that we have a Cahn transition, which is the subject of § 8.5.

We turn now to a further consideration of the structure of the $\alpha\gamma$ interface—particularly when β is not stable as a bulk phase, but when the two-phase states are in close proximity to three-phase states in which β is stable in bulk and spreads at the $\alpha\gamma$ interface.

8.4 The αγ interface

We know that when $\sigma^{\alpha\gamma} = \sigma^{\alpha\beta} + \sigma^{\beta\gamma}$ in three-phase equilibrium, there is a layer of bulk β at the $\alpha\gamma$ interface. But suppose now that β is not quite stable as a bulk phase; for example, suppose that $W(x^{\alpha}) = W(x^{\gamma}) = 0$ still, as in Fig. 8.5, but that now $W(x^{\beta}) = -w < 0$ with w small. Alternatively, referring to Fig. 8.6, suppose that $W(x^{\alpha}, y^{\alpha}) = W(x^{\gamma}, y^{\gamma}) = 0$ but that $W(x^{\beta}, y^{\beta}) = -w < 0$.

We show such a state, in which β is nearly but not quite in equilibrium as a bulk phase, as a point in the phase diagram of Fig. 8.7(a) or 8.7(b) for a system of c components. Figure 8.7(a) is in the space of two density or composition variables with $c - 1$ fields held fixed. The interior of the triangle is the three-phase region; adjoining each side of the triangle is a two-phase region, shown with some representative tielines; and these regions, in turn, are separated by one-phase regions, labelled α, β, and γ, at the vertices of the triangle. The state we contemplate is that marked by a cross in the $\alpha\gamma$ two-phase region, close to the $\alpha\gamma$ side of the three-phase triangle. Its distance from that side is proportional to w, the small quantity referred to above.[18] We see the same point also in Fig. 8.7(b), which is in the space of two fields, the remaining $c - 1$ fields being again fixed. The one-phase regions are there bounded and separated by two-phase lines that meet at the triple point. The state we contemplate is shown by the cross on the $\alpha\gamma$ equilibrium line, a small distance (proportional to w) from the triple point.

In such a state, then, what is the nature of the $\alpha\gamma$ interface? We may determine that from the van der Waals theory, based on an appropriate modification of the potential W in Fig. 8.5. The simplest $W(x)$ with the required properties is

$$W(x) = -W_0(a - x)^2(a + x)^2(x^2 + w/a^4W_0), \qquad (8.11)$$

in which w, a, and W_0 are positive parameters, with $w < \frac{1}{2}a^6W_0$. We show this $W(x)$ schematically in Fig. 8.8. We have stable phases α and γ and a metastable phase β, with $x^{\alpha} = -x^{\gamma} = -a$ and $x^{\beta} = 0$. The parameter w is a measure of how unstable β is with respect to α and γ.

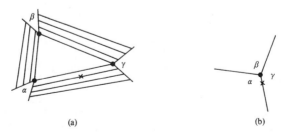

(a)   (b)

FIG. 8.7

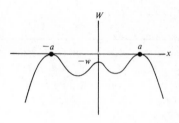

FIG. 8.8

When $w = 0$ the β phase can coexist stably with α and γ. In that limit $\sigma^{\alpha\beta} + \sigma^{\beta\gamma} = 2\sigma^{\alpha\beta} = 2\sigma^{\beta\gamma} = \sigma^{\alpha\gamma}$, by the obvious symmetry of the model. Let the common value of these four quantities be called σ. Then by (3.13), with m (for purposes of this illustration) taken to be another constant parameter,

$$\sigma = (\tfrac{1}{2}mW_0)^{\frac{1}{2}}a^4. \tag{8.12}$$

The composition profile of the $\alpha\gamma$ interface when $w > 0$ is found from (3.9) with (8.11). If we choose the direction of increasing distance z to be that of increasing x, and define $z = 0$ to be the point where $x = 0$, we find[18]

$$\frac{z}{\xi} = (1+\varepsilon)^{-\frac{1}{2}} \ln \left[\frac{\left[(1+\varepsilon)^{\frac{1}{2}}\left(\dfrac{x^2}{a^2}+\varepsilon\right)^{\frac{1}{2}} + \varepsilon + \dfrac{x}{a}\right]\left(1+\dfrac{x}{a}\right)}{\left[(1+\varepsilon)^{\frac{1}{2}}\left(\dfrac{x^2}{a^2}+\varepsilon\right)^{\frac{1}{2}} + \varepsilon - \dfrac{x}{a}\right]\left(1-\dfrac{x}{a}\right)} \right] \tag{8.13}$$

or the inverse[19]

$$\frac{x}{a} = \frac{\sinh[(1+\varepsilon)^{\frac{1}{2}}z/2\xi]}{\left(\dfrac{1}{\varepsilon}+\cosh^2[(1+\varepsilon)^{\frac{1}{2}}z/2\xi]\right)^{\frac{1}{2}}} \tag{8.14}$$

where ε and ξ are defined by

$$\varepsilon = \frac{w}{a^6 W_0}, \qquad \xi = \frac{1}{2a^2}\left(\frac{m}{2W_0}\right)^{\frac{1}{2}} = \frac{ma^2}{4\sigma}. \tag{8.15}$$

The second expression for ξ comes from (8.12), in which σ is the tension of the $\alpha\gamma$ interface in the limit $w = 0$. We see in (8.13) or (8.14) that ξ sets the scale of distance in the interface. The dimensionless parameter ε is an alternative to w as a measure of how far the state of the system is from one in which the β phase would be stable in bulk.

The profiles of the composition x and of the gradient dx/dz, in units of a and of $a/3\sqrt{3}\xi$, respectively, as derived from (8.14), are shown

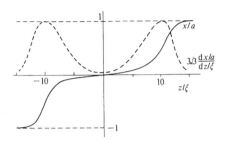

Fɪɢ. 8.9. Profile of composition x in units of a (solid curve) and profile of composition gradient dx/dz in units of $a/3\sqrt{3}\xi$ (dashed curve), plotted as functions of distance z measured in units of the correlation length ξ; from (8.14) with $\varepsilon = 10^{-4}$.

plotted in Fig. 8.9 for $\varepsilon = 10^{-4}$, the former as the solid curve and the latter as the dashed curve. The most noteworthy feature of the composition profile is the plateau at $x = x^{\beta}(=0)$, which for $\varepsilon = 10^{-4}$ extends over a range of z of about 10ξ. Thus, even though β is not stable as a bulk phase, there is still a layer of what is clearly identifiable as β phase at the $\alpha\gamma$ interface. That layer is now of microscopic rather than macroscopic thickness; but, as we shall see, that thickness increases with decreasing ε, and becomes infinite—the layer becomes macroscopically thick—in the limit $\varepsilon = 0$, where β becomes stable in bulk. We see that even before that limit is reached it is prefigured on a microscopic scale in the structure of the interface.

The smaller ε is, that is, the closer α and γ are to being in equilibrium with bulk β, the longer and more nearly horizontal is the plateau in the composition profile. At the midpoint, the gradient of the composition is found (most simply, directly from (3.9), (8.11), and (8.15)) to be

$$\left(\frac{dx/a}{dz/\xi}\right)_{z=0} = \tfrac{1}{2}\sqrt{\varepsilon} \tag{8.16}$$

so the slope of the plateau vanishes proportionally to $\sqrt{\varepsilon}$ as $\varepsilon \to 0$. The composition gradient is the velocity in the dynamical analogy, and so is greatest at the minima of $W(x)$. From (8.11), these are where $x/a = \pm[(1-2\varepsilon)/3]^{\frac{1}{2}}$. Then again from (3.9), (8.11), and (8.15), the maximum gradient is

$$\left(\frac{dx/a}{dz/\xi}\right)_{\max} = [(1+\varepsilon)/3]^{\frac{3}{2}}, \tag{8.17}$$

which for small ε is $\sim 1/3\sqrt{3}$, independently of ε. This is the reason we scale the gradient with $a/3\sqrt{3}\xi$ in making the plot in Fig. 8.9. From (8.13), the points z at which the maxima occur in the gradients are then $z \sim \pm\xi \ln(2/\varepsilon)$ for small ε. The distance between these points is a measure of the thickness of the interface, which we shall call h:

$$h \sim 2\xi \ln(2/\varepsilon) \quad \text{as} \quad \varepsilon \to 0. \qquad (8.18)$$

Thus, as the state of the system approaches that in which the third phase is stable in bulk, the interfacial thickness diverges as the logarithm of the distance from that limiting state.[8,18,20]

As we saw, the natural scale of distance in the composition profile is ξ, related to the model's parameters by (8.15). That basic distance is amplified by the logarithmic factor in (8.18), as, with decreasing ε, the $\alpha\gamma$ interface comes increasingly to resemble bulk β and its thickness becomes macroscopic. The reason we may identify the interface thickness h with the distance between the two maxima in the composition gradient when ε is small, is that, once past those maxima in its gradient, the composition goes exponentially rapidly—with the unamplified decay length ξ—to that of the bulk α and γ phases. We see that from (8.14): as $|z| \to \infty$,

$$|x|/a \sim 1 - 2(1 + 1/\varepsilon)\exp(-\sqrt{1+\varepsilon}\,|z|/\xi), \qquad (8.19)$$

the asserted exponential decay, with the decay length ξ when ε is small. Thus, if we identify the interface as that region in which the composition differs significantly from that of both bulk α and bulk γ, we see that the contributions to the interfacial thickness from the regions outside the two maxima in the composition gradient are only of order ξ, much less, for small ε, than the distance $2\xi \ln(2/\varepsilon)$ between the two maxima.

Because the logarithm is so slowly varying, it is only for very small ε that $\ln(2/\varepsilon)$ greatly exceeds 1. It is about 10 when $\varepsilon = 10^{-4}$; and, as we see in Fig. 8.9, the distance between the two maxima in the composition gradient is then indeed about 20ξ, as expected. But even for so small an ε as that, as we see in the figure, the distance between the maxima accounts for only about two-thirds of what we may reasonably identify as the total interface thickness; ε would have to be very much smaller still for $2\xi \ln(2/\varepsilon)$ to be entirely dominant.

These calculations of the structure of the incipient β layer at the $\alpha\gamma$ interface have been in the framework of the van der Waals theory, in which the composition of the interface approaches that of the bulk phases exponentially rapidly, with decay length ξ, the bulk-phase correlation length. But this β layer is subject to the same long-range London forces as is the wetting layer treated by de Gennes[16] (§ 8.3). If it should prove that the effects of those forces are not adequately expressed in a phenomenological excess free-energy density of the form (3.5), with W given by (8.11), the results of the present calculation would have to be modified quantitatively; but, we may reasonably assume, there would still be such a β layer, qualitatively like the one we have found here.

8.5 Phase transitions in interfaces. The Cahn transition

We shall first review very briefly the thermodynamics of first-order phase transitions between ordinary bulk phases at equilibrium. We shall then be able to describe in similar terms the closely analogous phase transitions in interfaces. Among these is the Cahn transition, our present subject.

In first-order transitions between bulk phases at given values of the fields $(\mu_1, \ldots, \mu_c, T) \equiv \mathbf{\mu}$, the thermodynamic potential $p(\mathbf{\mu})$, which is the pressure, is continuous, while the densities, which are the derivatives of p with respect to those fields, are discontinuous (except for occasional azeotropies; Appendix 1). Two phases that are at the same p with equal values of $\mathbf{\mu}$, but with unequal densities and hence with distinct molecular structures, are equally stable and may coexist in equilibrium.

The Gibbs–Duhem equation specifies how p varies with the $c + 1$ independent fields $\mathbf{\mu}$ of which it is a function; in the notation of § 2.3,

$$dp = \mathbf{\rho} \cdot d\mathbf{\mu}. \tag{8.20}$$

At two-phase equilibrium those $c + 1$ fields $\mathbf{\mu}$ are not all independent, for their possible variations are then restricted by the Clapeyron equation

$$(\mathbf{\rho}^\alpha - \mathbf{\rho}^\beta) \cdot d\mathbf{\mu} = 0. \tag{8.21}$$

This follows from the equilibrium condition $p^\alpha(\mathbf{\mu}) = p^\beta(\mathbf{\mu})$, by (8.20).

If we specify a set of values of the $c + 1$ fields appropriate to the coexistence of two phases, α and β, it still leaves indeterminate the relative amounts of the two phases that are present. That depends on otherwise irrelevant factors such as the nature of the walls of the containing vessel. If the walls favour one of the phases, only that phase will be present, for there is then no reason for the system to tolerate the excess free energy $\sigma^{\alpha\beta}A$ of an interface of area A between coexisting α and β phases. If, on the other hand, the $c + 1$ quantities we specify include at least one density, then some of both phases will be present, with an interface between them (provided the value we specify for any such density ρ_i is between ρ_i^α and ρ_i^β).

Analogously, in an interface we may have distinct surface phases with first-order transitions between them. That occurs when, for a given set of values of the thermodynamic fields, an interface has two alternative structures of equal tension. The two structures differ in their microscopic composition profiles, and therefore also (except for occasional azeotropies) in their macroscopic adsorptions (surface excesses). We may take as the surface thermodynamic potential the interfacial tension σ, which is analogous to $-p$ in our discussion of bulk-phase equilibrium; and the adsorptions are the surface densities, analogous to the earlier densities ρ_i. The surface analogue of the Gibbs–Duhem equation (8.20) is the Gibbs adsorption equation in the form (2.46),

$$d\sigma = -\sum_{i \neq j} \Gamma_{i(j)} \, d\mu_i, \qquad (8.22)$$

where the summation is over the c values of i not equal to the one arbitrarily chosen j; where the $\Gamma_{i(j)}$ are the c adsorptions relative to j; and where the μ_i are the c fields that remain independent after it is specified that two bulk phases (the two that are separated by the interface in question) are in equilibrium. We note that there is one field less in (8.22) than in (8.20), because of that condition of bulk-phase equilibrium.

If we call the two *surface* phases α_s and β_s, we have from $\sigma^{\alpha_s} = \sigma^{\beta_s}$, and (8.22), the surface analogue of the Clapeyron equation (8.21),

$$\sum_{i \neq j} [\Gamma_{i(j)}^{\alpha_s} - \Gamma_{i(j)}^{\beta_s}] \, d\mu_i = 0. \qquad (8.23)$$

This restricts the possible variations of the previously independent μ_i if the two surface phases are to remain in equilibrium. The condition of bulk-phase equilibrium, (8.21), had already decreased the number of independently variable fields from $c + 1$ to c, and now the additional condition of surface-phase equilibrium, (8.23), decreases it from c to $c - 1$.

Suppose we have two bulk phases coexisting, with an interface between them, and we then specify for the c fields that are still at our disposal a set of values that satisfy the further restriction for two surface phases to coexist. Then, as in the analogous circumstances in bulk-phase equilibrium, such a specification of fields alone still leaves indeterminate the relative amounts of (that is, the areas covered by) the two surface phases. Only if one or more of the c quantities that is specified is a density—here, one of the $\Gamma_{i(j)}$, specified to have, for the surface overall, some value between $\Gamma_{i(j)}^{\alpha_s}$ and $\Gamma_{i(j)}^{\beta_s}$—would the relative amounts of the two surface phases, as well as the intrinsic properties of each, be determined. That is what we do when we adjust the interfacial area in a Langmuir film balance.[21] When two such surface phases do coexist, they are separated by a macroscopically one-dimensional interface with a positive tension. That, too, is a line tension, but it is otherwise not the same thing as the tension of the line in which three bulk phases meet, which we discuss in § 8.6.

In bulk-phase equilibrium, coexisting phases may become gradually more alike as the thermodynamic parameters vary, ultimately becoming identical at a critical point. We may similarly have critical points of surface-phase equilibria, which are again the limit points of first-order transitions, at which the previously distinct phases become identical. Many kinds of higher-order phase transitions—at least as great in variety and complexity as those in bulk fluids—are also possible in interfaces, and are frequently reported.

The most familiar transitions between surface phases in fluid interfaces are those in the so-called insoluble monomolecular films that some higher alcohols and fatty acids form at a water–air interface.[21] Distinct surface phases and transitions between them are frequently observed also in monolayers of adsorbed gases on solid substrates,[22] and are the subject of an exuberant modern literature.[23]

First-order phase transitions in interfaces and the critical points associated with them may be described pictorially in the framework of the generalized van der Waals theory. We imagine the free-energy-density excess $-W$ as a function of the full set of densities $\rho_1, \ldots, \rho_{c+1}$ at given values of the μ_i appropriate to bulk two-phase equilibrium, and we consider the trajectory of a particle that moves from one to the other of the two equally high peaks of the potential-surface W, as in the two-dimensional example of Fig. 3.7. That trajectory and its time dependence then determine, in the way that is now familiar, the interfacial density profiles, and the dynamical action on it gives the interfacial tension. As in § 8.3 (but now for two-phase rather than three-phase equilibrium), we may have two distinct trajectories that are both locally of minimum action. Then the one of lower action determines the stable interfacial structure, the one of higher action a metastable structure. When the two distinct trajectories are of equal action they describe two equally stable but distinct surface phases that may coexist. The transitions between them are the interfacial phase transitions we have been discussing. As some thermodynamic parameter varies, the two trajectories of equal action may move closer together, until finally some critical point is reached at which they coincide. That is the critical point of the transition.

The Cahn transition is the particular case in which the transition is between the wetting and non-wetting of an $\alpha\gamma$ interface by β phase (§ 8.3)—or by incipient β phase if β is not stable in bulk (§ 8.4). It is thus the transition between two alternative structures of the $\alpha\gamma$ interface: one in which it consists of a macroscopic layer of bulk β (or a microscopic layer of incipient bulk β), and another in which it does not.

Such a transition as it occurs when β is stable in bulk was already illustrated in Fig. 8.4. There we see a discontinuity in the derivative of surface tension with respect to a field—hence, a discontinuous adsorption—while the surface tension itself is continuous. That, as we have seen, is the characteristic feature of a first-order interfacial phase transition. We have already described a general interfacial phase transition in the language of the extended van der Waals theory; we may apply that description particularly to the Cahn transition, referring now to Fig. 8.6. If, as the thermodynamic state changes, there is a point at which the trajectory that was formerly of higher action becomes the one of lower action, it marks a point of Cahn transition.

We may similarly describe the Cahn transition as it occurs when the wetting layer is a microscopically thick layer of an incipient β phase rather than a macroscopic layer of bulk β (§ 8.4). Again referring to Fig. 8.6, we imagine that the height of the β peak in the W surface is slightly less than the common height of the α and γ peaks. There is still one trajectory from α to γ that passes over or near the β peak; a particle following that trajectory would still move slowly near the β peak, though now not infinitely slowly. There may also, as in Fig. 8.6, be a competing $\alpha\gamma$ trajectory that does not pass near the β peak; and again, as in our description of the general interfacial phase transition, as the thermodynamic state varies there may be a transition in which first one and then the other of the two trajectories is of lower action. That is again a Cahn transition, now a more general one than that which occurs when β is present in bulk.

The existence of such a transition—both when β is and when it is not stable in bulk—was predicted by Cahn.[8] He assumed the proximity of a critical point of $\beta\gamma$ (or $\alpha\beta$) phase equilibrium, and then referred to the transition from non-wetting to wetting of the $\alpha\gamma$ interface by a bulk or incipient β phase as 'critical-point wetting'. We paraphrase here his argument that such a transition is to be expected in the neighbourhood of, say, the $\beta\gamma$ critical point.

As the critical point is approached $\sigma^{\beta\gamma}$ vanishes proportionally to $|T^c - T|^\mu$, where $|T^c - T|$ is the distance from that critical point measured as the difference between the temperature T and the critical temperature T^c, and where μ (not to be confused with chemical potential) is the critical-point surface-tension exponent with the universal value $\mu \simeq 1\cdot26$ (Chapter 9). In Fig. 8.10, also due to Cahn,[17] the ordinate of one of the curves is $\sigma^{\beta\gamma}$, shown vanishing at the origin with a power $\mu > 1$ of $|T^c - T|$. The ordinate of the other curve is the difference $\sigma_*^{\alpha\gamma} - \sigma^{\alpha\beta}$, where $\sigma^{\alpha\beta}$ is, in our usual notation, the tension of the equilibrium $\alpha\beta$ interface, while $\sigma_*^{\alpha\gamma}$ is the tension of the $\alpha\gamma$ interface when its structure is not that of a layer of β phase; that is, when its structure is that given by the trajectory in Fig. 8.6 that goes directly from α to γ, not via β. That may or may not be the trajectory of lower action, so $\sigma_*^{\alpha\gamma}$ may or may not be the equilibrium tension $\sigma^{\alpha\gamma}$. That interface of tension $\sigma_*^{\alpha\gamma}$, whether it is stable or metastable, has the normal structure of any simple, two-phase

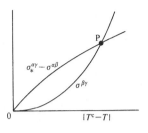

FIG. 8.10

interface, such as the $\alpha\beta$ interface; it does not have the layered structure that is associated with the other, indirect path from α to γ, via β. Near the $\beta\gamma$ critical point, where the β and γ phases are nearly identical, that unlayered $\alpha\gamma$ interface and the $\alpha\beta$ interface are very similar, and their tensions $\sigma_*^{\alpha\gamma}$ and $\sigma^{\alpha\beta}$ are nearly the same. Cahn supposes these to differ by an amount that is proportional to the difference in composition of the β and γ phases, which, in turn, is known to vanish proportionally to $|T^c - T|^\beta$, with a universal critical-point exponent $\beta \approx \frac{1}{3}$, as the critical point is approached (§ 9.3). Alternatively, $\sigma_*^{\alpha\gamma} - \sigma^{\alpha\beta}$ may be supposed to vanish proportionally to the difference between some field conjugate to the composition, and the critical value of that field; hence, as the first power of $T^c - T$. With either supposition, $\sigma_*^{\alpha\gamma} - \sigma^{\alpha\beta}$ vanishes less rapidly than $\sigma^{\beta\gamma}$ does as we approach the critical point, and, correspondingly, ultimately increases less rapidly than $\sigma^{\beta\gamma}$ does as we recede from the critical point. We therefore expect a crossing, as at P in Fig. 8.10. Further from the critical point than that crossing point, $\sigma_*^{\alpha\gamma} - \sigma^{\alpha\beta} < \sigma^{\beta\gamma}$. The tension $\sigma_*^{\alpha\gamma}$ of the normal, unlayered $\alpha\gamma$ interface is then less than the tension of the layered interface, which we know from § 8.3 is $\sigma^{\alpha\beta} + \sigma^{\beta\gamma}$. The equilibrium structure is that of the former, therefore, with the associated tension $\sigma^{\alpha\gamma} = \sigma_*^{\alpha\gamma} < \sigma^{\alpha\beta} + \sigma^{\beta\gamma}$. Nearer the critical point than the crossing point, $\sigma_*^{\alpha\gamma} - \sigma^{\alpha\beta} > \sigma^{\beta\gamma}$. The tension $\sigma_*^{\alpha\gamma}$ of the unlayered $\alpha\gamma$ interface is then greater than that of the layered interface, which is $\sigma^{\alpha\beta} + \sigma^{\beta\gamma}$; so the equilibrium structure is that of the latter, with the associated tension $\sigma^{\alpha\gamma} = \sigma^{\alpha\beta} + \sigma^{\beta\gamma}$. The crossing point P is thus a point of Cahn transition.

The important conclusion from this argument is that if a Cahn transition in the $\alpha\gamma$ interface occurs near a $\beta\gamma$ (or, equally well, an $\alpha\beta$) critical point, the states in which β spreads at the $\alpha\gamma$ interface are those which are nearer the critical point, while those in which β does not spread are those which are further. That was brilliantly verified in the experiments of Moldover and Cahn.[9] As a corollary of great practical importance, we note that sufficiently near to a $\beta\gamma$ or $\alpha\beta$ critical point, the interface of highest tension, which we have been calling the $\alpha\gamma$ interface, necessarily consists of a spread layer of β phase, and Antonow's rule, $\sigma^{\alpha\gamma} = \sigma^{\alpha\beta} + \sigma^{\beta\gamma}$, necessarily holds. (In § 8.3 we noted an obvious, though hypothetical, exception: two incompletely miscible liquids, β and γ, that are enantiomers, in equilibrium with their common vapour, α. By symmetry, $\sigma^{\alpha\gamma} = \sigma^{\alpha\beta} < \sigma^{\alpha\beta} + \sigma^{\beta\gamma}$, however close the system is to the critical mixing point of the two liquids. But in just such a case $\sigma^{\alpha\gamma}_* - \sigma^{\alpha\beta} = \sigma^{\alpha\gamma} - \sigma^{\alpha\beta} \equiv 0$, contradicting a basic premise of the earlier argument.)

In Fig. 8.11, which we have adapted from Cahn and from Teletzke *et al.*,[8] we show the temperature (T) vs composition (x) coexistence curve for the equilibrium of the phases β and γ (two liquids, say), while these are also in equilibrium with a third phase, α, which is not shown in the diagram (a vapour phase, say, or a solid boundary). The $\beta\gamma$ critical point is at C. The points marked γ and β and shown connected by a tieline are a general pair of equilibrium γ and β phases. The tieline labelled P marks the Cahn transition in the three-phase ($\alpha\beta\gamma$) region, and corresponds to P in Fig. 8.10. In the three-phase region above P, that is, closer to the critical point C, the $\alpha\gamma$ interface is wetted by β, below P it is not.

We saw that we may have a Cahn transition also in the two-phase ($\alpha\gamma$) region, where β is not stable in bulk. The curve P'C' in Fig. 8.11 is the locus of these transition points. On the side of P'C' that is toward C, the $\alpha\gamma$ interface consists of a layer of incipient β; on the other side of the locus it does not. It is at the coexistence curve that β becomes stable in bulk. As any point of the coexistence curve between P' and C is approached, the thickness of the β layer diverges, and does so proportionally to $\ln(1/\varepsilon)$, where ε is a measure of the distance from the coexistence curve. We saw that in § 8.4 and it is also as found by Cahn.[8]

Note the asymmetry: there is no locus corresponding to P'C' at the β side of the coexistence curve. That is because the high-tension interface

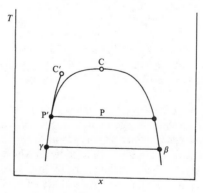

FIG. 8.11

(by our convention) is $\alpha\gamma$, so β may wet the $\alpha\gamma$ interface but γ does not wet the $\alpha\beta$ interface.

The point C' is the critical point of this interfacial phase transition, also predicted by Cahn.[8] Along P'C' the two alternative structures of the $\alpha\gamma$ interface are of equal tension. As C' is approached, those two equally stable but distinct structures become gradually more alike, in the way we saw in our description of the critical points of general first-order interfacial phase transitions. At C' they have become identical.

The theory of Cahn that led to Fig. 8.11 and to the foregoing interpretation was based on the van der Waals theory, but instead of describing the three-phase equilibrium as we do here, with a three-peak potential W, Cahn described it as a two-phase ($\beta\gamma$) equilibrium, with the third phase (α) introduced only via a boundary condition specifying the composition at the α surface. Minimizing the free energy then led either to only one possible value of that boundary composition, associated with the bulk composition x^β, or to two possible values of it, one associated with the bulk composition x^β and the other with the bulk composition x^γ. The former implied that there was an intruding β layer at the $\alpha\gamma$ interface, the latter that the α, β, and γ phases met at a three-phase contact line with contact angle $\beta > 0$.

Although eminently successful, the theory in that form is not fully satisfactory, for instead of treating the three phases on an equal footing it replaces one of them by an *ad hoc* boundary condition. Sullivan[8] and Teletzke *et al.*[8] have given alternative formulations. Another possible form of the theory would be based on a three-peak potential surface $W(x, y)$, on which the β peak is of height less than or equal to the common height of the α and γ peaks. The dynamical trajectories that connect the α and γ peaks must be determined, and the actions on those trajectories calculated and compared—all as functions of the thermodynamic fields that determine the W surface. Some results of this kind for a model surface with three peaks of equal height (thus, with β stable in bulk) have been obtained by Currie and Bishop,[24] and they did observe the Cahn transition. More extensive and systematic calculations, particularly in the regime in which β is not stable in bulk, would be welcome.

So far, the only Cahn transitions that have been seen with certainty in fluid interfaces are in three-phase systems,[9,25] where they are visible as transitions between spreading and non-spreading. The phenomenon may manifest itself as in Fig. 8.4, as a transition between a negative value of $\sigma^{\alpha\gamma} - (\sigma^{\alpha\beta} + \sigma^{\beta\gamma})$ and the value 0; or, equivalently, as a transition between a positive value of the contact angle β and the value 0, which is how Moldover and Cahn observed it;[9] or as a discontinuity in the derivative of $\sigma^{\alpha\gamma}$ with respect to a varying thermodynamic field. The more general Cahn transition that is predicted to occur in the two-phase region (at P'C' in Fig. 8.11) might have no immediately visible, macro-scopic manifestation; but it, too, would lead to a discontinuity in the derivative of $\sigma^{\alpha\gamma}$ with respect to a varying field, and it would also be detectable in principle as a discontinuous change in the microscopic structure—hence, in the optical properties—of the $\alpha\gamma$ interface. That more general Cahn transition, with its predicted critical point C', has not yet been observed with certainty in any fluid interface; it remains as an urgent challenge to experiment.

8.6   Three-phase line and line tension

We turn now to a consideration of the tension and microscopic structure of the line in which three phases meet—when there is such a line; that is, when the three interfacial tensions satisfy the Neumann-triangle conditions as expressed by the inequalities (8.7) or by Fig. 8.2.

The thermodynamics of the line tension may be developed in close analogy with that of surface tension, and we give it here in brief outline.[26-28] We may take the surfaces shown in Fig. 8.1 to be arbitrary dividing surfaces. Their intersection, which is the three-phase line, has then also an arbitrary location, determined by the locations of the surfaces. The dihedral angles between the surfaces are not arbitrary; they are determined by the interfacial tensions via (8.5). The orientation of the planes about the three-phase line is fixed by an external field such as gravity—just as the orientation of the single interface in two-phase equilibrium is determined by gravity. With fixed dihedral angles and fixed orientation, the location of the dividing surfaces is still arbitrary with respect to a parallel translation. Figure 8.12 shows two alternative and equally arbitrary choices, both viewed along the three-phase line (the perspective of Fig. 8.1(b)).

We make one such choice, and then isolate for consideration a right circular cylindrical sample of material, as in Fig. 8.13. (Alternatively, we may imagine the sample in the form of a triangular prism with walls perpendicular to the interfaces.[3,28] It is important only that the sides of

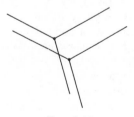

F<small>IG</small>. 8.12

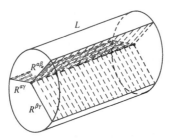

FIG. 8.13

the cylinder be or become perpendicular to the interfaces in the limit of
infinite sample size.) This sample is far from any container walls. The
surface of the cylinder is not itself a physical surface, only a mathematical
boundary outlining the region of the fluid that is under consideration. The
axis of the cylinder is parallel to the three-phase line. The cylinder is of
length L, and the lengths of the intersections of the three dividing
surfaces with the bases of the cylinder are $R^{\alpha\beta}$, $R^{\beta\gamma}$, and $R^{\alpha\gamma}$.

Let V^{α}, V^{β}, and V^{γ} be the volumes of the three phases in the
cylindrical sample; that is, the volumes of the sectors of the cylinder that
are within the respective dihedral angles α, β, and γ. We adopt the Gibbs
convention that the total volume V of the sample is the sum of the
volumes ascribed to the individual phases, so that there is no surface or
linear excess volume:

$$V = V^{\alpha} + V^{\beta} + V^{\gamma}, \tag{8.24}$$

as in (2.17). Let U^{α}, S^{α}, and $\mathbf{n}^{\alpha}$ be the energy, entropy, and masses or
molar amounts of the various constituents in a volume V^{α} of bulk α
phase; and similarly for the β and γ phases. Likewise, let $U^{\alpha\beta}$, $S^{\alpha\beta}$, and
$\mathbf{n}^{\alpha\beta}$ be the surface excesses of those quantities at the $\alpha\beta$ interface (§ 2.3)
when, hypothetically, the properties of that interface remain unchanged,
up to the three-phase line, from what they are far from that line; and
similarly for the $\beta\gamma$ and $\alpha\gamma$ interfaces. Then the corresponding linear
excesses U^{l}, S^{l}, and $\mathbf{n}^{l}$ at the three-phase line are

$$U^{l} = U - (U^{\alpha} + U^{\beta} + U^{\gamma}) - (U^{\alpha\beta} + U^{\beta\gamma} + U^{\alpha\gamma}), \quad \text{etc.} \tag{8.25}$$

where U (or S or $\mathbf{n}$) is the total energy (or entropy, etc.) in the whole
cylindrical sample. For macroscopic $R^{\alpha\beta}$, $R^{\beta\gamma}$, $R^{\alpha\gamma}$, and L, the linear
excesses (8.25) are independent of the Rs and proportional to L. Those
excesses depend only on the properties of the fluid in the microscopic
transition region in which the three phases meet, and on the location

(typically within and certainly parallel to the transition region, but otherwise arbitrary) of the three-phase line.

The linear excess Ω^l of the free energy Ω (the grand potential; cf. § 2.3) is defined as in (8.25), with

$$\Omega^\alpha = -pV^\alpha, \quad \text{etc.} \tag{8.26}$$

and

$$\Omega^{\alpha\beta} = \sigma^{\alpha\beta} R^{\alpha\beta} L, \quad \text{etc.} \tag{8.27}$$

from (2.25) and (2.26); where, as before, p is the uniform pressure in the bulk phases, and $\sigma^{\alpha\beta}$ is the tension of the $\alpha\beta$ interface. Then from (8.24), (8.26), and (8.27) we have

$$\Omega^l = \tau L = \Omega + pV - (\sigma^{\alpha\beta} R^{\alpha\beta} + \sigma^{\beta\gamma} R^{\beta\gamma} + \sigma^{\alpha\gamma} R^{\alpha\gamma})L \tag{8.28}$$

which defines the *line tension* τ, and is the analogue of (2.24) and (2.26).

We have here as the analogue of (2.22),

$$\tau L = \Omega^l = U^l - TS^l - \boldsymbol{\mu} \cdot \mathbf{n}^l. \tag{8.29}$$

We shall now see that, although U^l, S^l, and $\mathbf{n}^l$ individually depend on the location of the dividing surfaces, Ω^l does not; so that the line tension τ, like each of the surface tensions σ, is independent of that choice.[28]

From (8.28), any change $\Delta\tau$ in τ resulting from a change in the location of the dividing surfaces would be

$$\Delta\tau = -\sigma^{\alpha\beta} \Delta R^{\alpha\beta} - \sigma^{\beta\gamma} \Delta R^{\beta\gamma} - \sigma^{\alpha\gamma} \Delta R^{\alpha\gamma}, \tag{8.30}$$

where $\Delta R^{\alpha\beta}$, etc., are the corresponding changes in the Rs. In Fig. 8.14 we view the cylinder of Fig. 8.13 along the three-phase line. We see the base of the cylinder as a circle of radius R; we see its intersection with the $\alpha\beta$ dividing surface as a line segment of length $R^{\alpha\beta}$, as in Fig. 8.13; and we see the three-phase line displaced from the axis of the cylinder, to which it is parallel, by the vector $\mathbf{r}$, of length r. By the law of cosines

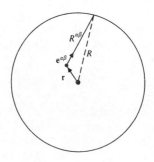

FIG. 8.14

applied to the triangle in Fig. 8.14,

$$R^2 = (R^{\alpha\beta})^2 + r^2 + 2\mathbf{e}^{\alpha\beta} \cdot \mathbf{r} R^{\alpha\beta}, \qquad (8.31)$$

where $\mathbf{e}^{\alpha\beta}$ is a unit vector parallel to the $\alpha\beta$ dividing surface and perpendicular to the three-phase line, directed in the sense shown in the figure. From (8.31)

$$R^{\alpha\beta} = R - \mathbf{e}^{\alpha\beta} \cdot \mathbf{r} + O(1/R). \qquad (8.32)$$

We now make a parallel displacement of the dividing surfaces (see Fig. 8.12), by $\Delta \mathbf{r}$, say. Then by (8.32) the accompanying change $\Delta R^{\alpha\beta}$ in the length $R^{\alpha\beta}$ is

$$\Delta R^{\alpha\beta} = -\mathbf{e}^{\alpha\beta} \cdot \Delta \mathbf{r} + O(1/R). \qquad (8.33)$$

This is the $\Delta R^{\alpha\beta}$ required in (8.30). If now $\mathbf{e}^{\beta\gamma}$ and $\mathbf{e}^{\alpha\gamma}$ are unit vectors related, respectively, to the $\beta\gamma$ and $\alpha\gamma$ dividing surfaces as $\mathbf{e}^{\alpha\beta}$ is to the $\alpha\beta$ surface, it follows from (8.30) and (8.33) that in the macroscopic limit $(R \to \infty)$,

$$\Delta\tau = (\sigma^{\alpha\beta}\mathbf{e}^{\alpha\beta} + \sigma^{\beta\gamma}\mathbf{e}^{\beta\gamma} + \sigma^{\alpha\gamma}\mathbf{e}^{\alpha\gamma}) \cdot \Delta\mathbf{r}. \qquad (8.34)$$

But the quantity in parentheses is the net force exerted on unit length of the three-phase line by the tensions in the three interfaces that meet at that line; and, at equilibrium, that force vanishes. (Indeed, the left-hand sides of the three relations (8.2) are just the scalar products of $\sigma^{\alpha\beta}\mathbf{e}^{\alpha\beta} + \sigma^{\beta\gamma}\mathbf{e}^{\beta\gamma} + \sigma^{\alpha\gamma}\mathbf{e}^{\alpha\gamma}$ with $\mathbf{e}^{\alpha\beta}$, $\mathbf{e}^{\beta\gamma}$, and $\mathbf{e}^{\alpha\gamma}$ in turn.) Thus, at equilibrium in a macroscopic system,

$$\Delta\tau = 0; \qquad (8.35)$$

that is, the line tension is independent of the arbitrary choice of location of the dividing surfaces, as we wished to show.

Besides (2.28) for the bulk phases and (2.29) for the interfaces, we have now the additional differential thermodynamic identity $dF^l = -S^l dT + \mathbf{\mu} \cdot \mathbf{dn}^l + \tau dL$; or, equivalently,

$$-dU^l + T dS^l + \mathbf{\mu} \cdot \mathbf{dn}^l + \tau dL = 0. \qquad (8.36)$$

We have continued to use the Gibbs convention $V^l = 0$; and we have assumed that we are in the macroscopic limit, in which the three-phase line may be treated as linear so that curvature terms[27] may be neglected in (8.29) and (8.36)—just as curvature terms were neglected in the analogous equations for the surface thermodynamic functions in § 2.3, where we assumed planar interfaces.

From (8.29) and (8.36) we find

$$L\, d\tau = -S^l dT - \mathbf{n}^l \cdot \mathbf{d\mu}. \qquad (8.37)$$

To express this in the compressed notation introduced in (2.35) (already used in this chapter in (8.20)–(8.23), but not yet in this section), we include S^l as a component of $\mathbf{n}^l$ and T as the corresponding component of $\boldsymbol{\mu}$, and we define the linear adsorptions $\boldsymbol{\Lambda}$,

$$\Lambda_i = n_i^l / L. \tag{8.38}$$

Then (8.37) becomes the fundamental equation of linear adsorption,

$$d\tau = -\boldsymbol{\Lambda} \cdot \mathbf{d}\boldsymbol{\mu}, \tag{8.39}$$

the linear analogue of the Gibbs adsorption equation (2.35). The individual adsorptions Λ_i, like the surface adsorptions Γ_i, depend on the locations of the dividing surfaces and three-phase line, but $\boldsymbol{\Lambda} \cdot \mathbf{d}\boldsymbol{\mu}$ does not, because, as we know, τ does not.

The Clapeyron equations of three-phase equilibrium,

$$\boldsymbol{\rho}^\alpha \cdot \mathbf{d}\boldsymbol{\mu} = \boldsymbol{\rho}^\beta \cdot \mathbf{d}\boldsymbol{\mu} = \boldsymbol{\rho}^\gamma \cdot \mathbf{d}\boldsymbol{\mu}, \tag{8.40}$$

are consequences of the Gibbs–Duhem equation (8.20) and the equilibrium conditions $p^\alpha(\boldsymbol{\mu}) = p^\beta(\boldsymbol{\mu}) = p^\gamma(\boldsymbol{\mu})$, as was the analogous (8.21) for two-phase equilibrium. In a c-component system there are $c + 1$ differentials $d\mu_i$ on the right-hand side of (8.39), but because of (8.40) only $c - 1$ of them are independent. In two-phase equilibrium there is a one-parameter infinity of possible locations of the dividing surface—it can be placed at any height z—and by choosing it to be at such a height that the adsorption Γ_j vanished, we were led to the relative adsorptions $\Gamma_{i(j)}$ and the form (2.46) of the adsorption equation. There is now a two-parameter infinity of possible locations of the three-phase line (Fig. 8.12), and we may correspondingly choose for it a location that makes two of the linear adsorptions, say Λ_j and Λ_k, vanish. Thus, defining the relative adsorption $\Lambda_{i(jk)}$ to be Λ_i when the three-phase line is so placed that $\Lambda_j = \Lambda_k = 0$, we obtain the adsorption equation, (8.39), in the form

$$d\tau = - \sum_{i \neq j,k} \Lambda_{i(jk)} \, d\mu_i \tag{8.41}$$

where the $c - 1$ infinitesimal changes $d\mu_i$ are all independent. This is the analogue of (2.46).

We shall see later that, both in theory and by experiment, the tension τ of a three-phase line may be of either sign. It is to be distinguished from the physically quite different tension of a one-dimensional interface between two surface phases, which we mentioned in passing in § 8.5, and which, like any two-phase boundary tension, is necessarily positive. The reason such a boundary tension must be positive at equilibrium is that if it were negative the interface between the phases (the two-dimensional interface in three dimensions or the one-dimensional interface in two dimensions) would spontaneously pucker, and become sponge-like, to increase its area (or length); and this would continue for as long as the interfacial tension still had its macroscopic, negative value; that is, it would continue until the two phases were molecularly dispersed in each other. Such a mutual dispersion on a molecular scale (of which a micro-emulsion, so-called, is an example) is a single, homogeneous phase, with no interface. Thus, if there is indeed a separation into two distinct, macroscopic phases at equilibrium, the tension of the boundary between them must be positive. This argument does not apply to the tension of the line in which three phases meet, which is what we are concerned with here: for in three-phase equilibrium the contact line cannot pucker to increase its length without at the same time changing the areas of the two-phase interfaces—which are of positive tension—in such a way as to increase the free energy of the whole system.

The tension of our three-phase line is different also from the quantity defined as line tension by Langmuir.[29] That, it is generally agreed,[30,31] arises from a misconception: an incorrect formula for the balance of surface forces in a sessile lens, which leads to deviations between the theoretical and true shape of the lens, the discrepancy being then corrected by the introduction of a spurious line tension.

A soap film and its Gibbs ring, or Plateau border,[31,32] meet in the configuration of Fig. 8.1(a). (We continue for the moment to neglect curvature of the surfaces and of the three-phase line.) The film itself constitutes the $\alpha\gamma$ interface, and the film's border is bulk β phase; see Fig. 8.15. The line in which the film meets its border is a three-phase line of the kind we have been considering, and the most reliable measurements we have of the line tension τ are those of the tension of such a line at the boundary of a soap film.[33] Though the film comes originally from the β phase, its structure, once it achieves its equilibrium thinness,[32] is so different from that of bulk β that the film tension $\sigma^{\alpha\gamma}$ does not equal, but is less than, $\sigma^{\alpha\beta} + \sigma^{\beta\gamma}$; so the contact angle β is positive.[31,33] The α and γ phases are identical here, so $\sigma^{\alpha\beta} = \sigma^{\beta\gamma}$; and then by (8.4) the contact angles α and γ are equal, as in Fig. 8.15.

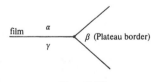

FIG. 8.15

Just as all the common measurements of surface tension σ depend on observing the effects of the pressure difference across an interface of known curvature, via Laplace's equation, (2.2), so also the measurement of line tension τ is made possible by the term τ/r which enters the equation for the balance of forces on unit length of a line[31,33] at a point where its radius of curvature is r.

As a first step towards a microscopic theory of the line tension τ we shall imagine, as in Chapter 3, that there is a local free-energy-density excess, $\Psi(\mathbf{r})$, due to the deviation of the fluid structure at $\mathbf{r}$ from the structure of any of the three bulk phases. Thus, Ψ vanishes in the interior of each phase but is non-vanishing within microscopic distances of the two-phase interfaces and of the three-phase line. We shall presently consider for $\Psi(\mathbf{r})$ a form similar to that in (3.5), but for now we leave it more general.

In our earlier theory of a planar interface we took Ψ to be independent of the position in any plane parallel to the interface, and to depend only on the distance z in the direction perpendicular to such planes. Now, instead, Ψ depends in an essential way on two coordinates, viz. those in any plane perpendicular to the three-phase line; but all such planes are equivalent, as Ψ is independent of position along the three-phase line. For example, let r, θ, x be cylindrical coordinates in the cylinder of Fig. 8.13, with x the axial coordinate, so that r, θ are plane polar coordinates in any plane perpendicular to the three-phase line. Since the axis of the cylinder is parallel to the three-phase line,

$$\Psi = \Psi(r, \theta). \tag{8.42}$$

Let $da(= r\, dr\, d\theta)$ be an element of area of any of the circular cross-sections of the cylinder in Fig. 8.13, and let A be the whole interior of that circle. Then the total excess free energy of the cylindrical sample due to the presence within it of the three interfaces and the three-phase line, per unit length in the direction of the three-phase line, is $\int_A \Psi\, da$. Since Ψ differs from 0 only within microscopic distances of the interfaces and the three-phase line, that excess free energy per unit length, for a macroscopic sample, is

$$\int_A \Psi\, da \sim \sigma^{\alpha\beta} R^{\alpha\beta} + \sigma^{\beta\gamma} R^{\beta\gamma} + \sigma^{\alpha\gamma} R^{\alpha\gamma} + \tau, \tag{8.43}$$

where $R^{\alpha\beta}$, etc., are as measured from any conveniently chosen origin within the circular cross-section. We already know that in the macroscopic limit the right-hand side of (8.43) is independent of the choice of origin. By (8.28) it is $\Omega + pV$ per unit length; that is, it is the excess of the grand potential Ω, per unit length, that is due to the inhomogeneities. (In a homogeneous fluid, $\Omega = -pV$.) We had already remarked following (3.5) that $\Psi(\mathbf{r})$ has the meaning of the local excess density of the free energy Ω.

If, as is convenient, we choose as origin the centre of the circular cross-section, so that $R^{\alpha\beta} = R^{\beta\gamma} = R^{\alpha\gamma} = R$, the radius of the cylindrical sample, then (8.43) becomes

$$\int_A \Psi \, da \sim (\sigma^{\alpha\beta} + \sigma^{\beta\gamma} + \sigma^{\alpha\gamma})R + \tau \qquad (R \to \infty). \tag{8.44}$$

The line tension τ associated with any known or postulated local excess free-energy density Ψ is thus found by integrating Ψ over the interior of a circle of radius R, finding the large-R asymptotic behaviour of the integral, and identifying τ as the constant term in the asymptotic expansion. The coefficient of the first power of R in the expansion is the sum of the three interfacial tensions.

Buff and Saltsburg[3] have given the analogue for the line tension τ of the formula (2.88) for surface tension. As in (2.88), one expresses the local excess free-energy density Ψ as the difference between the uniform, isotropic pressure in the interior of the bulk phases and an appropriately defined local stress (the latter now r- and θ-dependent rather than only z-dependent), and one then identifies τ as in (8.44). No progress has yet been made in evaluating τ this way, for we do not yet have *a priori* knowledge of the local stress in the neighbourhood of a three-phase line.

As a schematic illustration of the application of (8.44), let us imagine that in any plane perpendicular to the three-phase line $\Psi(r, \theta)$ is a constant Ψ_0 within an equilateral triangle, of side s, centred at the origin; another constant $\Psi_1 > 0$ through each of the three semi-infinite rectangular strips that terminate on the sides of the triangle; and 0 elsewhere (Fig. 8.16(a)). We infer that the three two-phase interfaces lie in directions parallel to the sides of the rectangles. The three contact angles are then $\alpha = \beta = \gamma = 2\pi/3$, and so from (8.4) the three interfacial tensions are equal. Let their common value be σ. From (3.3) applied far from the three-phase line,

$$\sigma^{\alpha\beta} = \sigma^{\beta\gamma} = \sigma^{\alpha\gamma} = \sigma = s\Psi_1. \tag{8.45}$$

The distance between the midpoint of an arc of a circle of radius R and

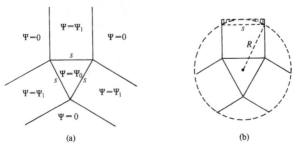

(a)   (b)

FIG. 8.16

the midpoint of a chord of length s subtending that arc is $R - \sqrt{(R^2 - s^2/4)}$. The shaded area in Fig. 8.16(b) is therefore less than $s(R - \sqrt{(R^2 - s^2/4)})$, which in turn vanishes as $s^3/8R$ when $R \to \infty$ for fixed s. Thus, for this model Ψ,

$$\int_A \Psi \, da \sim 3s\left(r - \frac{\sqrt{3}}{6} s\right)\Psi_1 + \frac{\sqrt{3}}{4} s^2\Psi_0 \qquad (8.46)$$

as $R \to \infty$. From (8.44)–(8.46) we then find

$$\tau = (\tfrac{1}{4}\sqrt{3})(\Psi_0 - 2\Psi_1)s^2. \qquad (8.47)$$

Ψ_1 and Ψ_0 are both positive but they may be of any magnitude relative to each other. Thus, as anticipated, line tensions in three-phase equilibrium may be of either sign; though they have a propensity to be negative: in this model, if Ψ_0 and Ψ_1 are the same, τ is $-\tfrac{1}{4}\sqrt{3}s\sigma$.

It had already been remarked by Gibbs[34] that the tension of the line in which three phases meet could be negative. Negative values of τ were calculated for soap films by de Feijter and Vrij.[31] They also remarked that the tension of the sharp edge (small contact angle β) of a liquid lens in air is probably negative, contrary to the theoretical estimate of Harkins,[35] which for 'a lens of an ordinary organic liquid in air' was $\tau = +10^{-11}$ to 10^{-10} N. The magnitude of Harkins's estimate, irrespective of its sign, is probably right, as confirmed also by Buff and Saltsburg.[3] The τ measured for soap films (Newton black films) by Platikanov et al.[33] range from about -10^{-9} N to $+10^{-9}$ N, depending on the chemical composition, and are typically of magnitude 10^{-10} N. We can understand this even from our crude model, which gave $|\tau| = \tfrac{1}{4}\sqrt{3}s\sigma$. The thickness s of a Newton black film[32] is about 5 nm, and typical surface and film tensions of soap solutions and films are about 0.05 N m^{-1}, so we would estimate $|\tau| \approx 10^{-10}$ N.

We may imagine a more detailed phenomenological theory along the lines of the van der Waals theory in Chapter 3. A local excess free-energy density Ψ that depends on the local composition and its gradients via some suitably generalized form of (3.5) or (3.16) may be postulated, and the equilibrium composition profiles then determined by minimizing $\int_A \Psi \, da$. The minimum value in question would have the asymptotic behaviour (8.43) or (8.44), with the line tension τ then identifiable as the constant term in the asymptotic expansion. A reasonable form to imagine for Ψ when the independent densities are $\rho_1, \rho_2, \ldots$, and their gradients are small, is

$$\Psi = -W(\rho_1, \rho_2, \ldots) + \tfrac{1}{2} \sum_{i,j} m_{ij}(\nabla\rho_i) \cdot (\nabla\rho_j), \qquad (8.48)$$

an obvious generalization of (3.16). The gradients are here two-dimensional: in the earlier language, each ρ_i is a function of r and θ but is independent of x. The Euler–Lagrange equations for the minimization of $\int_A \Psi \, da$ are then

$$-\partial W/\partial \rho_i = \tfrac{1}{2} \sum_j (m_{ij} + m_{ji}) \nabla^2 \rho_j \qquad (i = 1, 2, \ldots) \qquad (8.49)$$

with the two-dimensional Laplacian ∇^2. Because the gradients are now two-dimensional the spatial variables are no longer analogous to the time in a dynamical system, so there is no simple dynamical analogy as in the theory of the two-phase interface. There is correspondingly no analogue of the energy, and so no obvious first integral of (8.49) that would be analogous to (3.8).

We may guess the forms of the typical composition or density profiles that are likely to emerge from (8.43) [or (8.44)], (8.48), and (8.49). Non-monotonic composition profiles, analogous to the non-monotonic surface profiles in Fig. 3.6, may occur, but the simplest behaviour would be that illustrated in Figs 8.17 and 8.18. Figure 8.17 shows a representative one of the ρ_i, there called ρ, as a function of θ for some fixed r much greater than the radius of the microscopic, cylindrical, three-phase transition region. In (a), the angle θ is displayed linearly, in (b) around a circle; (b) is just (a) wrapped around a cylinder. The values of ρ in the bulk phases are shown as ρ^α, ρ^β, and ρ^γ. Figure 8.18 is a perspective sketch of the whole surface $\rho = \rho(r, \theta)$ plotted vertically above an imagined horizontal r, θ-plane. The regions of vertical shading are the nearly vertical parts of the surface, on which are seen the rapid transitions in ρ from ρ^α to ρ^β, etc., that occur in the two-phase interfaces; while the regions of horizontal shading are the plateaux on which $\rho = \rho^\alpha$, etc.

An example of an analytic $\rho(r, \theta)$ that has all the expected qualitative properties, for the special, symmetrical case in which the contact angles

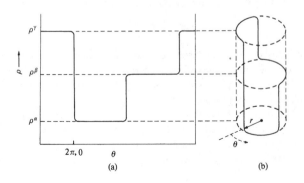

FIG. 8.17

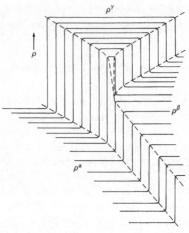

FIG. 8.18

are $\alpha = \beta = \gamma = 2\pi/3$ (so that the three interfacial tensions are equal), is

$$\rho(r, \theta) = \frac{\rho^\alpha g^\alpha(r, \theta) + \rho^\beta g^\beta(r, \theta) + \rho^\gamma g^\gamma(r, \theta)}{g^\alpha(r, \theta) + g^\beta(r, \theta) + g^\gamma(r, \theta)} \tag{8.50}$$

with

$$g^\alpha(r, \theta) = \{1 + [a + 2(1-a)\sin(\theta + \pi/6)]^{-r/\sqrt{3}(1-a)\xi}\}^{-1}$$
$$g^\beta(r, \theta) = g^\alpha(r, \theta - 2\pi/3) \tag{8.51}$$
$$g^\gamma(r, \theta) = g^\alpha(r, \theta - 4\pi/3),$$

in which a and ξ are constant parameters such that

$$\tfrac{2}{3} < a < 1, \qquad \xi > 0. \tag{8.52}$$

There is no need in a microscopic theory to define the two-phase dividing surfaces or the three-phase line in which they meet. It may be inferred from (8.50)–(8.52), as we shall soon see, that in this picture the bulk α, β, and γ phases lie within the dihedral angles $0 < \theta < 2\pi/3$, $2\pi/3 < \theta < 4\pi/3$, and $4\pi/3 < \theta < 2\pi$, respectively; the physical two-phase interfaces are planar regions that lie parallel to and within microscopic distances (distances of order ξ) of the planes $\theta = 0$, $2\pi/3$, and $4\pi/3$; and the physical three-phase line is a cylindrical region with axis $r = 0$ and radius of order ξ.

Some special values of this $\rho(r, \theta)$ are

$$\rho(0, \theta) = \tfrac{1}{3}(\rho^\alpha + \rho^\beta + \rho^\gamma) \qquad \text{(all } \theta\text{)}$$

$$\rho(\infty, \theta) = \begin{cases} \rho^\alpha & \text{when } 0 < \theta < 2\pi/3 \\ \rho^\beta & \text{when } 2\pi/3 < \theta < 4\pi/3 \\ \rho^\gamma & \text{when } 4\pi/3 < \theta < 2\pi \end{cases}$$

$$\rho(r, 0) \sim \tfrac{1}{2}(\rho^\alpha + \rho^\gamma) \qquad (r \to \infty)$$

$$\rho(r, 2\pi/3) \sim \tfrac{1}{2}(\rho^\alpha + \rho^\beta) \qquad (r \to \infty)$$

$$\rho(r, 4\pi/3) \sim \tfrac{1}{2}(\rho^\beta + \rho^\gamma) \qquad (r \to \infty).$$

$$(8.53)$$

In Fig. 8.19 we plot the $\rho(r, \theta)$ of (8.50) and (8.51) for $\rho^\alpha = -1$, $\rho^\beta = 0$,

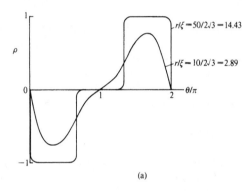

(a)

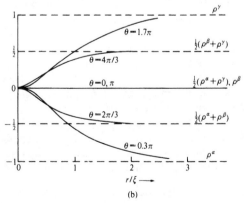

(b)

FIG. 8.19. (a) Profile of the density $\rho(r, \theta)$ as a function of θ for two values of r; from (8.50) and (8.51) with the parameters $\rho^\alpha = -1$, $\rho^\beta = 0$, $\rho^\gamma = 1$, and $a = \tfrac{5}{6}$. (b) Same $\rho(r, \theta)$, now as a function of r for several fixed θ.

$\rho^\gamma = 1$, and $a = \frac{5}{6}$, the latter satisfying (8.52). Figure 8.19(a) shows $\rho(r, \theta)$ as a function of θ for two fixed values of r, one $\approx 14 \cdot 4\xi$ and the other $\approx 2 \cdot 9\xi$. Except for the obvious displacement of the origin of the θ-scale along the horizontal axis, the former plot is essentially the same as that in the schematic Fig. 8.17(a), which was intended to represent $\rho(r, \theta)$ as a function of θ for fixed $r \gg \xi$. Figure 8.19(b) shows $\rho(r, \theta)$ as a function of r for several fixed θ, and is in accord with (8.53).

To see the behaviour of $\rho(r, \theta)$ at the $\alpha\gamma$ interface, say, far from the three-phase line, we let $r \to \infty$ and $\theta \to 0$ for fixed $r\theta$. This yields $\rho(r, \theta)$ at a point distant $z = r\theta$ from the plane $\theta = 0$, the mid-plane of the physical $\alpha\gamma$ interface. We find from (8.50) and (8.51) that as $r \to \infty$ for fixed z,

$$\rho(r, z/r) \sim \tfrac{1}{2}(\rho^\alpha + \rho^\gamma) + \tfrac{1}{2}(\rho^\alpha - \rho^\gamma)\tanh(z/2\xi). \qquad (8.54)$$

This is the historic hyperbolic tangent profile, implicit in van der Waals[36] and explicit in Landau and Lifshitz[37] and in Cahn and Hilliard,[38] of which we have already had an intimation in (5.103) and which we shall encounter again in Chapter 9. By the obvious symmetry of (8.50) and (8.51), the profile of the density ρ in the $\alpha\beta$ or $\beta\gamma$ interface far from the three-phase line is also given by (8.54), with ρ^γ or ρ^α replaced by ρ^β, and with z then the distance from the mid-plane of the $\alpha\beta$ or $\alpha\gamma$ interface.

The parameter ξ is the model's only microscopic length. As anticipated, and as we now see from (8.54) and Fig. 8.19, it measures the thickness of the three interfaces, and also the radius of the cylindrical region around the three-phase line in which the fluid is markedly inhomogeneous.

We should emphasize that we have not derived (8.50) and (8.51) [of which (8.53), (8.54), and Fig. 8.19 are consequences] from theory; we have constructed them to accord with our preconceptions.

In § 5.6 we studied the interface in a symmetric mixture model—the two-component, primitive version of the penetrable-sphere model—which we treated in a mean-field approximation. There is a three-component version of that model due to Guerrero et al.,[39] with which we may illustrate many of the ideas of this section. Again, like molecules do not interact, while unlike molecules repel each other as hard spheres. If the components are a, b, c, then in any homogeneous phase, in mean-field approximation, the densities ρ_a, etc., and activities ζ_a, etc., all in units of the volume v_0 of the exclusion sphere, are related by[39]

$$\rho_a/\zeta_a = \exp(-(\rho_b + \rho_c)), \quad \text{etc.} \qquad (8.55)$$

This is an obvious extension of the relation $\rho_a/\zeta_a = \exp(-\rho_b)$ which holds, in mean-field approximation, in any homogeneous phase of the two-component primitive model of § 5.6; cf. (5.128). Now the associated potential is $W(\rho_a, \rho_b, \rho_c; \zeta_a, \zeta_b, \zeta_c)$, given (in units of kT/v_0) by

$$-W = \rho_a\rho_b + \rho_b\rho_c + \rho_c\rho_a + \rho_a \ln(\rho_a/\zeta_a) + \rho_b \ln(\rho_b/\zeta_b) + \rho_c \ln(\rho_c/\zeta_c)$$
$$-\rho_a - \rho_b - \rho_c + (\rho_a + \rho_b + \rho_c + \rho_a\rho_b + \rho_b\rho_c + \rho_c\rho_a)^0, \quad (8.56)$$

where the superscript ^{0} means that the expression in parentheses is to be evaluated at the densities that are the equilibrium densities for the given activities; that is, at the densities that satisfy (8.55). This W has the required properties that W, $\partial W/\partial\rho_a$, $\partial W/\partial\rho_b$, and $\partial W/\partial\rho_c$ all vanish when $\rho_a = \rho_a^0$, $\rho_b = \rho_b^0$, and $\rho_c = \rho_c^0$. It is the counterpart for the present three-component model of (5.131) for the two-component model.

There is a range of thermodynamic states in which three phases, symmetrically related to each other, coexist. If those phases are α, β, γ, then by the symmetry of the model $\rho_a^\alpha = \rho_b^\beta = \rho_c^\gamma$ and $\rho_b^\alpha = \rho_c^\alpha = \rho_a^\beta = \rho_c^\beta = \rho_a^\gamma = \rho_b^\gamma$. (These are obvious extensions of the last two of (5.128).) Thus, in each phase one of the three components plays a distinguished role while the other two play roles symmetrical with each other, and the three phases differ only by permutation of the components. Also, in this symmetrical equilibrium,

$$\zeta_a = \zeta_b = \zeta_c = \zeta, \quad (8.57)$$

calling the common value ζ; this is the extension of the earlier $\zeta = \zeta_a = \zeta_b$ that appears immediately below (5.127). Because the phases are symmetrically related to each other they necessarily meet in a three-phase line with equal contact angles of $120°$ and with three equal interfacial tensions $\sigma^{\alpha\beta} = \sigma^{\beta\gamma} = \sigma^{\alpha\gamma}$, just as in the hypothetical picture discussed above. It is found[28] that such equilibria exist stably or metastably for all $\zeta \geqslant 4\cdot259$, or, equivalently, $\rho_a^\alpha \geqslant 2\cdot708$ and $\rho_b^\alpha \leqslant 0\cdot226$. According to the results of Guerrero et al.,[39] until ζ becomes as large as $4\cdot490$, where $\rho_a^\alpha = 3\cdot614$ and $\rho_b^\alpha = 0\cdot109$, these symmetrical, three-phase equilibria remain metastable relative to other, more stable phase equilibria; but at all higher ζ, where $\rho_a^\alpha > 3\cdot614$ and $\rho_b^\alpha < 0\cdot109$, these three-phase systems are fully stable.

Integral equations for the density profiles $\rho_a(r, \theta)$, etc., may be obtained in mean-field approximation by the same methods that led to (5.127) with (5.83) or (5.102). In this way it is found for $s = 3$, for example (s is the dimensionality of the potential: see the remarks below (5.133)), that[28]

$$\frac{\rho_a(r, \theta)}{\zeta_a} = \exp\left\{-\frac{3}{2\pi}\int_0^1 dt \int_0^{2\pi} d\phi \, t(1-t^2)^{\frac{1}{2}}[\rho_b(r', \theta') + \rho_c(r', \theta')]\right\} \text{ etc.} \quad (8.58)$$

where r', θ' are the functions of r, θ, ϕ, t defined by Fig. 8.20; or for $s = 1$,

$$\frac{\rho_a(r, \theta)}{\zeta_a} = \exp\left\{-\frac{1}{4}\int_{-1}^1 dx \int_{-1}^1 dy[\rho_b(r', \theta') + \rho_c(r', \theta')]\right\} \text{ etc.} \quad (8.59)$$

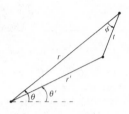

Fɪɢ. 8.20

where the integration variables x and y are the rectangular coordinates that are related to the polar coordinates ϕ and t of (8.58) and Fig. 8.20 by $x = -t \cos(\phi + \theta)$ and $y = -t \sin(\phi + \theta)$. Distances have all been scaled by the radius of the exclusion sphere, and are now dimensionless. The general integral equation, for any s, is also known,[28] as is the similar equation for the one-component penetrable-sphere or two-component primitive model.[40] When (8.58) or (8.59) is applied to the three-phase system of interest, we would set $\zeta_a = \zeta_b = \zeta_c = \zeta$.

When the fluid is uniform, so that ρ_b and ρ_c are constant, (8.58) and (8.59) reduce to (8.55), as they should. When $\rho_b(r, \theta)$ and $\rho_c(r, \theta)$ are functions only of $r \sin \theta$, as appropriate to a planar interface, (8.58) and (8.59) again reduce properly; when, in particular, $\rho_b(r, \theta) \equiv \rho_b(r \sin \theta)$, $\rho_c \equiv 0$, and $\zeta_a = \zeta_b = \zeta$, they become the integral equations for the density profile in the planar interface of the two-component primitive model, (5.127) with (5.83) or (5.102).

If the densities may be assumed to be slowly varying, the integral equations for the density profiles may be reduced to differential equations, just as (5.18) was reduced to (5.19), and just as (5.127), with (5.83) or (5.102) extended to general s, was reduced to (5.133). Here, the resulting differential equations for $\rho_a(r, \theta)$, etc., are[28]

$$-\frac{1}{2(s+2)} \nabla^2(\rho_b + \rho_c) = \rho_b + \rho_c + \ln(\rho_a/\zeta), \quad \text{etc.} \qquad (8.60)$$

where ∇^2 is the two-dimensional Laplacian operator (in dimensionless form, distances having been scaled with the radius of the exclusion sphere), and where ζ, defined by (8.57), is uniform throughout the otherwise inhomogeneous fluid. In terms of the analytic functions $\zeta_a(\rho_a, \rho_b, \rho_c)$, etc., defined by (8.55), and from (8.56) and (8.57), the equations (8.60) are

$$-\frac{1}{2(s+2)} \nabla^2(\rho_b + \rho_c) = \ln[\zeta_a(\rho_a, \rho_b, \rho_c)/\zeta] = -\partial W/\partial \rho_a, \quad \text{etc.} \quad (8.61)$$

These are close to (5.65) and (5.133) in form and origin. They are also of the anticipated form (8.49), with $m_{ii} = 0$, and $m_{ij} + m_{ji} = -1/(s+2)$ when $i \neq j$. The associated Ψ of (8.48) is now

$$\Psi = -W - \frac{1}{2(s+2)} [(\nabla \rho_a) \cdot (\nabla \rho_b) + (\nabla \rho_b) \cdot (\nabla \rho_c) + (\nabla \rho_a) \cdot (\nabla \rho_c)] \quad (8.62)$$

(in units of kT/v_0), with W given by (8.56) and (8.57), and with ∇ the two-dimensional gradient operator (in units of the reciprocal of the exclusion-sphere radius). This model treated in mean-field approximation thus provides a concrete realization of the phenomenological theory of the three-phase line.

Notes and references

1. Neumann, F. [E.] *Vorlesungen über die Theorie der Capillarität* (ed. A. Wangerin) Chapter 6, § 1, especially pp. 161–2. Teubner, Leipzig (1894).
2. Bakker, G. *Kapillarität und Oberflächenspannung*, Vol. 6 of *Handbuch der Experimentalphysik* (ed. W. Wien, F. Harms, and H. Lenz) p. 70. Akad. Verlags., Leipzig (1928).
3. Buff, F. P. and Saltsburg, H. *J. chem. Phys.* **26**, 23 (1957).
4. Buff, F. P. *Encyclopedia of Physics* (ed. S. Flügge) Vol. 10, § 7, pp. 298–9. Springer, Berlin (1960).
5. Antonow, G. N. *J. Chim. phys.* **5**, 372 (1907); *Kolloid-Zeit.* **59**, 7 (1932); **64**, 336 (1933); Adam, N. K. *The Physics and Chemistry of Surfaces*, 3rd ed. pp. 7, 214–5, Oxford University Press (1941). In his early writings his name was transliterated Antonow, later Antonoff. We have adopted the former.
6. One of the present authors (BW), in his first encounters with the question [*J. chem. Phys.* **62**, 1332 (1975); *Phys. Rev. Lett.* **34**, 999 (1975)], erroneously supposed (8.8) to be merely an asymptotic limit, not attainable while the three phases were all distinct.
7. Heady, R. B. and Cahn, J. W. *J. chem. Phys.* **58**, 896 (1973).
8. Cahn, J. W. *J. chem. Phys.* **66**, 3667 (1977). Following Cahn, there has been further theoretical work on the wetting transition, particularly at liquid–solid interfaces, by Sullivan, D. E. *J. chem. Phys.* **74**, 2604 (1981), and *Faraday Symp. chem. Soc.* **16** (1981) in press; and by Teletzke, G. F., Scriven, L. E. and Davis, H. T., preprint, 1981.
9. Moldover, M. R. and Cahn, J. W. *Science* **207**, 1073 (1980).
10. Lang, J. C., Lim, P. K. and Widom, B. *J. phys. Chem.* **80**, 1719 (1976).
11. Varea, C., Valderrama, A. and Robledo, A. *J. chem. Phys.* **73**, 6265 (1980).
12. The matter was essentially so summarized by Donahue, D. J. and Bartell, F. E. *J. phys. Chem.* **56**, 480 (1952), although they unnecessarily distinguished three cases instead of two.
13. Davis, H. T. and Scriven, L. E. *J. stat. Phys.* **24**, 243 (1981).
14. Harkins, W. D. *J. chem. Phys.* **9**, 552 (1941).
15. [Gibbs, J. W.] *The Collected Works of J. Willard Gibbs*, Vol. 1, p. 258, Longmans, Green, New York (1928). In the quotation the insertions in square brackets and the emphasis in italics are ours.
16. de Gennes, P. G. *J. Physique Lett.* **42**, L-377 (1981).
17. Cahn, J. W. privately communicated, 1975.
18. Widom, B. *J. chem. Phys.* **68**, 3878 (1978).

19. The possibility of an explicit analytical inversion of (8.13) was first recognized by Dr F. J. Ryan (1978, unpublished).
20. Lajzerowicz, J. *Ferroelectrics*, **35,** 219 (1981).
21. Adam, ref. 5, Chapter 2, particularly § 13, pp. 43–6.
22. Young, D. M. and Crowell, A. D. *Physical Adsorption of Gases*, pp. 117–32. Butterworth, London (1962).
23. Dash, J. G. and Ruvalds, J. (eds.) *Phase Transitions in Surface Films*, NATO Advanced Study Institute, Erice, Sicily, June 11–25, 1979, Plenum, New York (1980); Sinha, S. K. (ed.) *Ordering in Two Dimensions*, North-Holland, Amsterdam (1980).
24. Currie, J. F. and Bishop, A. R. *Can. J. Phys.* **57,** 890 (1979).
25. As a matter of historical interest, Buff and Saltsburg, in their paper on the theory of the three-phase contact line and Neumann's triangle (ref. 3), also reported, briefly and qualitatively, their observation that in the three-liquid system water–aniline–heptane, the middle, aniline-rich phase is transformed from a spread layer to a lens upon the addition of a detergent to the system. That was an early observation of a Cahn transition.
26. Gibbs, ref. 15, p. 288, footnote.
27. Boruvka, L. and Neumann, A. W. *J. chem. Phys.* **66,** 5464 (1977).
28. Kerins, J. E. Thesis, Cornell University (1982).
29. Langmuir, I. *J. chem. Phys.* **1,** 756 (1933).
30. Pujado, P. R. and Scriven, L. E. *J. Coll. Interf. Sci.* **40,** 82 (1972); Princen, H. M., in *Surface and Colloid Science* (ed. E. Matijević) Vol. 2, pp. 1–84, particularly p. 61, Wiley, New York (1969); and Pethica, B. A. *J. Coll. Interf. Sci.* **62,** 567 (1977).
31. de Feijter, J. A. and Vrij, A. *J. Electroanal. Chem.* **37,** 9 (1972).
32. Boys, C. V. *Soap-Bubbles: Their Colours and the Forces which Mold Them*, pp. 162–4, S.P.C.K., London (1890), reprinted by Dover Publications, 1958; C. Isenberg, *The Science of Soap Films and Soap Bubbles*, pp. 43–4, Tieto, Clevedon (1978).
33. Platikanov, D., Nedyalkov, M. and Scheludko, A. *J. Coll. Interf. Sci.* **75,** 612 (1980); Platikanov, D., Nedyalkov, M. and Nasteva, V. *J. Coll. Interf. Sci.* **75,** 620 (1980).
34. Gibbs, ref. 15, p. 296, footnote. See also Tarazona, P. and Navascués, G. *J. chem. Phys.* **75,** 3114 (1981).
35. Harkins, W. D. *J. chem. Phys.* **5,** 135 (1937).
36. van der Waals, J. D. *Zeit. phys. Chem.* **13,** 657 (1894); English translation, *J. stat. Phys.* **20,** 197 (1979).
37. Landau, L. and Lifshitz, E. *Phys. Zeit. Sowjetunion* **8,** 153 (1935), reprinted in English in *Collected Papers of L. D. Landau* (ed. D. ter Haar) p. 101, Pergamon, Oxford (1965). See also Landau, L. D. and Lifshitz, E. M. *Electrodynamics of Continuous Media*, § 39, p. 158, Pergamon, Oxford (1960).
38. Cahn, J. W. and Hilliard, J. E. *J. chem. Phys.* **28,** 258 (1958).
39. Guerrero, M. I., Rowlinson, J. S. and Morrison, G. *J. chem. Soc. Faraday Trans. II*, **72,** 1970 (1976); Rowlinson, J. S. *Adv. chem. Phys.* **41,** 1 (1980), particularly § VIA, pp. 31–6.
40. Leng, C. A., Rowlinson, J. S. and Thompson, S. M. *Proc. Roy. Soc. A* **352,** 1 (1976).

For further references, see p. 313.

INTERFACES NEAR CRITICAL POINTS

9.1 Introduction: mean-field approximation

The critical point of the equilibrium of two phases marks the limit of their coexistence. If the approach to the critical point is through the range of two-phase states, the interface between the phases disappears when that point is reached, and the previously distinct phases then become one. The tension of the interface, meanwhile, decreases as the critical point is approached and it vanishes at the moment that the interface itself disappears.

Historically, the critical point was a major focus of the theory of van der Waals, both in its original form[1] and as revived and reworked by Cahn and Hilliard.[2] One of the outstanding achievements of that theory was obtaining the form of the law by which the surface tension vanishes on approach to the critical point,

$$\sigma = \sigma_0(1 - T/T^c)^\mu; \qquad (9.1)$$

although, as we shall see, with the mean-field-theory value $\mu = \frac{3}{2}$ of that critical-point exponent instead of the more nearly correct $\mu \simeq 1\cdot26$. (We follow convention in using the symbol μ for this exponent, but it must not be confused with the chemical potential.)

Much of what has been done on the theory of the near-critical interface has been within the framework of the van der Waals theory of Chapter 3, so much of our present understanding of the properties of those interfaces comes from that theory or from some suitably modified or extended version of it. As we shall see, an interface thickens as its critical point is approached, and the gradients of density and composition in the interface are then small. Thus, the view that the interfacial region may be treated as matter in bulk, with a local free-energy density that is that of a hypothetically uniform fluid of composition equal to the local composition, with an additional term arising from the non-uniformity, and that the latter may be approximated by a gradient expansion, typically truncated in second order, is then most likely to be successful and perhaps even quantitatively accurate. In this section we shall see what the simplest theory of that kind—that which comes from treating simple models in mean-field approximation, as in Chapter 5—yields for the structure and tension of an interface near a critical point.

Anticipating that near the critical point the gradients of density or composition in the interface will prove to be small, we may take

$$m\, d^2\rho(z)/dz^2 = \mu[\rho(z), T] - \mu, \qquad (9.2)$$

as in (5.19) or in (5.36) [with (5.37)], to be the canonical form of the equation for the density or composition profile, $\rho(z)$, in mean-field approximation. Here, again, $\mu(\rho, T)$ is the chemical potential of the homogeneous fluid of density ρ in the mean-field approximation—it is the $M(\rho)$ of Chapter 3—while μ is the uniform value of the chemical potential in the two-phase system. The right-hand side of (9.2), as a function of ρ for fixed T, is as pictured in Fig. 3.4 (but with the baseline shifted vertically by the constant μ) and again in Fig. 5.1 (when $c\theta > 4$).

Near the critical point $\mu(\rho, T) - \mu$ may be expanded in powers of $\rho - \rho^c$ and $T - T^c$, the deviations of ρ and T from their critical values; thus,

$$\mu(\rho, T) - \mu = -A(T^c - T)(\rho - \rho^c) + B(\rho - \rho^c)^3 + \ldots \qquad (9.3)$$

with A and B two positive constants. The densities ρ^g and ρ^l of the bulk, equilibrium phases satisfy the equal-areas rule, and so, by symmetry, are the zeros, other than $\rho = \rho^c$, of $\mu(\rho, T) - \mu$. (We refer to liquid–gas equilibrium for definiteness, but the same theory, with obvious changes of notation, applies to the equilibrium of any two phases α and β near their critical point.) Near the critical point, then, from (9.3),

$$\rho^{l,g} \sim \rho^c \pm [A(T^c - T)/B]^{\frac{1}{2}} \qquad (T \to T^c) \qquad (9.4)$$

where $\rho^{l,g}$ means ρ^l when the $(+)$ sign is taken and ρ^g when the $(-)$ sign is taken. From the formula

$$\kappa^{-1} = \rho^2 (\partial\mu/\partial\rho)_T \qquad (9.5)$$

for the reciprocal of the isothermal compressibility κ [see (4.21) and just below it], and from (9.3) and (9.4), we see also that

$$(\kappa^l)^{-1} \sim (\kappa^g)^{-1} \sim 2B(\rho^c)^2(\rho^{l,g} - \rho^c)^2 \sim 2A(\rho^c)^2(T^c - T) \\ (T \to T^c) \quad (9.6)$$

in this mean-field approximation.

Both the profile and the surface tension are obtained most directly from the function $W(\rho)$ of Chapter 3, which is such that $W(\rho^l) = W(\rho^g) = 0$ and $-dW(\rho)/d\rho = \mu(\rho, T) - \mu$. From (9.3) and (9.4),

$$-W(\rho) = \tfrac{1}{4}B(\rho^l - \rho)^2(\rho - \rho^g)^2 \qquad (9.7)$$

near the critical point. From (3.9), with (9.4) and (9.7), we have for the profile near the critical point,

$$z = \left(\frac{2m}{B}\right)^{\frac{1}{2}} \frac{1}{\rho^l - \rho^g} \ln\left(\frac{\rho - \rho^g}{\rho^l - \rho}\right) \qquad (9.8)$$

where we have taken the sign in (3.9) to be such that $\rho \to \rho^g$ as $z \to -\infty$ and $\rho \to \rho^l$ as $z \to +\infty$, and the arbitrary constant to be such that $z = 0$ where $\rho = \rho^c$. From (3.13) with (9.7) we likewise have for the surface tension near the critical point,

$$\sigma = \frac{1}{6} \left(\frac{mB}{2}\right)^{\frac{1}{2}} (\rho^l - \rho^g)^3. \tag{9.9}$$

Inverting (9.8) while taking account of (9.4) and (9.6), we have for the profile

$$\rho(z) - \rho^c = \tfrac{1}{2}(\rho^l - \rho^g)\tanh(z/2\xi) \tag{9.10}$$

where the length ξ is

$$\xi = (2m/B)^{\frac{1}{2}}(\rho^l - \rho^g)^{-1} = [m/2A(T^c - T)]^{\frac{1}{2}} = \rho^c(m\kappa)^{\frac{1}{2}} \tag{9.11}$$

with κ the asymptotically common value of κ^l and κ^g. In (9.10) we have the classical hyperbolic tangent profile which we have encountered before [(8.54); cf. (5.103)]. It has the anticipated shape—that of the profiles in Figs 5.2 and 5.5 and of curve (c) of Fig. 3.3. Almost all of the variation from ρ^l to ρ^g through the interface occurs over a distance of a few ξ, so ξ, or any prescribed multiple of it, may be taken to be a measure of the interfacial thickness. From (9.11), that thickness diverges at the critical point as $(T^c - T)^{-\frac{1}{2}}$. At the same time the total change in density through the interface, $\rho^l - \rho^g$, vanishes at the critical point, and does so proportionally to $(T^c - T)^{\frac{1}{2}}$, according to (9.4). The maximum density gradient in the interface, according to (9.10), is $(\rho^l - \rho^g)/4\xi$, which then vanishes as the first power of $T^c - T$; so the gradients do become small as the critical point is approached.

As we shall see in § 9.2, the length ξ is also the coherence length of density (or composition) fluctuations in the bulk phases, which was discussed in § 3.5.

We now return to the surface tension, given in the present approximation by (9.9). The latter is like Macleod's relation[3], $\sigma \sim (\rho^l - \rho^g)^4$, though here, in mean-field theory, the exponent is 3 instead of the empirical 4. (An even more accurate value of this exponent is 3.88, as we shall see in § 9.3.) Equation (9.9), with (9.4), (9.6), and (9.11), yields the alternative expressions

$$\sigma = \sqrt{m}[2A(T^c - T)]^{\frac{3}{2}}/3B \tag{9.12}$$

$$= m(\rho^l - \rho^g)^2/6\xi \tag{9.13}$$

$$= \xi(\rho^l - \rho^g)^2/6(\rho^c)^2\kappa \tag{9.14}$$

$$= \sqrt{m/\kappa}(\rho^l - \rho^g)^2/6\rho^c. \tag{9.15}$$

Of these, the first is of the form (9.1) with $\mu = \frac{3}{2}$, as anticipated; it is the classical result of van der Waals. The remaining three are noteworthy in that, unlike (9.9) and (9.12), they contain neither A nor B and so make no explicit reference to the equation of state.

Both the profile, as given by (9.10), and the surface tension, as given by (9.9) or any of (9.12)–(9.15), have some elements of universality in their form but reflect the mean-field approximation in their details. We shall see in § 9.4 that near the critical point the profile is indeed of the form

$$\rho(z) - \rho^c = \frac{1}{2}(\rho^l - \rho^g) f(z/2\xi) \tag{9.16}$$

where the function f is an odd function of its argument and $f(\infty) = 1$; but only in mean-field theory is $f(x) = \tanh x$. Likewise, the amplitude $\rho^l - \rho^g$ of the step in the profile, and the interfacial thickness ξ, do generally vanish and diverge, respectively, proportionally to characteristic powers of $T^c - T$; but only in the mean-field approximation are those powers $\frac{1}{2}$ and $-\frac{1}{2}$, as by (9.4) and (9.11). We already remarked that (9.9) and (9.12) are of the right form, but that in real fluids the exponents are 3.88 and 1.26, respectively, instead of 3 and $\frac{3}{2}$, as they are in mean-field approximation. The forms of (9.13)–(9.15) will also prove to be generally correct; even the numerical coefficient $\frac{1}{6}$ may be nearly right, but it is otherwise approximation-dependent, and is exactly $\frac{1}{6}$ only in mean-field theory.

An interesting point of principle, briefly alluded to in § 3.3, is that near a critical point there is no distinction between the one- and two-density van der Waals theories, and therefore no improvement of the theory on allowing a second density to vary independently of the first one. This is most readily understood and illustrated for a classical critical point, and so is discussed here; although the argument is in fact of great generality[4] and depends only on the homogeneity of form of the free-energy density $-W$, which that function retains even when corrected for non-classical critical-point behaviour (§ 9.4).

Suppose ρ_1 and ρ_2 are two independent densities. In a binary mixture they might be the densities of the two chemical components; in a one-component fluid they might be the densities of matter and entropy, as in the example in § 3.3; etc. In Fig. 9.1 we see the coexistence curve in the ρ_1, ρ_2-plane (at fixed values of the fields $\mu_3, \ldots, \mu_{c+1}$, if this is a system of $c \geq 2$ components). We define a coordinate system x, y with origin at the critical point, as in the figure; the x-axis is tangent to the coexistence curve at the critical point while the y-axis is in any other direction. The equation of the y-axis is $x = 0$; so, the direction of the y-axis being almost arbitrary, x may be chosen to be an almost arbitrary linear combination of $\rho_1 - \rho_1^c$ and $\rho_2 - \rho_2^c$, the deviations of the densities from their values at the critical point. The equation of the x-axis is $y = 0$; so, the direction of the x-axis in the ρ_1, ρ_2 space being unique, y is a

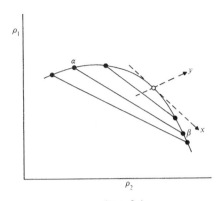

FIG. 9.1

linear combination of $\rho_1 - \rho_1^c$ and $\rho_2 - \rho_2^c$ that is uniquely determined up to a common constant factor in its two terms.

In the two-phase region near the critical point the two-density version of (9.7) is of the form

$$-W(x, y) = (B/2a^4)[(a^2x^2 + y)^2 + (y + t)^2] \tag{9.17}$$

where $t > 0$ is the one independently variable field, say $t = \text{const.}\ (T^c - T)$, while a^2 and B are positive constants. Contours of constant W in the x, y or ρ_1, ρ_2-plane are like those in Fig. 3.5 or 3.7. The equation of the coexistence curve in the x, y-plane (the ρ_1, ρ_2-plane) is $y = -a^2x^2$, while in the t, x-plane it is $t = a^2x^2$. One of the two phases, β, say, is at $y = -t$, $x = \sqrt{t}/a$, the other, α, is at $y = -t$, $x = -\sqrt{t}/a$. This is as in Fig. 9.1, asymptotically near the critical point, where the coexistence curve in the figure becomes parabolic with tielines parallel to the x-axis. The one-density $-W$ of (9.7) would follow from the two-density $-W$ of (9.17) upon replacing y by that function of x which it would be if it varied with x, at fixed values of the fields, as in a bulk phase; that is, so as to satisfy $-\partial W/\partial y = 0$ [cf. (3.19)], which here means $y = -\frac{1}{2}(a^2x^2 + t)$. Making this replacement, and the further identification $x = \rho - \rho^c$ (which we are free to make because of the arbitrariness in x), we need then only recall that $t = a^2x^2$ on the coexistence curve, to obtain (9.7) from (9.17). We may thus view the one-density $-W(x)$ as obtained from the two-density $-W(x, y)$ by minimizing the latter over y at fixed x and fixed values of the fields.

The equations that determine the profiles $x(z)$ and $y(z)$ through the $\alpha\beta$ interface are of the form of those in (3.20):

$$m_{xx}x''(z) + m_{xy}y''(z) = -\partial W/\partial x, \tag{9.18}$$

$$m_{xy}x''(z) + m_{yy}y''(z) = -\partial W/\partial y. \tag{9.19}$$

We suppose the coefficients m to have finite, non-vanishing limiting values at the critical point. Then with $-t$ and $\pm\sqrt{t}/a$ the values of y and x, respectively, in the bulk phases, and with ξ the thickness of the interface, the typical magnitudes of $x''(z)$ and $y''(z)$ in the interface are $\sqrt{t}/\xi^2$ and t/ξ^2, respectively, of which the second is negligible compared with the first. At the same time $-W$, by (9.17), is typically of order t^2, so $-\partial W/\partial x$ for typical x and y is of order $t^{\frac{3}{2}}$ and $-\partial W/\partial y$ would be of order t. Then asymptotically, on approach to the critical point, (9.18) becomes

$$m_{xx}x''(z) = -\partial W/\partial x, \qquad (9.20)$$

from which we conclude that $\sqrt{t}/\xi^2 \sim t^{\frac{3}{2}}$, or $\xi \sim t^{-\frac{1}{2}}$, as in (9.11). At the same time the left-hand side of (9.19) is also of order $\sqrt{t}/\xi^2$, hence of order $t^{\frac{3}{2}}$, but the magnitude of $-\partial W/\partial y$ for typical x and y, we saw, is t, which is much greater. Thus, the trajectory in the x, y-plane is precisely determined to be that on which the largest contributions to $-\partial W/\partial y$, which are of order t, occur with such signs and amplitudes as to cancel each other exactly, leaving residuals that are much smaller; that is, to be that on which, to a close approximation (cf. (3.19) and (5.146))

$$-\partial W(x, y)/\partial y = 0. \qquad (9.21)$$

This, we noted before, is just the condition that y should vary with x through the interface as in a bulk phase; that is, in such a way that $-W$ is minimized with respect to y for every x. Since on the path determined by (9.21) the one-density $W(x)$, related to $W(x, y)$ by

$$-W(x) = \min_{y}[-W(x, y)], \qquad (9.22)$$

has the property $dW(x)/dx = \partial W(x, y)/\partial x$, the right-hand side of (9.20) is now just $-dW(x)/dx$, so the two-density theory has reduced exactly to the one-density theory.

Near the critical point the surface tension, too, as calculated by the two-density theory becomes the same as in the one-density theory. In the present example,

$$\sigma = \int_{-\infty}^{\infty} [-W(x, y) + \tfrac{1}{2}m_{xx}x'(z)^2 + m_{xy}x'(z)y'(z) + \tfrac{1}{2}m_{yy}y'(z)^2]\,dz.$$

$$(9.23)$$

By the same estimates as before, $y'(z)$ is asymptotically negligible compared with $x'(z)$, while the trajectory, we know, is such that $W(x, y)$ on it becomes the one-density $W(x)$ of (9.22); so σ is asymptotically that of the one-density theory,

$$\sigma = \int_{-\infty}^{\infty} \left[-W(x) + \tfrac{1}{2}m_{xx}x'(z)^2\right] dz, \tag{9.24}$$

as asserted.

In the next section we shall see how very close the van der Waals theory of the density profile is to the Ornstein–Zernike theory of the pair-correlation function. The anticipated connection between the interfacial width and the correlation length near the critical point will follow when we observe that the latter is also asymptotically equal to $\rho^c\sqrt{(m\kappa)}$, as in (9.11). We shall find again in the Ornstein–Zernike theory that the parameter m is related generally to the second moment of the direct correlation function $c(r)$ by (4.162). It is characteristic of the van der Waals or Ornstein–Zernike theories that $c(r)$ is assumed to be so short-ranged that its second moment is finite, even at the critical point. We shall remark in §§ 9.2 and 9.3 that in reality m diverges as the critical point is approached; but so slowly—that is, with so small a negative power of $|T - T^c|$—that the divergence may be of little practical consequence; so that the van der Waals, Ornstein–Zernike assumption that m has a finite, positive limit at the critical point may not be seriously misleading.

While § 9.2 is a necessary digression on the theory of the correlation function near the critical point, § 9.3 is an equally necessary digression on critical-point exponents: what their non-classical (non-mean-field-theory) values are, and how, according to the ideas of critical-point scaling and homogeneity, they must be related to each other. Then in § 9.4 we shall see how one incorporates non-classical exponents into the framework of the van der Waals theory of the interface.

In § 9.5 we turn again to the problem of the interfaces in three-phase equilibrium, as in Chapter 8, but now the three phases are near that limit of their coexistence at which they become one: the tricritical point. In § 9.6 we treat still another problem related to three-phase equilibrium: the interface between a non-critical phase α and another fluid phase that is itself near the critical point of separation into two phases β and γ. An example is the interface between a liquid mixture and its equilibrium vapour when the former is near its consolute point. Finally, in § 9.7, we sketch the applications to interfaces of the ideas and methods of the modern renormalization-group theory of critical phenomena.

9.2 Digression on the Ornstein–Zernike theory of the pair-correlation function

We shall see from the Ornstein–Zernike theory of the pair-correlation function $h(r)$ in a uniform fluid, that near the critical point and at large distances r, where $h(r)$ is slowly varying, it satisfies a differential equation remarkably like (9.2). The theory of the asymptotic $(r \to \infty)$ form of $h(r)$ will then illuminate for us many aspects of the van der Waals theory of interfacial structure, particularly near the critical point, where the gradients of density or composition in the interface are small.

The function $h(r)$ is defined by (4.23), specialized now to a uniform, isotropic fluid, so that $h(\mathbf{r}_1, \mathbf{r}_2) = h(|\mathbf{r}_1 - \mathbf{r}_2|) = h(r_{12})$. The essence of the Ornstein–Zernike theory is the observation (which we shall make presently) that near the critical point the total correlation function $h(r)$ is much longer ranged than the direct correlation function $c(r)$, to which it is related by (4.25):

$$h(r_{12}) = c(r_{12}) + \rho \int c(r_{13}) h(r_{32}) \, d\mathbf{r}_3, \qquad (9.25)$$

where now ρ is what in (4.25) we called ρ_u, the uniform density of the homogeneous fluid.

The integrals over all space of $h(r)$ and $c(r)$, by (4.22) and (4.27), are

$$\int h(r) \, d\mathbf{r} = kT\kappa - 1/\rho \qquad (9.26)$$

$$\int c(r) \, d\mathbf{r} = (1 - 1/\rho kT\kappa)/\rho, \qquad (9.27)$$

of which the first diverges proportionally to κ as the critical point is approached while the second approaches the finite limit $1/\rho^c$ at the critical point. The range, ξ, of $h(r)$, is the coherence length of density or composition fluctuations. It is such that $h(r)$ vanishes rapidly with increasing r when $r \gg \xi$, so that while a major part of the integral in (9.26) comes from the range $r \lesssim \xi$, only a negligible part comes from $r \gg \xi$. The value of the integral near the critical point is roughly $\xi^d h(\xi)$, where d is the dimensionality of space (which we leave general for now). It is because the range ξ of $h(r)$ is infinite at the critical point that the integral (9.26) diverges there, and the divergences of ξ and κ are then clearly related: to within constant factors that approach finite limits at the critical point,

$$\xi^d h(\xi) \approx \kappa. \qquad (9.28)$$

By contrast, the convergence of (9.27) at the critical point shows that the range of $c(r)$ remains finite, at least in that sense; and that $c(r)$ is thus much shorter-ranged than $h(r)$ near the critical point, as we wished to observe. It was supposed further by Ornstein and Zernike (and, implicitly, by van der Waals, too) that $c(r)$ has the range of the intermolecu-

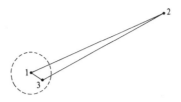

FIG. 9.2.

lar forces, and is thus negligible when $r > b$ where b is some fixed length comparable with that range. We shall for now make that same assumption, although we shall later have to qualify it.

We may visualize as in Fig. 9.2 the points $\mathbf{r}_1, \mathbf{r}_2,$ and $\mathbf{r}_3$ in (9.25): $\mathbf{r}_1$ and $\mathbf{r}_2$ are fixed, while $\mathbf{r}_3$ is the point that varies in the integration. The sphere that is shown centred at $\mathbf{r}_1$ is of radius b, so that $c(r_{13})$ may be taken to be zero when $\mathbf{r}_3$ lies outside that sphere. Then only those $\mathbf{r}_3$ that lie within the sphere contribute to the integral in (9.25). We are here interested in the behaviour of $h(r_{12})$ only for large r_{12}, by which we shall mean $r_{12} \gg b$, though it may still be of any magnitude relative to ξ. For such r_{12}, and near the critical point, where $\xi \gg b$, we shall find that $h(r_{32})$ varies by only a small fraction of itself as $\mathbf{r}_3$ varies through the interior of the sphere of radius b; that is,

$$-d \ln h(r)/dr \ll 1/b \qquad (9.29)$$

when $r \gg b$ and $\xi \gg b$. Since, as we noted, the only contribution to the convolution in (9.25) comes from $\mathbf{r}_3$ within the sphere of radius b about $\mathbf{r}_1$, and since we shall later verify (9.29), which says that $h(r_{32})$ varies little as $\mathbf{r}_3$ varies through that sphere, we may expand $h(r_{32})$ about the centre of the sphere, $\mathbf{r}_3 = \mathbf{r}_1$, and truncate the expansion after the second-derivative term:

$$\int c(r_{13})h(r_{32})d\mathbf{r}_3 \sim h(r_{12})\int c(r_{13})\,d\mathbf{r}_3 + \frac{1}{2d}\nabla_{12}^2 h(r_{12})\int r_{13}^2 c(r_{13})\,d\mathbf{r}_3.$$
$$(9.30)$$

Here d in the coefficient $1/(2d)$ is again the dimensionality of the space. The coefficient of the first-derivative term in the expansion (9.30) vanishes, by the spherical symmetry of $c(r)$ and $h(r)$. We shall refer to (9.30) as the Ornstein–Zernike approximation.

Since we are considering only $r_{12} > b$ we ignore the term $c(r_{12})$ on the right-hand side of (9.25). Then (9.25), with (9.27) and (9.30), becomes

$$\nabla^2 h(r) = \frac{1}{\xi^2}h(r) \qquad (r > b) \qquad (9.31)$$

where

$$\xi = \rho\sqrt{(m\kappa)}, \qquad m = \frac{kT}{2d}\int r^2 c(r)\,d\mathbf{r}. \tag{9.32}$$

With $d = 3$, this m is that of (4.162), where it arose in a closely analogous argument.

We see from (9.31) that ξ sets the scale of length for $h(r)$, and we shall see shortly that the dominant factor in determining the attenuation of $h(r)$ with increasing r when $r \gg \xi$ is $\exp(-r/\xi)$; so that the ξ of (9.32) is just that correlation (coherence) length, or range of $h(r)$, to which we have often referred and for which we have used the same symbol ξ. We see also, as anticipated in §9.1, that it is the same as the interface thickness ξ in the van der Waals theory: compare (9.32) with the last of the alternative forms of ξ in (9.11). Since the arguments that led to (9.31) with (9.32) can hold only in the neighbourhood of the critical point, the ρ in (9.32) is just ρ^c, as in (9.11); and the κ in (9.11) is that of either of the two homogeneous phases (we saw that they are asymptotically equal), so it is the same as its counterpart in (9.32).

In d dimensions,

$$\nabla^2 h(r) = \frac{1}{r^{d-1}}\frac{d}{dr}\left(r^{d-1}\frac{d}{dr}h(r)\right); \tag{9.33}$$

so the solution $h(r)$ of (9.31) that vanishes as $r \to \infty$ is

$$h(r) = Br^{-\frac{1}{2}(d-2)}K_{\frac{1}{2}(d-2)}(r/\xi), \tag{9.34}$$

where we use standard notation[5] for the modified Bessel function $K_\nu(z)$. The constant coefficient B is not determined by (9.31), which is homogeneous in h, but may be calculated from (9.26). To be sure, (9.34) gives $h(r)$ only for $r \gg b$, with b the range of the direct correlation function, while the integral in (9.26) requires $h(r)$ for all r; but (9.31) and (9.34) require also that we be close to the critical point, where $\xi \gg b$; and near the critical point, as we have seen, the major part of the integral (9.26) comes from large r. Thus, asymptotically, near the critical point, we may use (9.34) for $h(r)$ in (9.26), and so find B from the known[5] value of $\int_0^\infty x^{\nu+1}K_\nu(x)\,dx$. In this way we complete the formula (9.34),

$$h(r) = kT^c\kappa(2\pi)^{-d/2}\xi^{-\frac{1}{2}(d+2)}r^{-\frac{1}{2}(d-2)}K_{\frac{1}{2}(d-2)}(r/\xi) \tag{9.35}$$

with ξ still related to κ by (9.32). We note in passing that the expectation (9.28) is fulfilled.

While the formula (9.35) for $h(r)$ can hold only when r and ξ are both large compared with b, they may, as already mentioned, be of any magnitude relative to each other. We have from (9.32), (9.35), and the known properties of the Bessel function,[5,6] with $\rho \sim \rho^c$,

$$h(r) \sim kT^c \kappa (\tfrac{1}{2}\pi)^{\frac{1}{2}} (2\pi)^{-d/2} \xi^{-\frac{1}{2}(d+1)} r^{-\frac{1}{2}(d-1)} e^{-r/\xi} \qquad (r/\xi \gg 1) \qquad (9.36)$$

and

$$h(r) \sim \frac{1}{4\pi^{d/2}} \Gamma\left(\frac{d-2}{2}\right) \frac{kT^c}{(\rho^c)^2 m} r^{-(d-2)} \qquad (r/\xi \ll 1). \qquad (9.37)$$

In (9.37) we assumed $d \neq 2$; if $d = 2$, the $-(d-2)$ power of r is replaced by $\ln r$.

From (9.36) we see that as $r \to \infty$ in a fixed thermodynamic state, however close to the critical point, $h(r)$ decays proportionally to $r^{-\frac{1}{2}(d-1)} \exp(-r/\xi)$. The exponential factor is as anticipated.

According to (9.37), $h(r)$ at the critical point is asymptotically proportional to $r^{-(d-2)}$ (or to $\ln r$ in $d = 2$) as $r \to \infty$. In $d = 1$ or 2, then, $h(r)$ at a critical point would increase with increasing r at large r, which is impossible; so the Ornstein–Zernike approximation (9.30) is incompatible with the existence of a critical point in $d = 1$ or 2. That is a virtue of the theory in $d = 1$, where it is known that there can be no phase transition, and so no critical point, when the intermolecular forces are of short range;[7] but in $d = 2$, where there can be a critical point, it is a defect of the theory,[8] for which we shall ultimately have to account.

Now that we have the explicit asymptotic formulae (9.36) and (9.37) we may verify (9.29), which we required as justification for the Ornstein–Zernike approximation (9.30). For the exponential factor $\exp(-r/\xi)$ we have $-d \ln \exp(-r/\xi)/dr = 1/\xi$, which is indeed much less than $1/b$ near the critical point, where $\xi \gg b$; and for any negative power of r, say r^{-p}, we have $-d \ln r^{-p}/dr = p/r$, which for not-too-large p is indeed much less than $1/b$ when $r \gg b$.

We shall now see that (9.31) also follows from a relation analogous to (9.2). In the same mean-field and small-gradient approximation that led to (9.2), the local density $\rho(r)$ at a distance r from a chosen molecule in an isotropic, homogeneous fluid would satisfy

$$m\nabla^2 \rho(r) = \mu[\rho(r), T] - \mu. \qquad (9.38)$$

Now ρ is r-dependent, rather than z-dependent as it was in (9.2), and the relevant solution is that which for large r is asymptotic to the mean density $\bar{\rho}$ of the homogeneous fluid; while the uniform μ here is $\mu(\bar{\rho}, T)$. We have the mean local density $\rho(r)$ related to the pair-correlation function $h(r)$ by

$$\rho(r) = [1 + h(r)]\bar{\rho}, \qquad (9.39)$$

according to (4.23), so (9.38) is an equation for $h(r)$. The true $h(r)$ varies rapidly with r when r is small, so the gradients of h are then too large for (9.38) to hold; but near the critical point, and for large r, the correlation function varies slowly, as we have seen; so in that regime (9.38) would be expected to determine $h(r)$ as accurately as (9.2) determines the interfacial profile. Now, $h(r)$ is small at large r; so (9.38) with (9.39) and (9.5) is then precisely (9.31) with (9.32). This quasi-thermodynamic theory of the pair-correlation function[9] is thus equivalent to the Ornstein–Zernike theory, which further confirms how intimate the connection is between the Ornstein–Zernike theory of $h(r)$ and the van der Waals theory of $\rho(z)$.

The central point in the Ornstein–Zernike theory of $h(r)$ is the short-rangedness of $c(r)$ in comparison with $h(r)$, and, in particular, the assumption that the range b of $c(r)$ is finite even at the critical point, in contrast to the range ξ of $h(r)$. Certainly, as we saw from (9.27), the range of c is finite at the critical point in the sense that $\int c(r)\,d\mathbf{r}$ is convergent there; but the Ornstein–Zernike and van der Waals theories require also that m have a finite, limiting value at the critical point; that is, by (9.32), that $c(r)$ be so short-ranged as even to have a finite second moment at the critical point. Fisher[10] has shown that this is not so, in general; that, specifically, $c(r)$ at the critical point may decay as $r^{-(d+2-\eta)}$ with some $\eta > 0$. As a consequence, $h(r)$ at the critical point would decay as $r^{-(d-2+\eta)}$ rather than as in (9.37). The Ornstein–Zernike theory is inconsistent with the existence of a critical point in $d = 2$ because it yields an $h(r)$ that does not decay at long distances; but when $\eta > 0$, the correlation function does decay at large r when $d = 2$, and there is no inconsistency. Indeed, the exact result of Kaufman and Onsager[11] for the correlation function of the two-dimensional Ising model at its critical point implies[8,10] $\eta = \frac{1}{4}$.

We see, then, that (9.37) is incorrect in two ways: it supposes m to be finite, and it has $h(r)$ decaying too slowly at large r. These are connected, in the sense that each is a symptom of the other: the vanishing of the coefficient $1/m$ means that the leading term in $h(r)$ at the critical point is not $r^{-(d-2)}$, but one that vanishes more rapidly. The expansion (9.30) is of the wrong form at the critical point. The behaviour (9.36), on the other hand, is correct; so, near but not at the critical point, where ξ is large but finite, the Ornstein–Zernike approximation (9.30) holds for $r \gg \xi$.

The mere existence of a critical point does not entail a positive η in $d = 3$ as it does in $d = 2$; we could well have imagined that the Ornstein–Zernike theory, with $\eta = 0$ and with m finite at the critical point, might be correct in $d = 3$. It is therefore of interest that in model fluids with short-ranged forces, which we believe portray faithfully the behaviour of real fluids at their critical points, η is found to be positive, albeit extremely small, in three dimensions:[12] $\eta \simeq 0\cdot03$–$0\cdot04$. Thus, strictly, the Ornstein–Zernike theory, and (9.37) in particular, fails in $d = 3$; but the failure is so inconspicuous that we may be making a negligible error if we ignore it.

Still, insofar as (9.37) is not strictly correct, the formula (9.35) for $h(r)$ is not. In § 9.3 we shall see what replaces (9.35) in the modern theory of critical points. The alternative formula will again yield (9.36) when $r/\xi \gg 1$; but, when $r/\xi \ll 1$, it will yield $h(r) \sim r^{-(d-2+\eta)}$ with a finite, positive coefficient, instead of (9.37).

The exponent η, along with the exponents we already took note of in § 9.1 and others that we shall introduce, describe the analytic form of thermodynamic functions and correlation functions near the critical point, and, in particular, index the critical-point singularities of those functions. In § 9.3 we shall see how the many critical-point exponents are related to each other, and what their values are, both in the classical, mean-field theories and in reality.

9.3 Digression on critical-point exponents

In §§ 9.1 and 9.2 we saw thermodynamic functions and parameters in correlation functions vanishing or diverging at a critical point proportionally to some power of the distance—often measured as $(T - T^c)$—from that point. Those powers are the critical-point exponents, central to any discussion of critical phenomena.[13] Here we define and discuss those that are most frequently referred to, and to which we shall ourselves refer in the remaining sections of this chapter.

There are four important exponents, conventionally called α, β, γ, and δ, that describe the thermodynamics of the bulk phases near a critical point.

The difference in densities, $\rho^l - \rho^g$, of coexisting liquid and vapour near the critical point of a one-component fluid vanishes as

$$\rho^l - \rho^g \sim (T^c - T)^\beta. \tag{9.40}$$

This defines the exponent β, and is a necessary generalization of the mean-field theory (classical) result $\beta = \frac{1}{2}$ found in (9.4). The same power law (with, it is believed, the same numerical value of β) describes, similarly, the difference in composition of the two liquid phases on approach to the critical solution point of a binary liquid mixture, or the analogous quantities (each sometimes called an 'order parameter') at critical points in still other physical contexts.[13]

The generalization of (9.6) for the compressibility defines the exponent γ:

$$\kappa \sim |T - T^c|^{-\gamma} \tag{9.41}$$

with $\gamma = 1$ classically. This might be the compressibility of either of the two phases that are in equilibrium at $T < T^c$, in which case, as in (9.6), the respective coefficients of proportionality are asymptotically equal; or (with a different coefficient of proportionality but the same value of γ) it might be that of the single fluid phase when the critical point is approached from above T^c, with, for example, ρ fixed at $\rho.^c$ The same exponent γ determines the rate of divergence of the osmotic compressibility, $(\partial x/\partial \mu)_T$, at the critical solution point of a liquid mixture in which μ is the chemical potential of one of the components and x is the composition.

According to any of the classical equations of state, $\mu(\rho, T)$ in the one-phase region near the critical point has the form (9.3), with the second μ in $\mu(\rho, T) - \mu$ now understood to be $\mu(\rho^c, T)$, the chemical potential on the critical isochore. Then the critical isotherm in the μ, ρ-plane would be given by $\mu - \mu^c \sim (\rho - \rho^c)^3$ near the critical point. More generally, it is of the form

$$\mu - \mu^c \sim (\rho - \rho^c) |\rho - \rho^c|^\alpha \cdot \tag{9.42}$$

with $\delta = 3$ the classical value of this exponent. The critical isotherm in the pressure-volume or pressure-density plane has the same exponent: $p - p^c \sim -(v - v^c) |v - v^c|^{\delta - 1}$. It is the same δ because at fixed temperature $\mathrm{d}p = \rho \, \mathrm{d}\mu$, while $\rho = \rho^c + (\rho - \rho^c)$ and $\rho - \rho^c \sim -(\rho^c)^2(v - v^c)$.

The remaining bulk thermodynamic exponent α determines the rate of divergence of the constant-volume heat capacity C_V at the liquid–vapour critical point of a one-component fluid [or that of the constant-pressure heat capacity C_p, or of the literal mechanical (rather than osmotic) compressibility κ, or of the coefficient of thermal expansion, of a liquid mixture near its consolute point].[13] At fixed $\rho = \rho^c$, the heat capacity C_V as a function of temperature has the shape shown schematically in Fig. 9.3(a) with

$$C_V \sim A_\pm |T - T^c|^{-\alpha} \tag{9.43}$$

where $A_\pm$ means some coefficient A_+ for $T > T^c$ (the one-phase region) and some generally different coefficient A_- for $T < T^c$ (the two-phase region). Analogous behaviour found at the critical points of order–disorder transitions and at the superfluid transition in liquid helium led to their being called lambda points, because of a fancied resemblance[14] of Fig. 9.3(a) to the letter λ. In the classical, mean-field theories C_V at fixed $\rho = \rho^c$ is as in Fig. 9.3(b): finite but discontinuous at $T = T^c$. This corresponds to $\alpha = 0$. The most general meaning of $\alpha = 0$ in (9.43) is a symmetric logarithmic divergence superimposed on such a finite discontinuity, for (9.43) may be understood as[15]

$$C_V \sim A\alpha^{-1}\{[|T - T^c|\exp(-B_\pm/A)]^{-\alpha} - 1\}. \tag{9.44}$$

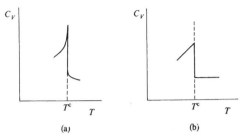

FIG. 9.3. Constant-volume heat capacity at a fixed density equal to the critical density, as a function of the temperature near the critical temperature T^c. (a) Non-classical critical point (λ-point), as in a real fluid; (b) classical critical point, as in mean-field theory.

The $A_\pm$ in (9.43) has here been re-expressed as $(A/\alpha)\exp(\alpha B_\pm/A)$, while the added constant $-A/\alpha$ is just part of the analytic background which is in any case present in C_V and which for any fixed, positive α is negligible compared with the singular term $A_\pm |T - T^c|^{-\alpha}$ as $T \to T^c$. From (9.44) in the limit $\alpha \to 0$,

$$C_V \sim A \ln|T - T^c|^{-1} + B_\pm \qquad (9.45)$$

which is the aforementioned symmetric logarithm superimposed on a finite discontinuity. In the two-dimensional Ising model[16] $B_+ = B_-$, so there is no discontinuity, while in the mean-field theories $A = 0$, so there is no divergence; but $\alpha = 0$ for both.

Two further exponents ν and η are defined by the properties of the correlation function $h(r)$. The dominant factor in the decay of $h(r)$ as $r \to \infty$ is $\exp(-r/\xi)$, as in (9.36). On approach to the critical point the correlation length ξ diverges as

$$\xi \sim |T - T^c|^{-\nu} \qquad (9.46)$$

defining the exponent ν. At the critical point, as we saw in § 9.2, $h(r)$ decays at large r as

$$h(r) \sim r^{-(d-2+\eta)} \qquad (9.47)$$

with some finite, positive constant of proportionality. This defines η.

The remaining exponent of importance to us is μ, defined by (9.1), which determines the rate at which the interfacial tension vanishes as the critical point is approached.

All the other quantities that have arisen in § 9.1 and § 9.2, or that will arise in the later sections of this chapter, are characterized by exponents that are expressible in terms of these. For example, the van der Waals parameter m, which by the second of (9.32) is a measure of the second moment of the direct correlation function $c(r)$, and which was supposed by van der Waals and by Ornstein and Zernike to have a finite, positive value at the critical point, in fact behaves as

$$m \sim |T - T^c|^{\gamma - 2\nu}, \tag{9.48}$$

according to the first of (9.32) and from (9.41) and (9.46). We shall see later in (9.57) that $\gamma - 2\nu = -\eta\nu$; so that (as anticipated in § 9.2) m diverges when $\eta > 0$. For another example, we conclude from (9.1) and (9.40) that the exponent in the Macleod relation (9.9) is μ/β. The value 3 of this exponent in (9.9) is its mean-field-theory value, while 4 is the empirical value in the relation as originally proposed.[3] From the more accurate values of the exponents, which we are about to quote, we calculate $\mu/\beta = 3.88$.

The distance $z_i - z_j$ between the $\Gamma_i = 0$ and $\Gamma_j = 0$ dividing surfaces, given by (2.49), also shows interesting critical behaviour describable by the exponents already introduced. This distance, by (2.49), is $\Gamma_{i(j)}/\Delta\rho_i$. The relative adsorption $\Gamma_{i(j)}$, by (2.47), is the rate at which the interfacial tension varies with the thermodynamic field μ_i. The temperature is representative of such a field, so, with distance from the critical point measured by $T^c - T$, say, $\Gamma_{i(j)}$ is seen to vanish proportionally to $(T^c - T)^{\mu - 1}$ as the critical point is approached. At the same time the density difference $\Delta\rho_i$ vanishes proportionally to $(T^c - T)^\beta$; with the result, then, that[17]

$$|z_i - z_j| \sim (T^c - T)^{\mu - 1 - \beta}. \tag{9.49}$$

In mean-field theory, with $\mu = \frac{3}{2}$ and $\beta = \frac{1}{2}$, the distance between dividing surfaces thus approaches a finite limiting value at the critical point. In reality, as we may see from the values of the exponents we are about to quote, $\mu - 1 - \beta$ is slightly negative ($\simeq -0.06$ or -0.07 in $d = 3$), so $|z_i - z_j|$ diverges. The quantity $(-\mu + 1 + \beta)$ is only about one-tenth of ν ($\simeq 0.63$ in $d = 3$), so the length $|z_i - z_j|$ diverges much less rapidly than the coherence length ξ. The latter, as we saw in § 9.2 and shall see again in § 9.4, is a measure of the interfacial thickness; so $|z_i - z_j|$, although indeed a fundamental length in the interface, is not the thickness.

We list in Table 9.1 the values of the critical-point exponents α, β, γ, δ, ν, η, and μ. In the first column are the classical, mean-field-theory values, in the second column the values for the two-dimensional Ising model (lattice gas), which are also known exactly, and in the third column those for real, three-dimensional fluids as determined from present-day critical-point theory,[13,18] from experiment,[13] or from both. As theoretical values the latter are probably correct to ± 0.01, but as experimental values their uncertainties may be two or three times that.

Table 9.1

	Classical	$d = 2$	$d = 3$
α	0	0	0·11
β	$\frac{1}{2}$	$\frac{1}{8}$	0·32$_5$
γ	1	$\frac{7}{4}$	1·24
δ	3	15	4·81
ν	$\frac{1}{2}$	1	0·63
η	0	$\frac{1}{4}$	0·03
μ	$\frac{3}{2}$	1	1·26

The exponents are not independent of each other. We remarked earlier that by any of the classical equations of state $\mu(\rho, T)$ in the one-phase region near the critical point is of the form (9.3), with the μ in $\mu(\rho, T) - \mu$ taken to be $\mu(\rho^c, T)$. The appropriate generalization of this form, incorporating non-classical critical-point exponents, is[15]

$$\mu(\rho, T) - \mu(\rho^c, T) = (\rho - \rho^c) |\rho - \rho^c|^{\delta - 1} j\left(\frac{T - T^c}{T^c - \tau(\rho)}\right) \qquad (9.50)$$

where $T = \tau(\rho)$ is the equation of the coexistence curve in the T, ρ-plane, so that for ρ near ρ^c,

$$T^c - \tau(\rho) \sim \text{const.} |\rho - \rho^c|^{1/\beta}. \qquad (9.51)$$

The function $j(x)$ in (9.50) is a smooth function, analytic in x for all finite $x > -1$. The coexistence curve, $T = \tau(\rho)$, corresponds to $x = -1$; the critical isotherm, $T = T^c$, to $x = 0$; and the critical isochore in the one-phase region, $\rho = \rho^c$ at $T > T^c$, to $x = +\infty$. On the critical isotherm (9.50) becomes (9.42), with coefficient of proportionality $j(0)$. The compressibility κ may be calculated from (9.50) and (9.51) by the identity $\kappa^{-1} = \rho^2(\partial\mu/\partial\rho)_T$, evaluated at the coexistence curve, $T = \tau(\rho) < T^c$, or on the critical isochore in the one-phase region, $\tau(\rho) = T^c < T$, and then compared with the definition (9.41); whereupon it is seen[15] that

$$\delta = 1 + \gamma/\beta, \qquad (9.52)$$

one of two relations among the thermodynamic exponents that are implied by (9.50) and (9.51). The second relation follows upon calculating C_V on the critical isochore from (9.50) and (9.51). The result[15] is that C_V is of the form (9.45) when $2\beta + \gamma = 2$ [with γ related to β and δ by (9.52)], while in general it is of the form (9.43) with

$$\alpha + 2\beta + \gamma = 2. \qquad (9.53)$$

Both exponent relations (9.52) and (9.53) are seen to be satisfied by the entries in each column of Table 9.1.

Equation (9.50) with (9.51) asserts that $[\mu(\rho, T) - \mu(\rho^c, T)]/(\rho - \rho^c)$ is a homogeneous function of the variables $T - T^c$ and $T^c - \tau(\rho)$. The correlation function $h(r)$, likewise, has an asymptotic homogeneity of form[10,19] for large r and ξ, which is a generalization of that in (9.35). For simplicity, and because it is sufficient for the purpose at hand, we suppose there to be only one thermodynamic variable determining the distance from the critical point; for example, we may specify a state on the critical isochore, $\rho = \rho^c$, at some $T > T^c$, so that κ, ξ, $h(r)$, etc., depend only on $T - T^c$. Then[10]

$$h(r) = r^{-(d-2+\eta)} H(r/\xi), \tag{9.54}$$

where the function $H(x)$ is such that, to within finite positive constants of proportionality,

$$H(x) \sim \begin{cases} x^{\frac{1}{2}(d-3)+\eta} e^{-x}, & x \to \infty \\ 1, & x \to 0. \end{cases} \tag{9.55}$$

(9.55) is the generalization of (9.36) and (9.37). Near the critical point, as we know, the major contribution to the integral in (9.26) comes from large r, where (9.54) is accurate, so from (9.26) and (9.54) we have the asymptotic proportionality

$$\kappa \sim \xi^{2-\eta}. \tag{9.56}$$

From this, and the definitions of γ and ν in (9.41) and (9.46), we find Fisher's relation[10]

$$(2 - \eta)\nu = \gamma. \tag{9.57}$$

This, too, is satisfied by the entries in each column of Table 9.1.

There are further relations among critical-point exponents in which the dimensionality d appears explicitly. The equilibrium fluctuations of the density about its average value are coherent (that is, are of one sign) over distances of order ξ. These fluctuations that thus occur spontaneously in regions of linear dimension ξ are the elementary density fluctuations with each of which is associated a free energy of order $kT \simeq kT^c$. The free-energy density associated with equilibrium density (or composition) fluctuations near a critical point is thus of order kT^c/ξ^d. But thermodynamically, the part of the free-energy density that is responsible for the heat-capacity singularity $|T - T^c|^{-\alpha}$ is proportional to $|T - T^c|^{2-\alpha}$. If there is only one free-energy density of significance near the critical point, these two would be essentially the same, so that

$$d\nu = 2 - \alpha. \tag{9.58}$$

From (9.52), (9.53), (9.57), and (9.58), we find another d-dependent relation[20] equivalent to (9.58):

$$2 - \eta = d(\delta - 1)/(\delta + 1). \tag{9.59}$$

The mean-square fluctuation in the density in a volume V at given μ and T, according to (4.20) and (4.21), is given by

$$\overline{(\rho - \bar{\rho})^2}/\bar{\rho}^2 = kT\kappa/V. \tag{9.60}$$

For this to hold accurately V must be macroscopic, that is, $V \gg \xi^d$; but it may be supposed qualitatively correct, that is, correct to within a factor of order unity, even when $V = \xi^d$. Then by the definitions of the exponents γ and ν in (9.41) and (9.46), and by the exponent relations (9.53) and (9.58), we conclude from (9.60) that at a distance from the critical point measured by $|T - T^c|$ the mean-square fluctuation in the density in the elementary volume ξ^d is of order $|T - T^c|^{2\beta}$. Thus, by definition of the exponent β in (9.40), the root-mean-square density fluctuations in volumes of size ξ^d are of the magnitude of the difference in the bulk densities of the distinct phases that would be in equilibrium at a comparable distance from the critical point. We may therefore think of the region of size ξ^d as being in some ways like a droplet of one phase in another. The free energy associated with such an elementary in-homogeneity, the magnitude of which we previously identified as $kT \; (\simeq kT^c$ near the critical point), we may now identify alternatively as $\sigma\xi^{d-1}$, with σ the surface tension between bulk phases at a comparable distance from the critical point. Thus, $\sigma\xi^{d-1} \sim kT^c$; so from the definitions of the exponents μ and ν in (9.1) and (9.46),

$$\mu = (d - 1)\nu. \tag{9.61}$$

From (9.53), (9.58), and (9.61) we find for the sum of the surface-tension exponent μ and correlation-length exponent ν the d-independent relations

$$\mu + \nu = 2 - \alpha = \gamma + 2\beta, \tag{9.62}$$

which are also seen to be satisfied by the entries in each column of Table 9.1.

Exponent relations such as those given here in (9.52), (9.53), (9.57), (9.58), (9.59), (9.61), and (9.62) are often called 'scaling laws' in the literature on critical phenomena. Those that refer explicitly to the dimensionality, d, here (9.58), (9.59), and (9.61), are sometimes distinguished from the others by being called 'hyperscaling' relations. It is seen in Table 9.1 that these hold in $d = 2$, and that to the precision with which the exponents are known they hold also in $d = 3$. [In $d = 3$ it is really only (9.58) or the equivalent (9.59), among the d-dependent relations, not

(9.61), that is verified by the entries in Table 9.1. There is no independent theoretical estimate of μ in $d = 3$; the entry in the table was calculated from the quoted ν assuming (9.61). Experimentally, $\mu = 1 \cdot 28 \pm 0 \cdot 05$, so (9.61) is indeed verified, but not with as high precision as (9.58). The entries under $d = 3$ in Table 9.1 are intended to be the most probable values of those exponents, taking account of the full body of present-day theory, including the hyperscaling relations.] These d-dependent relations do not hold in the mean-field theories, however, as we see from the table, except at the single dimensionality $d = 4$; and it is that which distinguishes them from the d-independent relations, which, as we have seen, are universal.

The picture of the exponents that has emerged from the modern theory of critical points is that they are dimensionality dependent and different from their classical, mean-field-theory counterparts up to $d = 4$, while for all $d \geqslant 4$ they are independent of dimensionality and coincide with their mean-field-theory values. For $d < 4$, when they are still d-dependent, they satisfy the d-dependent exponent relations such as (9.58), (9.59), and (9.61); while for $d > 4$, where the values of the exponents are fixed at those of the mean-field theories, the d-dependent relations no longer hold. It is not that the singularities represented by the non-classical, d-dependent exponents have disappeared in $d > 4$; it is that they are no longer the exponents of the leading terms in the quantities with which they are associated. The terms with the d-dependent exponents are simply overtaken by other terms with fixed exponents when, with increasing d, the values of the d-dependent exponents pass by those of the fixed ones. It is as though classical and non-classical critical behaviour were in competition, the latter dominating when $d < 4$, the former when $d > 4$.

A simple example of this, which makes clear the mathematical mechanism by which it occurs, is seen in the ideal Bose gas, which provides an exactly soluble model of a critical point.[21] The number of single-particle states with energy in the range ε to $\varepsilon + d\varepsilon$ is proportional to $\varepsilon^{\frac{1}{2}d-1} d\varepsilon$ in d dimensions, while the occupancy of a state of energy ε is $\{\exp[(\varepsilon - \mu)/kT] - 1\}^{-1}$, with $\mu \leqslant 0$, by the Bose–Einstein distribution law; so the density of the gas at given μ and T is expressible in terms of the integral

$$I(d, y) = \int_0^\infty \frac{x^{\frac{1}{2}d-1}}{e^{x+y} - 1} \, dx \qquad (9.63)$$

with $y \geqslant 0$. The critical point,[21] when there is one, is at $y = 0$. The integral $I(d, y)$ consists of a part that is analytic in y at $y = 0$, and is thus expandable in the non-negative integer powers $y^0, y^1, y^2, \ldots$, and a part that is singular at $y = 0$, and that is of order $y^{\frac{1}{2}d-1}$ when d is not an even integer and of order $y^{\frac{1}{2}d-1} \ln(y^{-1})$ when d is an even integer. Thus, with increasing d the singular term of order $y^{\frac{1}{2}d-1}$ moves along the line of integer powers $y^0, y^1, y^2, \ldots$, passing each one in turn, and having the additional factor $\ln(y^{-1})$ attached to it each time it is itself momentarily an integer power. As long as $d \leq 2$ that singular term is infinite at $y = 0$, so there is no critical point at finite density. That is because $\eta = 0$ for the Bose gas,[21] which we know to be incompatible with the existence of a critical point in $d \leq 2$. When $2 < d < 4$ the integral $I(d, 0)$ is finite, and the singular term $y^{\frac{1}{2}d-1}$ is the leading term in $I(d, y) - I(d, 0)$ at small y. But the power of y in the leading term in $I(d, y) - I(d, 0)$ is $2/(\delta - 1)$ for the Bose gas,[21] so when $2 < d < 4$ the exponent δ is d-dependent: $\delta = (d + 2)/(d - 2)$. While it is thus non-classical ($\delta > 3$) and d-dependent it satisfies the d-dependent exponent relation (9.59) (since $\eta = 0$ for all d). When $d > 4$ the leading term in $I(d, y) - I(d, 0)$ is no longer the singular term, which is of order $y^{\frac{1}{2}d-1}$, but is the term with the fixed, integer power 1. Thus, for all $d > 4$ the exponent $2/(\delta - 1)$ remains fixed at 1; that is, δ remains fixed at its classical value $\delta = 3$. At the borderline dimensionality, $d = 4$, the singular term in $I(d, y) - I(d, 0)$ is still the leading term, by virtue of the factor $\ln(y^{-1})$ that then accompanies the power y^1. Because of the logarithm the critical behaviour is still non-classical; but, since a logarithm varies more slowly than any power, the formal value of the exponent at $d = 4$ is its classical value: $2/(\delta - 1) = 1$, or $\delta = 3$.

The same phenomenon is seen also in the modern renormalization-group theory of critical points,[22] though in a different guise. There, two fixed points of the renormalization-group equations—the so-called Gaussian and Ising fixed points—are in competition. When the critical-point behaviour is determined by the former, as it is when $d > 4$, the exponents are d-independent and classical; when by the latter, as it is when $d < 4$, the exponents are non-classical and d-dependent. The borderline is again $d = 4$, where the two competing fixed points coincide.

The borderline dimensionality, d^*, is the highest at which the d-dependent exponent relations, such as (9.58), (9.59), and (9.61), hold, and the lowest at which the exponents have their classical values. Thus, d^* can always be found as the solution d of the algebraic equation that results from setting the exponents in any of the d-dependent exponent relations equal to their classical values. For example, setting $\nu = \frac{1}{2}$ and $\alpha = 0$ in (9.58), or $\eta = 0$ and $\delta = 3$ in (9.59), or $\nu = \frac{1}{2}$ and $\mu = \frac{3}{2}$ in (9.61), yields $d^* = 4$. Since $\eta = 0$ classically, it follows from (9.59) that the borderline dimensionality d^* is related to the classical value δ_{cl} of the critical-isotherm exponent δ by

$$d^* = 2(\delta_{cl} + 1)/(\delta_{cl} - 1). \tag{9.64}$$

Setting $\delta_{cl} = 3$ in (9.64) then gives $d^* = 4$, as before; but (9.64) also allows us to obtain d^* for higher-order critical points, where the classical value of η is still zero but where $\delta_{cl} \neq 3$.

A critical point of order p is a thermodynamic state in which p phases become identical: $p = 2$ for ordinary critical points, of the kind we have been discussing; $p = 3$ for tricritical points, the subject of § 9.5; etc. The mean-field-theory value δ_{cl} of the exponent δ for a critical point of order p may be inferred from Fig. 1.8 or 3.4 and their obvious generalizations. When $p = 2$, as in Fig. 1.8 or 3.4, every isotherm at $T < T^c$ makes three intersections with the tieline in the equal-areas construction; at $T = T^c$ those previously distinct intersections coalesce in a single, three-fold root, which makes the critical isotherm cubic at the critical point: $\delta_{cl} = 3$. For p-phase equilibrium, the analogue of the curve in Fig. 1.8 or 3.4 has $2p - 2$ loops; the horizontal line that determines the equilibrium pressure or chemical potential makes $2p - 1$ intersections with the curve, in such a way that the areas of the $2p - 2$ loops are equal in pairs; and at $T = T^c$ the $2p - 1$ roots coalesce, so that the critical isotherm is of degree $\delta_{cl} = 2p - 1$ ($= 5$ for a tricritical point). Then from (9.64), the borderline dimensionality for a critical point of order p is

$$d^* = 2p/(p - 1). \tag{9.65}$$

For $p = 2$ this again gives $d^* = 4$. For $p = 3$ it gives $d^* = 3$. Thus, in three dimensions the exponents associated with tricritical points, unlike those of ordinary, two-phase critical points, are given correctly by a mean-field theory. That is not to say that the mean-field theory of tricritical points is fully correct in $d = 3$; on the contrary, since $d = 3$ is borderline, there are in reality non-classical logarithmic factors attached to the classical powers,[23] as in the example we saw earlier. But, as we also remarked earlier, those factors do not alter the formal values of the exponents, which remain classical. We shall need to recall this in § 9.5.

We turn next to the question of how to modify the mean-field theory of the near-critical interface that was outlined in § 9.1, so as to incorporate in it the correct, non-classical values of the critical-point exponents.

9.4 Van der Waals theory with non-classical exponents

We need non-classical versions of the mean-field-theory formulae (9.4) and (9.7) from which to calculate the surface tension and the density or composition profile—as in (9.8)–(9.15), but corrected for non-classical critical-point behaviour.

The appropriate generalization of (9.4) is (9.40), which it will be convenient to take in the form (9.51). To replace (9.7) we seek a $W(\rho)$ that is such that $W(\rho^g) = W(\rho^l) = 0$ and $-dW(\rho)/d\rho = \mu(\rho, T) - \mu$, with an appropriate generalization of the $\mu(\rho, T) - \mu$ in (9.3). We know that, classically, $\mu(\rho, T) - \mu(\rho^c, T)$ in the one-phase region near the critical point is also given by the right-hand side of (9.3), and that its non-classical generalization is (9.50). Thus, we may assume[24] the required non-classical $\mu(\rho, T) - \mu$ in the two-phase region to be of the form

$$\mu(\rho, T) - \mu = (\rho - \rho^c) \, |\rho - \rho^c|^{\delta-1} \, j^{\dagger} \left(\frac{T - T^c}{T^c - \tau(\rho)} \right) \qquad (9.66)$$

with $T^c - \tau(\rho)$ as in (9.51), and with some function $j^{\dagger}(x)$ that is a smooth extension of the $j(x)$ of (9.50) from the one-phase region $(x > -1)$ to the two-phase region $(x < -1)$. The $[\mu(\rho, T) - \mu]/(\rho - \rho^c)$ of (9.66), like the $[\mu(\rho, T) - \mu(\rho^c, T)]/(\rho - \rho^c)$ of (9.50), is homogeneous in $T - T^c$ and $T^c - \tau(\rho)$. The degree of homogeneity* is $\beta(\delta - 1)$, which, by (9.52), is γ. Since $-dW(\rho)/d\rho = \mu(\rho, T) - \mu$, the $-W(\rho)$ we seek is then also homogeneous in $T - T^c$ and $T^c - \tau(\rho)$, and of degree $\beta(\delta + 1)$; hence, of the form

$$-W(\rho) = |\rho - \rho^c|^{\delta+1} u \left(\frac{T^c - \tau(\rho)}{T^c - T} \right). \qquad (9.67)$$

By (9.52), the degree of homogeneity $\beta(\delta + 1)$ is $\gamma + 2\beta$, which in turn, by (9.53), is $2 - \alpha$, as is expected of a free-energy density. Since $W(\rho^{g,l}) = 0$ and $W'(\rho^{g,l}) = 0$ (where the prime means differentiation with respect to the indicated argument), $u(x)$ must be such that

$$u(1) = 0, \qquad u'(1) = 0. \qquad (9.68)$$

Again because $-W'(\rho) = \mu(\rho, T) - \mu$, the function $u(x)$ is related to $j^{\dagger}(-1/x)$ by a differential equation of the first order, which may be thought of as having to be solved subject to either of the two conditions (9.68); the other of the two would then hold automatically, since $j^{\dagger}(-1) = 0$, by virtue of $\mu(\rho^{g,l}, T) - \mu = 0$.

The analytical forms of the thermodynamic functions at a critical point are universal, the same for all substances. That includes the critical-point exponents and the functional forms, up to multiplicative constants, of functions such as $u(x)$ or $T^c - \tau(\rho)$; while the locations of the critical points (the values of T^c, ρ^c, etc.) and the amplitudes of the scaling functions (the multiplicative constants, such as that in (9.51)) are non-universal, varying from substance to substance. The principle of universality goes beyond the principle of corresponding states, for it does not require a universal form of the intermolecular potentials; but it would be implied by the principle of corresponding states and is in practice difficult to distinguish from it.

With $u(x)$ universal up to a multiplicative constant, the function $v(x)$ defined by

$$v(x) = u(x)/\tfrac{1}{2}u''(1) \qquad (9.69)$$

*The power n is said to be the degree of the homogeneous function $x^n f(y/x)$.

is universal, the same for all substances;[24] and we conclude from (9.68) and (9.69) that

$$v(1) = 0, \qquad v'(1) = 0, \qquad \tfrac{1}{2}v''(1) = 1. \tag{9.70}$$

The classical $v(x)$, from (9.7), (9.51), (9.67), and (9.69), with $\delta = 3$ and $\beta = \tfrac{1}{2}$, is just

$$v(x) = \left(\frac{1}{x} - 1\right)^2 \quad \text{(classical)}, \tag{9.71}$$

with $\tfrac{1}{2}u''(1)$ in that case what in (9.7) was called $\tfrac{1}{4}B$.

We take the van der Waals, Ornstein–Zernike parameter m to be independent of ρ in (3.8) and (3.13) although we recognize from (9.48) that it may depend on T and may even diverge (albeit extremely slowly, with exponent $-\eta\nu \simeq -0{\cdot}02$) as $T \to T^c$. We shall first calculate the surface tension, then the density profile.[24]

From (3.13) with (9.51), (9.67), and (9.69),

$$\sigma = 2\beta[mu''(1)]^{\frac{1}{2}}\left(\frac{T^c - T}{C}\right)^{\frac{1}{2}\beta(\delta+3)}\int_0^1 x^{\frac{1}{2}\beta(\delta+3)-1}[v(x)]^{\frac{1}{2}}\,dx, \tag{9.72}$$

where C [called B/A in (9.4)] is the constant coefficient of proportionality in (9.51). The integral is a pure number that is universal, and that we shall evaluate later for an assumed $v(x)$. Equation (9.72) is the non-classical generalization of (9.12) and reduces to it when $\beta = \tfrac{1}{2}$ and $\delta = 3$, when $v(x)$ has its classical form (9.71), and with $\tfrac{1}{2}u''(1) = \tfrac{1}{4}B$. We see from (9.48) and (9.52) that (9.72) implies $\sigma \sim (T^c - T)^{-\nu+\gamma+2\beta}$; so $\mu = -\nu + \gamma + 2\beta$, the exponent relation already found in (9.62) and verified by the entries in Table 9.1 (§ 9.3).

The profile is obtained from (3.9) with (9.67) and (9.51). If as in (9.8) we take z to be 0 where $\rho = \rho^c$, and if we also take account of (9.52) and (9.69), we have

$$|z| = \beta\left(\frac{m}{u''(1)}\right)^{\frac{1}{2}}\left(\frac{T^c - T}{C}\right)^{-\frac{1}{2}\gamma}\int_0^{x^*} x^{-1-\frac{1}{2}\gamma}[v(x)]^{-\frac{1}{2}}\,dx, \qquad x^* = \frac{T^c - \tau(\rho)}{T^c - T}. \tag{9.73}$$

If we then define the universal function $X(y)$ for all $y \geqslant 0$ by

$$y = \int_0^{[X(y)]^{1/\beta}} x^{-1-\frac{1}{2}\gamma}[v(x)]^{-\frac{1}{2}}\,dx, \tag{9.74}$$

(9.73) with (9.51) becomes the formula for the profile $\rho(z)$,

$$|\rho(z) - \rho^c| = \tfrac{1}{2}(\rho^l - \rho^g)\,X\!\left[\left(\frac{T^c - T}{C}\right)^{\frac{1}{2}\gamma}\frac{1}{\beta}\left(\frac{u''(1)}{m}\right)^{\frac{1}{2}}|z|\right]. \tag{9.75}$$

This is explicit once $v(x)$ is specified, for then the function $X(y)$ is known from (9.74).

The coefficient of $|z|$ in the argument of the function X in (9.75) will now be seen to be ξ^{-1}, the reciprocal of the correlation length in either of the two bulk phases. From (9.5) and $-dW(\rho)/d\rho = \mu(\rho, T) - \mu$, the compressibility κ is related to $W(\rho)$ by $\kappa^{-1} = -\rho^2 d^2W/d\rho^2$. A typical $W(\rho)$ is as in Fig. 3.2. We assume it and its first and second derivatives to be continuous at ρ^α and ρ^β: that, specifically, is what was meant above by the statement that $j^\dagger(x)$ is a smooth extension of $j(x)$ from the one-phase to the two-phase region. Thus, although (9.67) refers to the two-phase region, if we use it to evaluate $-\rho^2 d^2W/d\rho^2$ at ρ^g and ρ^l the result may be identified with $(\kappa^g)^{-1}$ and $(\kappa^l)^{-1}$, the reciprocals of the compressibilities of the separate, bulk phases. We then find from (9.51), (9.52), (9.67), and (9.68), that

$$\frac{1}{(\rho^g)^2 \kappa^g} = \frac{1}{(\rho^l)^2 \kappa^l} = \frac{u''(1)}{\beta^2}\left(\frac{T^c - T}{C}\right)^\gamma. \tag{9.76}$$

Then from the first of (9.32) we have for the correlation length $\xi^{g,l}$ in either of the two bulk phases (we shall call it simply ξ),

$$\xi = \beta\left(\frac{m}{u''(1)}\right)^{\frac{1}{2}}\left(\frac{T^c - T}{C}\right)^{-\frac{1}{2}\gamma}. \tag{9.77}$$

We may now rewrite the formula (9.75) for the density profile as

$$|\rho(z) - \rho^c| = \tfrac{1}{2}(\rho^l - \rho^g)X(|z|/\xi). \tag{9.78}$$

The function $v(x)$ is implicitly contained in this, for it determines the form of the function $X(y)$ by (9.74). For the classical $v(x)$ as given by (9.71), and with the classical $\beta = \frac{1}{2}$ and $\gamma = 1$ as well, (9.74) gives $X(y) = \tanh(y/2)$; so (9.78) then becomes (9.10) (if we again adopt the convention that z increases in the direction of increasing ρ), as it should. More generally, (9.78) is of the form anticipated in (9.16). Since $X(y)$ is universal, the same for all fluids, (9.78) asserts that when z is scaled by ξ and $\rho(z) - \rho^c$ by $\rho^l - \rho^g$, the resulting profile is universal.

The deviation of the profile $\rho(z)$ from its limit ρ^g or ρ^l as $|z| \to \infty$ has a universal exponential form in which the exponential is the same as that in the pair-correlation function $h(r)$ at large r. We have $v(x) \sim (1-x)^2$ near $x = 1$, because of (9.70). Then the universal constant q defined by

$$q = \int_0^1 \left[\frac{1}{x^{1+\frac{1}{2}\gamma}[v(x)]^{\frac{1}{2}}} - \frac{1}{1-x}\right]dx \tag{9.79}$$

is a convergent definite integral. We have $X(y) \to 1$ as $y \to \infty$ in (9.74); then

$$y \equiv \int_0^{[X(y)]^{1/\beta}}\frac{dx}{1-x} + \int_0^{[X(y)]^{1/\beta}}\left[\frac{1}{x^{1+\frac{1}{2}\gamma}[v(x)]^{\frac{1}{2}}} - \frac{1}{1-x}\right]dx$$

$$\sim -\ln\{1 - [X(y)]^{1/\beta}\} + q + o(1), \tag{9.80}$$

where the terms represented as $o(1)$ in this asymptotic formula approach 0 as $y \to \infty$ $(X \to 1)$. From (9.78) and (9.80), and the symmetry $\rho^{\mathrm{l}} - \rho^{\mathrm{c}} = \rho^{\mathrm{c}} - \rho^{\mathrm{g}}$, we find the anticipated exponential decay,

$$|\rho(z) - \rho^{\mathrm{g,l}}| \sim \tfrac{1}{2} \beta e^q (\rho^{\mathrm{l}} - \rho^{\mathrm{g}}) e^{-|z|/\xi} \qquad (|z| \to \infty). \tag{9.81}$$

The $|z|$-dependence in this asymptotic formula is the same as that of the pair-correlation function $h(|z|)$ at large $|z|$ in a one-dimensional fluid in which the correlation length is ξ [cf. (9.36)]. The decay (9.81) is characteristically one-dimensional because the density gradient in the interface is one-dimensional. In mean-field theory, with $\gamma = 1$ and with $v(x)$ given by (9.71), we find $q = \ln 4$ from (9.79), which we may verify from (9.10) and (9.81), with $\beta = \tfrac{1}{2}$, to be the right result.

From (9.51), (9.52), (9.76), and (9.77), we may rewrite the formula (9.72) for the surface tension in the alternative forms

$$\sigma = Km(\rho^{\mathrm{l}} - \rho^{\mathrm{g}})^2 / \xi \tag{9.82}$$

$$= K\xi(\rho^{\mathrm{l}} - \rho^{\mathrm{g}})^2 / (\rho^{\mathrm{c}})^2 \kappa \tag{9.83}$$

$$= K\sqrt{m}(\rho^{\mathrm{l}} - \rho^{\mathrm{g}})^2 / \rho^{\mathrm{c}} \sqrt{\kappa} \tag{9.84}$$

where K is the universal dimensionless constant

$$K = \tfrac{1}{2}\beta^2 \int_0^1 x^{\frac{1}{2}\beta(\delta+3)-1} [v(x)]^{\frac{1}{2}} \, dx \tag{9.85}$$

and κ is κ^{g} or κ^{l}, these being asymptotically equal, along with $\rho^{\mathrm{g}} \sim \rho^{\mathrm{l}} \sim \rho^{\mathrm{c}}$ [see (9.76)]. The formulae (9.82)–(9.84) are the non-classical generalizations of the mean-field-theory results (9.13)–(9.15), respectively. We see directly from (9.83) that $\mu = -\nu + \gamma + 2\beta$, as found before, less directly, from (9.72). When $\beta = \tfrac{1}{2}$ and $\delta = 3$, and $v(x)$ has its classical form (9.71), we have from (9.85) the classical

$$K = \tfrac{1}{6} \quad \text{(classical)}, \tag{9.86}$$

as in (9.13)–(9.15).

To complete the foregoing formulae and make them fully explicit, we must derive or devise an appropriate non-classical $v(x)$. The classical $-W(\rho)$ of (9.7) may be written

$$-W(\rho) = \tfrac{1}{4}B[(\rho - \rho^{\mathrm{c}})^4 + (\rho^{\mathrm{g,l}} - \rho^{\mathrm{c}})^4 - 2(\rho^{\mathrm{g,l}} - \rho^{\mathrm{c}})^2(\rho - \rho^{\mathrm{c}})^2] \quad \text{(classical)}. \tag{9.87}$$

A suitable[25] (although not unique) non-classical generalization of this $-W(\rho)$ in the two-phase region is

$$-W(\rho) = \frac{B}{\delta+1}\left[|\rho - \rho^{\mathrm{c}}|^{\delta+1} + \frac{\delta-1}{2}|\rho^{\mathrm{g,l}} - \rho^{\mathrm{c}}|^{\delta+1} \right.$$
$$\left. - \frac{\delta+1}{2}|\rho^{\mathrm{g,l}} - \rho^{\mathrm{c}}|^{\delta-1}(\rho - \rho^{\mathrm{c}})^2 \right] \tag{9.88}$$

which with (9.51)–(9.53), (9.67), and (9.69) implies

$$v(x) = 4 \frac{1 + \dfrac{\delta - 1}{2} x^{-2+\alpha} - \dfrac{\delta + 1}{2} x^{-\gamma}}{(\delta - 1)(2 - \alpha)(3 - \alpha) - (\delta + 1)\gamma(\gamma + 1)}. \tag{9.89}$$

This satisfies (9.70) [taking account of the exponent relations (9.52) and (9.53)], and reduces to the classical (9.71) when $\alpha = 0$, $\gamma = 1$, $\delta = 3$.

As a convenient and numerically accurate approximation in three dimensions we shall take $\eta = 0$, as in the Ornstein–Zernike theory. From (9.57) this implies $\nu = \frac{1}{2}\gamma$, and from (9.59) and (9.61) in $d = 3$ it implies $\delta = 5$ and $\mu = \gamma$. Combined with the additional numerically accurate approximation $\beta = \frac{1}{3}$, which we shall also adopt, it yields the full set of exponents

$$\begin{aligned} \alpha = 0, \qquad & \beta = \tfrac{1}{3}, \qquad \gamma = \tfrac{4}{3}, \qquad \delta = 5, \\ \nu = \tfrac{2}{3}, \qquad & \eta = 0, \qquad \mu = \tfrac{4}{3} \end{aligned} \tag{9.90}$$

which satisfies all the exponent relations of § 9.3 for $d = 3$ and is close enough for all practical purposes to the accurate values in Table 9.1. (An even more accurate self-consistent set of exponents for $d = 3$ with $\eta = 0$ has $\alpha = \frac{1}{8}$, $\beta = \frac{5}{16}$, $\gamma = \frac{5}{4}$, $\delta = 5$, $\nu = \frac{5}{8}$, and $\mu = \frac{5}{4}$, but these are more awkward to use in analytical theories.) From (9.89) and (9.90), the approximation to the universal $v(x)$ that we adopt[24–26] in our further calculations is

$$v(x) = \tfrac{3}{4}(1 - 3x^{-\frac{4}{3}} + 2x^{-2}) \equiv \tfrac{3}{4}(x^{-\frac{2}{3}} - 1)^2(2x^{-\frac{2}{3}} + 1). \tag{9.91}$$

We shall now estimate the universal constant K in the surface tension (9.82)–(9.84) and the universal function $X(y)$ in the profile (9.78), from (9.85) and (9.74), respectively, using the $v(x)$ of (9.91) and the exponents (9.90).

From (9.85) and (9.91), with $\beta = \frac{1}{3}$ and $\delta = 5$, we calculate

$$K = \frac{\sqrt{3}}{8} \ln\left(\frac{1 + \sqrt{3}}{\sqrt{2}}\right) = 0 \cdot 1426 \simeq \tfrac{1}{7}. \tag{9.92}$$

This is very close to the classical $K = \frac{1}{6} = 0 \cdot 1667$ [see (9.86)]. Thus, although the non-classical behaviour of the separate quantities σ, $\rho^l - \rho^g$, ξ, κ, and perhaps even m, is significantly different from the classical, the relations that connect σ to the others are given nearly correctly by the mean-field theory, the only modification required being a slight decrease in the coefficient of proportionality. Indeed, since the form of $v(x)$ in (9.91) is to some extent conjectural, it is not even certain that $K \simeq \frac{1}{7}$ is an improvement over $K = \frac{1}{6}$ (although it is confirmed in the renormalization-group theory; see § 9.7). Unfortunately, κ and ξ are not yet known precisely enough in the two-phase region near the critical point to test

(9.83) and (9.92) rigorously, although the results of the attempts that have been made are generally satisfactory.[27]

From (9.74) and (9.91), with $\beta = \frac{1}{3}$ and $\gamma = \frac{4}{3}$, we find

$$X(y) = \sqrt{2} \tanh(\tfrac{1}{2}y)[3\text{-}\tanh^2(\tfrac{1}{2}y)]^{-\frac{1}{2}} \tag{9.93}$$

as the functional form of the universal profile in (9.78). This was found by Webb and his co-workers[28] to fit their profile measurements perceptibly better than the classical $X(y) = \tanh(y/2)$.

We turn next to the theory of the interfaces, and particularly their tensions, near the tricritical point of three-phase equilibrium.

9.5 Tricritical points

The theoretical possibility of tricritical points—thermodynamic states in which three or more previously distinct phases become simultaneously identical—was known to van der Waals,[29] but it is only in recent times that such states have been recognized experimentally.[30,31] We shall here outline the mean-field theory of such points, due to Griffiths;[32] and then, with Griffiths's potential W as the essential ingredient, and within the framework of the van der Waals theory, we shall calculate the tensions of the interfaces as the tricritical point is approached through the three-phase region.

We call the three phases α, β, and γ, as in Chapter 8. There are states in which two of them, say α and β, are at a critical point of their own phase equilibrium while they are still in equilibrium with the distinct third phase γ. Such a state is called a critical endpoint—an $\alpha\beta$ critical endpoint in that example. The tricritical point is that state in which the $\alpha\beta$ and $\beta\gamma$ critical endpoints coincide. That is shown schematically in Fig. 9.4(a) in the plane of two independent fields. The choice of the two fields is arbitrary; in the figure they are the temperature T and the chemical potential μ_1 of one of the components, just for illustration. The three-phase states are in the region that is shown shaded. That region is bounded by the loci of $\alpha\beta$ and $\beta\gamma$ critical endpoints. The tricritical point is the point of confluence of those two loci. By varying μ_1 at fixed T, say, we may traverse the three-phase region from an $\alpha\beta$ critical endpoint to a $\beta\gamma$ critical endpoint.

Figure 9.4(a) shows the system to have at least two thermodynamic degrees of freedom in the three-phase region. If it has more than two, say f, then the additional $f-2$ fields are being held fixed in the figure. For a system of $p = 3$ phases to have $f \geq 2$ degrees of freedom requires that it consist of $c \geq 3$ independent components, since $c = f + p - 2$ by the phase rule. The condition $c \geq 3$ for a tricritical point was known to van der Waals.[29]

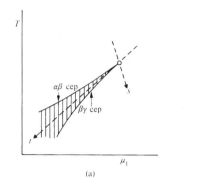
               (a)   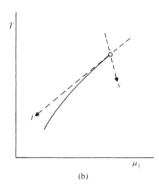
                                                        (b)

FIG. 9.4

When there are only three components, which is the minimum number for a tricritical point to be realizable, the system is invariant at that point, so the temperature, pressure, and chemical composition are all uniquely determined there. That is the case, for example, in the acetic acid–water–n-butane system, which was one of the first mixtures in which a tricritical point was found (Krichevskii et al.[31]). We may think of the $\alpha\beta$ critical endpoints there as liquid–gas critical points, and of the $\beta\gamma$ critical endpoints as liquid–liquid consolute points; but we should keep in mind that in such a system, near the $\alpha\beta$ critical endpoints generally and near the tricritical point particularly, there is no fundamental distinction between liquid and gas.

If there were an additional component, $c = 4$, we could have a tricritical endpoint, at which three phases were at their tricritical point while in equilibrium with a fourth, distinct phase. In this way we could have a liquid–liquid–liquid tricritical point, in which the tricritical fluid was an ordinary, dense liquid in equilibrium with its vapour. The first reported tricritical point, in the system benzene–ethanol–water–ammonium sulphate, was of this kind (Radyshevskaya et al.[30]). Here, the $\alpha\beta$ critical endpoints arise from, and may be likened to, the critical solution point in the ternary system benzene–ethanol–water, in which the ethanol plays the role of an amphiphile that increases the mutual solubility of benzene and water; while the $\beta\gamma$ critical endpoints arise from the critical solution point that is associated with the salting-out of ethanol from ethanol–water solutions by a salt such as ammonium sulphate, where it is the water that is the common solvent for the otherwise mutually insoluble alcohol and salt.[33] But the closer the system is to its tricritical point the less distinction is there between the $\alpha\beta$ and $\beta\gamma$ critical endpoints, and the less does either one, in the four-component mixture, continue to resemble the critical solution point in the three-component mixture from which it arose.

Such four-component liquid mixtures are closely analogous[34] to the hydrocarbon–surfactant-brine mixtures in which the middle phase is often described as a microemulsion and in which ultra-low interfacial tensions are found.[35] If in the systems we treat here we traverse the three-phase region from an $\alpha\beta$ to a $\beta\gamma$ critical endpoint (by varying the composition at fixed temperature, for example), the three tensions $\sigma^{\alpha\beta}$, $\sigma^{\beta\gamma}$, and $\sigma^{\alpha\gamma}$ are all low if we are near the tricritical point, while $\sigma^{\alpha\beta}$ and $\sigma^{\beta\gamma}$, respectively, vanish at the two limits.

Returning to Fig. 9.4(a), we define two fields, $t(>0)$ and s, as linear combinations of the deviations of the two laboratory field variables (T and μ_1 in the figure) from their values at the tricritical point. We take the t-axis in the direction of the common tangent to the two loci of critical endpoints at the tricritical point, and the s-axis in any other direction. Since $t = 0$ is the equation of the s-axis, t is an almost arbitrary linear combination of the deviations of the two laboratory fields from tricritical. In practice it is most often taken to be just $T^t - T$, the deviation of the temperature from its value T^t at the tricritical point, and we may so think of it. Since $s = 0$ is the equation of the t-axis, s is unique to within a scale factor.

Figure 9.4(a) may be compared with the simpler and more familiar Fig. 9.4(b) that describes two-phase equilibrium and an ordinary critical point. Again, T and μ_1 have been chosen arbitrarily to be the two field variables for this representation, while the remaining $c - 1$ independent fields, say $\mu_2, \ldots, \mu_c$, are held fixed. Now $f = 2$ in the one-phase region, $f = 1$ in the two-phase region (i.e. on the equilibrium curve, which is that shown), and $f = 0$ at the critical point. Here, too, we introduce new variables $t(>0)$ and s, the former almost arbitrary and usually chosen to be $T^c - T$, the latter unique. Now $s = 0$ in the two-phase region near the critical point, and only t varies; whereas in the three-phase equilibrium depicted in Fig. 9.4(a) both s and t are variable.

The mean-field $-W$ appropriate to the two-phase equilibrium in Fig. 9.4(b) near the critical point is that in (9.7), which, with (9.4), with $x = \rho - \rho^c$, $x^\alpha = \rho^g - \rho^c$, and $x^\beta = \rho^l - \rho^c$, and with $x^\alpha = -x^\beta = -\sqrt{t}/a$ as in (9.17), may be written

$$-W(x) = \tfrac{1}{4}B(x - x^\alpha)^2(x^\beta - x)^2 = \tfrac{1}{4}B(x^2 - t/a^2)^2. \qquad (9.94)$$

This is the one-density $-W(x)$ associated with the two-density $-W(x, y)$ of (9.17), as we saw there. It is shown in Fig. 9.5(a) (cf. Fig. 3.2).

Now in the three-phase system, if x is some density or composition variable that vanishes at the tricritical point, the mean-field $-W(x)$ appropriate to the phase equilibrium (Fig. 9.4(a)) is the extension of

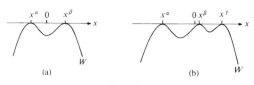

FIG. 9.5

(9.94), due to Griffiths,[32]

$$-W(x) = \tfrac{1}{6}B(x - x^\alpha)^2(x - x^\beta)^2(x - x^\gamma)^2 \qquad (9.95)$$

where x^α, x^β, and x^γ, the respective values of the composition variable x in the three bulk phases, are the roots of the cubic equation

$$x^3 - 3tx + 2s = 0 \qquad (9.96)$$

with t and s as in the figure. [We have now fixed the otherwise arbitrary scale factors in t and s—the counterparts of the a^2 in (9.94)—by having chosen the coefficients in (9.96) as we did.] This $-W(x)$ is as pictured in Fig. 9.5(b) (cf. Fig. 8.5).

From (9.96), the roots x^α, x^β, and x^γ must be such that

$$x^\alpha + x^\beta + x^\gamma = 0, \qquad x^\alpha x^\beta + x^\beta x^\gamma + x^\alpha x^\gamma = -3t,$$
$$x^\alpha x^\beta x^\gamma = -2s. \qquad (9.97)$$

By the convention in Fig. 9.5(b) we have $x^\alpha \leq x^\beta \leq x^\gamma$. Then from (9.97) we see that x^α is always negative, x^γ is always positive, and x^β has the sign of s. They are given explicitly as functions of the fields t and s by

$$x^\alpha = -\sqrt{t}(\sqrt{3}\cos\phi + \sin\phi), \qquad x^\beta = 2\sqrt{t}\sin\phi,$$
$$x^\gamma = \sqrt{t}(\sqrt{3}\cos\phi - \sin\phi) \qquad (9.98)$$

where the angle ϕ is

$$\phi = \tfrac{1}{3}\arcsin(s/t^{\frac{3}{2}}) \qquad \left(-\frac{\pi}{6} \leq \phi \leq \frac{\pi}{6}\right). \qquad (9.99)$$

The critical endpoints are at

$$x^\alpha = x^\beta = -\tfrac{1}{2}x^\gamma = -\sqrt{t}, \qquad s = -t^{\frac{3}{2}} \quad (\alpha\beta \text{ cep})$$
$$x^\beta = x^\gamma = -\tfrac{1}{2}x^\alpha = \sqrt{t}, \qquad s = t^{\frac{3}{2}} \quad (\beta\gamma \text{ cep}) \qquad (9.100)$$

and the tricritical point is at $x^\alpha = x^\beta = x^\gamma = t^{\frac{1}{2}} = s^{\frac{1}{3}} = 0$.

Since $c \geq 3$ for a tricritical point, the full density space would be at least four-dimensional. The $W(x)$ in (9.95) is the one-density potential to which a hypothetical four-density $W(x_1, x_2, x_3, x_4)$ would reduce (with

$x \equiv x_1$, say) if x_2, x_3, and x_4 were taken to be related to x_1 by $\partial W/\partial x_2 = \partial W/\partial x_3 = \partial W/\partial x_4 = 0$. Just as in the representation of the two-phase equilibrium in the two-dimensional space of Fig. 9.1, where $y \sim x^2$ at coexistence, so also here, in the three-phase equilibrium represented in a four-dimensional space, the additional densities x_n could be chosen so as to form with x_1 a hierarchy in which, at coexistence, $x_n \sim x_1^n$. Let us suppose the x_n to have been so chosen. With $x_1 \sim \sqrt{t}$, the density x_n in any of the coexistent phases is then typically of magnitude $t^{n/2}$; so as $t \to 0$ on approach to the tricritical point the region of three-phase coexistence in the four-dimensional space shrinks non-uniformly: proportionally to t^2 in one direction, to $t^{\frac{3}{2}}$ in a second direction, etc.

It shrinks most slowly—proportionally to $t^{\frac{1}{2}}$—in the x_1 direction. Thus, near the tricritical point, the three points that represent the three coexisting phases in the four-dimensional density or composition space come asymptotically to lie on a line: that of the x_1-axis itself ($x_2 = x_3 = x_4 = 0$). There is then always a distinguished phase, the β phase, which in its chemical composition and all other properties is intermediate between the α and γ phases. (In particular, it is intermediate in its density; hence, in the field of gravity, it is the middle phase.) Even in the full four-dimensional space, then, there can be no trajectory from α to γ other than via β, and the three interfacial tensions necessarily satisfy Antonow's rule, (8.8). At the same time and by the same critical-point scaling that reduced the two-density theory of (9.18), (9.19), and (9.23) to the one-density theory of (9.20) and (9.24), the four-density theory in the present case reduces to the associated one-density theory.[4] That alone could have told us that the tensions would be related by (8.8), for we saw in § 8.3 that only in a two- (or more-) density theory could (8.8) be violated. Thus, we may conclude from critical-point scaling alone that when a three-phase system is close to its tricritical point the β phase perfectly wets (spreads at) the $\alpha\gamma$ interface, and that the latter then consists, in its equilibrium structure, of a macroscopically thick layer of β phase, as in Figs 8.3 and 8.5.

We may thus calculate the composition profiles and tensions of the several interfaces by the prescriptions of the one-density van der Waals theory, with $-W$ as in (9.95). We take the van der Waals, Ornstein–Zernike parameter m to be independent of x, and then find any of the interfacial profiles from (3.9); here,

$$z = \pm (\tfrac{1}{2}m)^{\frac{1}{2}} \int_{x_0}^{x} [-W(x')]^{-\frac{1}{2}} \, dx' \qquad (9.101)$$

with z having been chosen to be 0 at some $x = x_0$. The resulting profiles[36] are qualitatively like those in Fig. 8.5, where the part of the curve between $x = x^\alpha$ and $x = x^\beta$ is the profile of the $\alpha\beta$ interface, that between x^β and x^γ is the $\beta\gamma$ profile, and the whole curve, including the step of infinite (that is, macroscopic) height at $x = x^\beta$, is the $\alpha\gamma$ profile. Near a critical endpoint the composition profile of the near-critical interface is the classical hyperbolic tangent, as in (9.10).

The tensions, from (3.13), are

$$\sigma^{\alpha\beta} = (2m)^{\frac{1}{2}} \int_{x^\alpha}^{x^\beta} [-W(x)]^{\frac{1}{2}} dx, \qquad \sigma^{\beta\gamma} = (2m)^{\frac{1}{2}} \int_{x^\beta}^{x^\gamma} [-W(x)]^{\frac{1}{2}} dx,$$

(9.102)

with $\sigma^{\alpha\gamma}$ given in terms of these by (8.8). With the $-W(x)$ of (9.95), and with account taken of the first of (9.97), these become

$$\sigma^{\alpha\beta} = \frac{1}{4} \left(\frac{mB}{3}\right)^{\frac{1}{2}} (x^\beta - x^\alpha)^3 x^\gamma$$

(9.103)

$$\sigma^{\beta\gamma} = \frac{1}{4} \left(\frac{mB}{3}\right)^{\frac{1}{2}} (x^\gamma - x^\beta)^3 (-x^\alpha).$$

(Recall that $x^\gamma > 0$, $x^\alpha < 0$, and $x^\alpha < x^\beta < x^\gamma$.)

From (9.98), (9.99), and (9.103), both $\sigma^{\alpha\beta}$ and $\sigma^{\beta\gamma}$ (hence also $\sigma^{\alpha\gamma}$, which is their sum) are homogeneous of degree $\frac{4}{3}$ in the fields $t^{\frac{3}{2}}$ and s; that is, they are of the form

$$\sigma^{\alpha\beta} = t^2 f^{\alpha\beta}(s/t^{\frac{3}{2}}), \quad \text{etc.,}$$

(9.104)

with functions $f^{\alpha\beta}$, etc., that are known explicitly. The function $f^{\alpha\beta}(y)$ is positive and finite for all y in the interval $-1 < y \leq 1$, while, as $y \to -1$,

$$f^{\alpha\beta}(y) \sim \frac{8}{9}(2mB)^{\frac{1}{2}}(1+y)^{\frac{3}{2}} \quad (y \to -1).$$

(9.105)

The function $f^{\beta\gamma}(y)$ behaves analogously as $y \to +1$; and $f^{\alpha\gamma}(y) = f^{\alpha\beta}(y) + f^{\beta\gamma}(y)$.

As any one of the $\alpha\beta$ critical endpoints is approached through the three-phase region, $\sigma^{\alpha\beta}$, according to (9.104) and (9.105), vanishes proportionally to the $\frac{3}{2}$ power of the distance from that critical endpoint in the space of the thermodynamic fields. That is just the reappearance of the classical critical-point surface-tension exponent $\mu = \frac{3}{2}$, arising here because a critical endpoint is an ordinary critical point of two-phase coexistence, the presence of the additional phase (the γ phase in this instance) being irrelevant to the critical behaviour. But if, instead, the tricritical point is approached, and the approach is along a path through the three-phase region on which $s/t^{\frac{3}{2}}$ takes on some limiting value between -1 and 1 (the path $s = 0$, for example), then, according to (9.104), the tensions vanish proportionally to t^2; so the classical tricritical-point surface-tension exponent is $\mu = 2$, in contrast to the $\mu = \frac{3}{2}$ that is associated with an ordinary, two-phase critical point. The respective values $\mu = \frac{3}{2}$ and $\mu = 2$ at a critical endpoint and at the tricritical point are direct reflections of the structure of (9.103): at an $\alpha\beta$ critical endpoint $x^\beta - x^\alpha$ vanishes with the classical exponent $\beta = \frac{1}{2}$, while x^γ approaches a non-zero limit, so $\sigma^{\alpha\beta}$ vanishes with exponent $\frac{3}{2}$; while at the tricritical point each of x^α, x^β, and x^γ vanishes proportionally to $\sqrt{t}$ (thus, again with the classical exponent $\beta = \frac{1}{2}$), so $\sigma^{\alpha\beta} \sim t^2$.

This dependence of the rate of vanishing of the surface tension on the character of the critical point manifests itself in an interesting 'cross-over' phenomenon that occurs when a critical endpoint that lies very near the tricritical point is approached. Suppose, for example, that the $\alpha\beta$ critical endpoint at $t^* = (-s^*)^{\frac{2}{3}}$ is approached along a path of constant, very slightly negative $s = s^*$ (Fig. 9.4(a)). Then over most of the way t is much greater than $(-s)^{\frac{2}{3}}$, and so from (9.104) the tension $\sigma^{\alpha\beta}$ appears to be vanishing with decreasing t as $f^{\alpha\beta}(0)t^2$, as is characteristic of approach to a tricritical point; but when t becomes so small as to be comparable with the very small $t^* = (-s^*)^{\frac{2}{3}}$—three or four times it, say—deviations from tricritical behaviour begin to be noticeable; and, finally, when t is so close to t^* that $0 < t/t^* - 1 \ll 1$, then, from (9.104) and (9.105), $\sigma^{\alpha\beta} \sim \frac{4}{3}(3mBt^*)^{\frac{1}{2}}(t - t^*)^{\frac{3}{2}}$, the tension thus vanishing as at an ordinary critical point, with $\mu = \frac{3}{2}$.

Both values of the exponent μ, namely $\frac{3}{2}$ at a critical endpoint and 2 at a tricritical point, are the classical, mean-field-theory values. We know that for ordinary critical points, among which are critical endpoints, the classical $\mu = \frac{3}{2}$ is wrong, the true value being close to $\mu = 1\cdot26$. We find classical behaviour here because (9.95) with (9.96) describes the critical endpoints, as well as the tricritical point, classically. But we saw at the end of § 9.3 that $d = 3$ is the borderline dimensionality for tricritical points, so that in $d = 3$ the mean-field tricritical exponents are expected to be correct (although the behaviour would not be fully classical because of the additional logarithmic factors expected at the borderline d). Thus, the phenomenon we illustrated above would in reality be a crossover between $\mu = 2$ and $1\cdot26$, not between 2 and $\frac{3}{2}$. But until we can correctly describe the crossover between non-classical critical-endpoint and classical tricritical-point behaviour of the thermodynamic properties of the bulk phases (perhaps along the lines suggested by Fox[37]), we shall not be able to give a more detailed account of it in the interfacial tensions.

That the classical tricritical-point exponent μ is 2 can be understood in the following way. As in § 9.4, the reciprocal of the compressibility (here, the osmotic compressibility) κ^{-1} in any of the equilibrium phases is proportional to $-W''(x)$ evaluated in that phase. Thus, from (9.95), the osmotic compressibilities of the several phases are proportional to (or, if appropriately defined, may be taken to be)

$$\kappa^\alpha = 3/B(x^\beta - x^\alpha)^2(x^\gamma - x^\alpha)^2, \quad \text{etc.} \tag{9.106}$$

Since x^α, x^β, and x^γ, according to (9.98), all vanish proportionally to $\sqrt{t}$ when the tricritical point is approached along any path through the three-phase region on which $s/t^{\frac{3}{2}}$ takes on some limiting value between -1 and 1 (the path $s = 0$, for example), the classical tricritical-point exponent β is $\frac{1}{2}$ (as remarked earlier), the same as at an ordinary critical point; and, by (9.106), the osmotic compressibilities diverge proportionally to t^{-2}; that is, the classical tricritical-point exponent γ is 2. With the van der Waals, Ornstein–Zernike assumption of a finite limiting m, we have $\nu = \frac{1}{2}\gamma$ from (9.32) [and so also, as we know, $\eta = 0$, by (9.56) or (9.57)]; so that the classical tricritical-point exponent ν is 1. Thus, $\beta = \frac{1}{2}$, $\gamma = 2$, and $\nu = 1$; so $\mu = 2$, from (9.62). Alternatively, since the d-dependent exponent relations still hold at the borderline d (although that is the highest d in which they do), we find $\mu = 2$ from (9.61) with $d = 3$ and $\nu = 1$. Note that the relations among exponents, even if not their values, are the same at higher-order critical points as at ordinary critical points.

The theoretical prediction $\mu = 2$ has never been tested by experiment in multicomponent fluids of the kind we have been considering, but it has been experimentally confirmed in a different system that is supposed, theoretically, to be equivalent to such a mixture. It was remarked by Griffiths[38] that the consolute point that is found at $0 \cdot 9$ K in mixtures of ^{3}He and ^{4}He is thermodynamically equivalent to a tricritical point because it is also the terminus of the line of λ-points associated with the superfluid transition. There are special symmetries in this system that make a tricritical point attainable with only two components in the mixture. The tension of the interface between the two liquid phases (of which the one richer in ^{4}He is superfluid) as this tricritical point is approached from lower temperatures through the two-liquid-phase region has been measured by Leiderer *et al.*,[39] who find $\mu = 2 \cdot 0 \pm 0 \cdot 1$. (It was in the context of this tricritical point that the theoretical prediction $\mu = 2$ was originally made, by Papoular,[40] by a critical-point scaling argument like that outlined above. It was only later[41] that the same result was obtained, by similar arguments, for the more conventional tricritical point in ordinary fluid mixtures of three or more components.) We may take comfort in the experimental verification of the theoretical prediction at the ^{3}He–^{4}He tricritical point, but we must not consider it a substitute for a direct experimental test in ordinary fluid mixtures.

Ramos Gómez[42] has pointed out an interesting connection between the tensions and the osmotic compressibilities. From (9.103), the several relations (9.106), and the first of (9.97), we find

$$\frac{\sigma^{\alpha\beta}}{\sigma^{\beta\gamma}} = \frac{1/\kappa^{\alpha} - 1/\kappa^{\beta}}{1/\kappa^{\gamma} - 1/\kappa^{\beta}}.\tag{9.107}$$

The compressibilities are experimentally accessible by light scattering[43] or by the direct thermodynamic measurement of the rate of change of chemical potential or osmotic pressure with composition. Thus, the two sides of (9.107) are obtained from two different kinds of measurements. Experimental verification of (9.107) would be a significant confirmation of this whole body of theory.

At a fixed distance from the tricritical point measured by t ($= T^t - T$, say) there is still a one-parameter infinity of possible three-phase mixtures (Fig. 9.4(a)), each with a different value of s in the range $-t^{\frac{3}{2}} \leqslant s \leqslant t^{\frac{3}{2}}$ or of ϕ in the range $-\pi/6 \leqslant \phi \leqslant \pi/6$. In practice[30,31,44] one varies s (or, equivalently, ϕ) at fixed t, and sweeps through the three-phase region from an $\alpha\beta$ critical endpoint to a $\beta\gamma$ critical endpoint, by varying the composition (or, when $c = 3$, sometimes just the pressure) of the three-phase mixture at fixed temperature. The three tensions, from (9.98), (9.99), and (9.103), with $\sigma^{\alpha\gamma} = \sigma^{\alpha\beta} + \sigma^{\beta\gamma}$, then vary as in Fig. 9.6. The qualitative features of Fig. 9.6, including the convexity toward the origin of the plot of $\sigma^{\beta\gamma}$ vs $\sigma^{\alpha\beta}$ and the minimum in the $\sigma^{\alpha\gamma}$ curve, are verified by experiment,[45] as is the quantitative relation $\sigma^{\alpha\gamma} = \sigma^{\alpha\beta} + \sigma^{\beta\gamma}$. The tangencies seen in Fig. 9.6 at the critical endpoints are all $\frac{3}{2}$-power tangencies, reflecting the classical $\mu = \frac{3}{2}$, but would in reality be given by the power 1.26.

We may inquire more generally how the tensions vary in the neighbourhood of a critical endpoint, both in the three- and in the two-phase region. (In Fig. 9.6 we see only the three-phase region.) We might, for

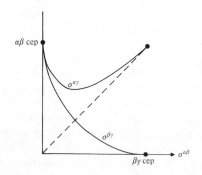

FIG. 9.6. Interfacial tensions $\sigma^{\alpha\gamma}$ and $\sigma^{\beta\gamma}$ as functions of $\sigma^{\alpha\beta}$ at a fixed temperature in the three-phase region between the $\alpha\beta$ and $\beta\gamma$ critical endpoints; from (9.98), (9.99), and (9.103). The tensions are such that $\sigma^{\alpha\gamma} = \sigma^{\alpha\beta} + \sigma^{\beta\gamma}$.

example, consider the liquid–vapour surface tensions of the one or two liquid phases that are present above or below the consolute point of the liquids. The liquid–liquid interfacial tension vanishes with critical-point exponent $\mu = 1\cdot26$ at that point. The liquid–vapour tensions do not vanish, but, it will appear, reflect the critical point in more subtle ways. That is the subject of the next section.

9.6 Non-critical interface near a critical endpoint

We had occasion in § 9.5 (in Fig. 9.6, for example) to consider how interfacial tensions vary as a critical endpoint is approached, but we have not yet reflected on how special must be the interface between the non-critical and critical phases at such a point. At a $\beta\gamma$ critical endpoint, for example, we have a critical phase $\beta\gamma$ in which the coherence length ξ of composition fluctuations is infinite (i.e. macroscopic), in equilibrium with and distinct from a non-critical phase α in which ξ is finite. Since ξ is also a measure of how far from the interface we must penetrate into a phase before the chemical composition is sensibly that of the bulk, the interface separating the critical and non-critical phases at a critical endpoint must then have a very asymmetric composition profile, approaching the bulk composition infinitely slowly on the critical side and very rapidly—over microscopic distances of just a few ångstroms—on the non-critical side.[46,47]

This is shown schematically in Fig. 9.7, where we see a density or composition variable x varying with distance z through the interface, between the non-critical α phase and the critical $\beta\gamma$ phase. Our calculations later will show that x approaches the bulk x^α exponentially rapidly, with a microscopic decay length ξ; while it approaches the bulk $x^{\beta\gamma}$ very slowly, roughly as $1/\sqrt{z}$, as indicated in the figure. The adsorption of the

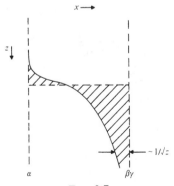

Fig. 9.7

quantity whose density is x is then infinite for any choice of dividing surface. The divergence of the adsorption as the critical endpoint is approached is reflected, we shall see, in singular behaviour of the tension of the interface.

We may, as suggested at the end of § 9.5, think of a binary liquid mixture in equilibrium with its vapour near its critical solution point as a system prototype. Because of the presence of the vapour, phase α, that critical solution point, at which the distinct liquid phases β and γ become the single liquid phase $\beta\gamma$, is a $\beta\gamma$ critical endpoint. The tension $\sigma^{\beta\gamma}$ of the liquid–liquid interface vanishes at that point, and does so, we know, proportionally to $|T^c - T|^\mu$, with $\mu = \frac{3}{2}$ in mean-field approximation or with $\mu = 1 \cdot 26$ in reality. That will prove to play an important role in this story, but it is not otherwise the object of our present inquiry. We are concerned now, instead, with the non-critical interfaces—the liquid–vapour interfaces in this example—and their tensions, $\sigma^{\alpha,\beta\gamma}$ when there is only the one liquid phase, or $\sigma^{\alpha\beta}$ and $\sigma^{\alpha\gamma}$ when there are two. These do not vanish at the $\beta\gamma$ critical endpoint for α and $\beta\gamma$ remain distinct phases there. Rather, they approach some common limiting value $\sigma^c > 0$. The infinite adsorption we remarked upon above then reflects itself in critical-point singularities, with characteristic critical-point exponents, in the temperature- and composition-dependence of $\sigma^{\alpha,\beta\gamma} - \sigma^c$, or in the temperature-dependence of $\sigma^{\alpha\beta} - \sigma^c$ and of $\sigma^{\alpha\gamma} - \sigma^c$. It is these that we wish now to study.

The analysis here will go beyond the treatment of critical endpoints in § 9.5 in three ways: we shall be interested in the structure (Fig. 9.7) as well as in the tension; we shall (as already mentioned at the end of § 9.5) be interested in the behaviour of the tensions in the two-phase region near the critical endpoint as well as in their behaviour in the three-phase region; and we shall recognize that, a critical endpoint being an ordinary critical point of two-phase equilibrium, the exponents that characterize it are not given correctly by the mean-field approximation but have non-classical values.

Let us represent by ω the deviation of some density or composition variable (like the earlier ρ or x) from its value in the critical phase at the $\beta\gamma$ critical endpoint. Had no α phase been present, the free-energy density $-W(\omega)$ associated with the critical $\beta\gamma$ phase, in which $\omega = 0$, would by (9.88) have been proportional to $|\omega|^{\delta+1}$. With the α phase present $-W(\omega)$ would still vanish proportionally to $|\omega|^{\delta+1}$ near $\omega = 0$, but would also have a factor $(\omega - \omega^{\alpha,c})^2$ associated with the coexisting α phase in which the value of ω at the $\beta\gamma$ critical endpoint is some $\omega^{\alpha,c}$. Such a $-W(\omega)$, shown in Fig. 9.8(a), may now be taken to be

$$-W(\omega) = W_0(\omega - \omega^{\alpha,c})^2 \omega^6 \qquad (9.108)$$

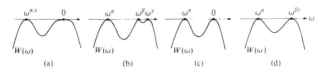

FIG. 9.8

with a positive coefficient W_0. We have here explicitly taken the dimensionality to be 3 and made the Ornstein–Zernike approximation $\eta = 0$, which we know (§ 9.4) together entail $\delta = 5$.

By (3.9) and (9.108) the composition profile of the α, $\beta\gamma$ interface at the $\beta\gamma$ critical endpoint is

$$\omega - \omega^{\alpha,c} \sim \text{const. } e^{z/\xi} \qquad (z \to -\infty) \qquad (9.109)$$

deep in the α phase, where $\omega \sim \omega^{\alpha,c}$; and

$$\omega \sim \omega^{\alpha,c}(\xi/2z)^{\frac{1}{2}} \qquad (z \to \infty) \qquad (9.110)$$

deep in the critical $\beta\gamma$ phase, where $\omega \sim 0$. [We have arbitrarily chosen $\omega^{\alpha,c} < 0$, as in Fig. 9.8(a), and have chosen the sign in (3.9) so that z increases in the direction of increasing ω.] In both these formulae ξ is the correlation length in the bulk α phase,

$$\xi = [m/2(-\omega^{\alpha,c})^6 W_0]^{\frac{1}{2}} = [m/-W''(\omega^{\alpha,c})]^{\frac{1}{2}}. \qquad (9.111)$$

In the α phase, as we anticipated, and as we now see in (9.109), the decay is exponential; while in the critical $\beta\gamma$ phase, according to (9.110), it is much slower, being proportional to $1/\sqrt{z}$, as indicated in Fig. 9.7.

By (3.13) (with m assumed constant) and (9.108), the tension of that α, $\beta\gamma$ interface at the $\beta\gamma$ critical endpoint, which we called σ^c before, is

$$\sigma^c = \tfrac{1}{20}(2mW_0)^{\frac{1}{2}}(-\omega^{\alpha,c})^5. \qquad (9.112)$$

We wish now to see how the tensions of the several non-critical interfaces, that is, $\sigma^{\alpha,\beta\gamma}$ in the two-phase region or $\sigma^{\alpha\beta}$ and $\sigma^{\alpha\gamma}$ in the three-phase region, approach σ^c as the critical endpoint is approached by varying the temperature or composition.[25,47] We require a generalization of (9.108) appropriate to states that are near but not at the $\beta\gamma$ critical endpoint. We may obtain it by simply appending a factor $(\omega - \omega^\alpha)^2$ to the $-W(\omega)$ that would otherwise describe the one ($\beta\gamma$) or two (β and γ) phases of a fluid near its critical point when no additional phase α is present. A possible form of such a $-W(\omega)$ below the critical point, where there are distinct β and γ phases, is that in (9.88); so with a coexisting α phase also present, we may take[25]

$$-W(\omega) = W_0(\omega - \omega^\alpha)^2\{b_2[\omega^2 - (\omega^{\beta,\gamma})^2] + \omega^6 - (\omega^{\beta,\gamma})^6\}$$

$$\text{(3-phase)} \qquad (9.113)$$

as the required generalization of (9.108). Here, $b_2 < 0$ is a thermodynamic field (the subscript 2 distinguishes it from a different field b_1 to be introduced later), like $-\frac{1}{2}(\delta + 1)|\rho^{g,1} - \rho^c|^{\delta - 1}$, with $\delta = 5$, in (9.88); and $\omega^{\beta,\gamma}$ means either ω^β or ω^γ,

$$-\omega^\beta = \omega^\gamma = (-b_2/3)^{\frac{1}{4}}. \tag{9.114}$$

The conditions (9.114) ensure $W'(\omega^\beta) = W'(\omega^\gamma) = 0$. The $-W(\omega)$ of (9.113), with (9.114), is shown schematically in Fig. 9.8(b). When β and γ are two liquid phases in equilibrium with their vapour α at a distance $t \sim (T^c - T) > 0$ below an upper critical solution temperature T^c (or a distance $T - T^c$ above a lower critical solution temperature T^c), we have from (9.114), from the definition of the critical-point exponent β, and from the exponent relation (9.52) [the exponent $\frac{1}{4}$ in (9.114) is understood from the remark above (9.114) more generally to be $1/(\delta - 1)$],

$$b_2 \sim -B_2 t |t|^{\gamma - 1} \tag{9.115}$$

with γ the compressibility (or osmotic compressibility) exponent in conventional notation, and with $B_2 > 0$ a constant parameter.

Equations (9.113)–(9.115) with $t > 0$ and $b_2 < 0$ are appropriate to the three-phase region. In the two-phase region, where the non-critical α phase is in equilibrium with a single near-critical phase $\beta\gamma$, there is another degree of freedom in addition to the field b_2; but if for now we restrict consideration to states in which $\omega^{\beta\gamma} = 0$ then[25] the appropriate $-W(\omega)$ is still of the form (9.113), but with $\omega^{\beta,\gamma}$ replaced by $\omega^{\beta\gamma} = 0$, and with b_2 still related to t by (9.115), but with $t < 0$ and $b_2 > 0$. Thus,

$$-W(\omega) = W_0(\omega - \omega^\alpha)^2(b_2\omega^2 + \omega^6) \quad \text{(2-phase, } \omega^{\beta\gamma} = 0). \tag{9.116}$$

This $-W(\omega)$ is shown schematically in Fig. 9.8(c). In our example of a binary liquid mixture near its consolute point, in equilibrium with vapour, the condition $\omega^{\beta\gamma} = 0$ is the condition that the composition of the single-phase liquid be the same as at the consolute point, say $x = x^c$. This is shown in Fig. 9.9 as the locus of states marked (i), intersecting the temperature–composition coexistence curve at the consolute point.

The locus marked (ii) in Fig. 9.9 is also in that range of two-phase states in which a homogeneous liquid, $\beta\gamma$, is in equilibrium with its vapour, α. This locus is the critical isotherm $b_2 = 0$; that is, $t = 0$ or $T = T^c$. Here, too, there is only a single $\omega^{\beta\gamma}$; it varies along the isotherm and is zero only at the critical point. In these states the appropriate form of $-W(\omega)$ with $\delta = 5$ is[25]

$$-W(\omega) = W_0(\omega - \omega^\alpha)^2[b_1(\omega - \omega^{\beta\gamma}) + \omega^6 - (\omega^{\beta\gamma})^6] \tag{9.117}$$

where the thermodynamic field b_1, distinct from b_2 (which is now 0), is

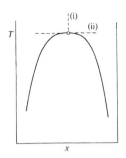

FIG. 9.9

related to the composition $\omega^{\beta\gamma}$ by

$$b_1 = -6(\omega^{\beta\gamma})^5. \tag{9.118}$$

The condition (9.118) ensures $W'(\omega^{\beta\gamma}) = 0$. The $-W(\omega)$ of (9.117) with (9.118), is shown schematically in Fig. 9.8(d).

Our object now is to calculate the several interfacial tensions for states in the three-phase region (those within the coexistence curve in Fig. 9.9), from (9.113) and (9.114); for states in the two-phase region on the locus (i), from (9.116); and for states in the two-phase region on the locus (ii), from (9.117) and (9.118). In § 9.5 (Fig. 9.6) we did only the first of these calculations and then only in mean-field approximation. Here we have been treating the critical endpoint as non-classical with $\delta = 5$.

By (3.13) with m constant, and from (9.112)–(9.114), we calculate that in the three-phase region

$$\sigma^{\beta\gamma} = 20 \ln\left(\frac{1+\sqrt{3}}{\sqrt{2}}\right)\left[\frac{b_2}{(\omega^{\alpha,c})^5}\right](-\omega^\alpha)\sigma^c \tag{9.119}$$

$$\sigma^{\alpha\beta} = 20(-\omega^{\alpha,c})^{-5}\left\{\tfrac{3}{2}\omega^\alpha(\omega^\beta)^4 \ln\left[\frac{-(\sqrt{3}-1)\omega^\beta}{\omega^\alpha + [(\omega^\alpha)^2 + 2(\omega^\beta)^2]^{\frac{1}{2}}}\right]\right.$$

$$-\frac{6\sqrt{3}}{5}(-\omega^\beta)^5 - [(\omega^\alpha)^2 + 2(\omega^\beta)^2]^{\frac{1}{2}}[\tfrac{1}{20}(\omega^\alpha\omega^\beta)^2$$

$$\left. -\tfrac{6}{5}(\omega^\beta)^4 - \tfrac{1}{20}(\omega^\alpha)^4]\right\}\sigma^c \tag{9.120}$$

$$\sigma^{\alpha\gamma} = \sigma^{\alpha\beta} + \sigma^{\beta\gamma} \tag{9.121}$$

with ω^β in (9.120) related to b_2, and thence to t, by (9.114) and (9.115). On the locus (i) we have only the single tension $\sigma^{\alpha,\beta\gamma}$, which we calculate from (3.13) and (9.116) to be

$$\sigma^{\alpha,\beta\gamma} = 20\left(\frac{\omega^\alpha}{\omega^{\alpha,c}}\right)^5 \sigma^c \int_0^1 x(1-x)[x^4 + b_2/(\omega^\alpha)^4]^{\frac{1}{2}}\,dx. \tag{9.122}$$

It is expressible in elliptic integrals or is readily evaluated numerically. On the critical isotherm, locus (ii), we also have only the single $\sigma^{\alpha,\beta\gamma}$, which from (3.13) and (9.117) with (9.118) we calculate to be[25]

$$\sigma^{\alpha,\beta\gamma} = [1 + j^{\mp}\omega^{\beta\gamma}|\omega^{\beta\gamma}|^3/(\omega^{\alpha,c})^4 + \ldots]\sigma^c \qquad (9.123)$$

to order $(\omega^{\beta\gamma})^4$; where the coefficient $j^{\mp}$ is one constant j^- when $\omega^{\beta\gamma} > 0$ and a different constant j^+ when $\omega^{\beta\gamma} < 0$, these constants being given by

$$j^{\mp} = 5 + 20\int_{\mp\infty}^{1} |x|^3\{[1 - 6x/|x|^6 + 5/|x|^6]^{\frac{1}{2}} - 1\}\,dx \qquad (9.124)$$

$$j^- \simeq 148, \qquad j^+ \simeq 42, \qquad j^-/j^+ \simeq 3\cdot5.$$

In (9.123) we have identified ω^α with $\omega^{\alpha,c}$. While ω^α will in general vary along the critical isotherm, it may be expected to do so as a smooth function of the field b_1, and since b_1, by (9.118), is of order $(\omega^{\beta\gamma})^5$, the variation of ω^α would appear only in the higher order terms in (9.123).

In (9.119)–(9.122) we need to know how ω^α, or the composition of the α phase, varies with the field b_2 or equivalently [see (9.115)] with t. We may take ω^α to be a smooth function of t,

$$\omega^\alpha = \omega^{\alpha,c}(1 + gt + \ldots) \qquad (9.125)$$

with some constant coefficient g. If the α phase were a nearly perfect gas in equilibrium with the liquid solution $\beta\gamma$ (or with the separate liquid phases β and γ), for example, and ω^α a measure of its chemical composition, then ω^α would be a smooth function of the chemical potentials or partial pressures of the constituents. But the latter, in turn, like the vapour pressure of a pure liquid near its critical point, would vary linearly with t, to a first approximation, with a correction term[15] proportional to $|t|^{2-\alpha}$, where $\alpha(\simeq 0\cdot1$ in $d = 3)$ is the conventional heat-capacity exponent. Such a term $|t|^{2-\alpha}$ in (9.125) would be of higher order than the singular terms that will be of interest in (9.119)–(9.122), which will prove to be of order $|t|^\gamma$ or $|t|^\mu$, with $\gamma \simeq \mu \simeq 1\cdot25$; so we may truncate (9.125) at the linear term and we shall still determine correctly the leading singular behaviour of (9.119)–(9.122).

We shall now see what the results (9.119)–(9.123) imply for the behaviour of the interfacial tensions near a critical endpoint. At this $\beta\gamma$ critical endpoint β and γ are the critical phases, so in the three-phase region, of the three tensions in (9.119)–(9.121) it is $\sigma^{\beta\gamma}$ that should vanish while $\sigma^{\alpha\beta}$ and $\sigma^{\alpha\gamma}$ should approach the common limiting value σ^c. From (9.119) and (9.125) we see that $\sigma^{\beta\gamma}$ does indeed vanish, and does so proportionally to b_2, or, by (9.115), to t^γ. In deriving (9.119) we explicitly took $d = 3$ and $\eta = 0$, which implies $\delta = 5$ and also, as we know (§ 9.4), $\mu = \gamma$. Thus, $\sigma^{\beta\gamma}$ has here been found to vanish as t^μ, as expected for a critical tension. When we do not make the explicit assumptions $d = 3$ and $\eta = 0$, but instead derive the analogue of (9.119) from a form of (9.113) with a general δ, we do indeed find $\sigma^{\beta\gamma} \sim t^{\mu}$ as the general result.[25]

Since $\omega^{\beta} = 0$ at the critical endpoint, where $b_2 = 0$ [see (9.114)], and since $\omega^{\alpha} = \omega^{\alpha,c}$ and $\sigma^{\beta\gamma} = 0$ there, we confirm from (9.120)–(9.122) that $\sigma^{\alpha\beta}$ and $\sigma^{\alpha\gamma}$ in the three-phase region and $\sigma^{\alpha,\beta\gamma}$ on locus (i) in the two-phase region all approach σ^c. To leading order in b_2, we have from (9.114), (9.120), (9.122), and (9.125),

$$\sigma^{\alpha\beta} = [1 + 5gt + \tfrac{5}{2}(\omega^{\alpha,c})^{-4}b_2 \ln(1/-b_2) + \dots]\sigma^c \qquad (9.126)$$

in the three-phase region, where $b_2 < 0$, and

$$\sigma^{\alpha,\beta\gamma} = [1 + 5gt + \tfrac{5}{2}(\omega^{\alpha,c})^{-4}b_2 \ln(1/b_2) + \dots]\sigma^c \qquad (9.127)$$

on locus (i) in the two-phase region, where $b_2 > 0$. By (9.115) the leading singular term in $\sigma^{\alpha\beta}$, or in $\sigma^{\alpha,\beta\gamma}$ on locus (i), is of order $|t|^{\gamma} \ln|t|$, or equivalently $|t|^{\mu} \ln|t|$ because $\gamma = \mu$ here. The logarithm appears because $\delta = 5$ is a borderline case.[25] When $\delta \neq 5$, terms both of order $|t|^{\gamma}$ and $|t|^{\mu}$ (without logarithmic factors) appear[25] in the analogues of (9.126) and (9.127). In $d = 2$, where $\gamma = \tfrac{7}{4}$ and $\mu = 1$, the term t^{μ} joins the term gt in the analytic background and the leading singular term is then of order $|t|^{\gamma}$. In mean-field approximation, where $\gamma = 1$ and $\mu = \tfrac{3}{2}$, it is the term t^{γ} that becomes part of the analytic background while the leading singular term is then of order $|t|^{\mu} = |t|^{\tfrac{3}{2}}$.

The present results, which are for $\delta = 5$ and $\mu = \gamma$, are shown schematically in Fig. 9.10. Here T^c has been taken to be an upper critical

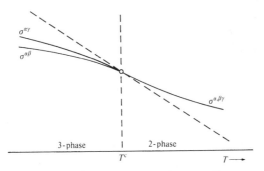

FIG. 9.10. Surface tension $\sigma^{\alpha,\beta\gamma}$ of the liquid phase $\beta\gamma$ of critical composition for $T > T^c$, and surface tensions $\sigma^{\alpha\beta}$ and $\sigma^{\alpha\gamma}$ of the two liquid phases β and γ for $T < T^c$, as functions of T, when critical exponents are $\delta = 5$, $\mu = \gamma$. The dashed line is the common tangent to all three curves at $T = T^c$; the small circle marks the critical point. This is an enlarged view (schematic) of the immediate neighbourhood of $T = T^c$ in Fig. 9.11; here, $\sigma^{\alpha\gamma}$ still lies below the tangent.

solution temperature so that $-t$ and b_2 have the sign of $T - T^c$, and the parameter g has been taken to be positive so that the tensions decrease with increasing T, as is typical. The difference between $\sigma^{\alpha\gamma}$ and $\sigma^{\alpha\beta}$ in the three-phase region is the critical $\sigma^{\beta\gamma} \sim t^\mu$, which is smaller in magnitude than $t^\mu \ln t$, so the curves of $\sigma^{\alpha\beta}$ and $\sigma^{\alpha\gamma}$ lie on the same side of their common tangent, as shown. The confluence of $\sigma^{\alpha\beta}$ and $\sigma^{\alpha\gamma}$ at the $\beta\gamma$ critical endpoint shown here in Fig. 9.10 is qualitatively the same as the confluence of $\sigma^{\beta\gamma}$ and $\sigma^{\alpha\gamma}$ at the $\alpha\beta$ critical endpoint depicted in Fig. 9.6. Further from the critical point it need no longer be true that $\sigma^{\alpha\beta}$ and $\sigma^{\alpha\gamma}$ lie on the same side of their common tangent, and Fig. 9.11 shows those same tensions on a scale covering a much greater range of temperature than that in Fig. 9.10. These were calculated[25] from (9.119)–(9.122) [still assuming the truncation in (9.125)], for ranges of the variables comparable with those in typical experiments. The curve of $\sigma^{\alpha\gamma}$, which in the immediate neighbourhood of the critical point lies below the tangent line (Fig. 9.10), here mostly lies above it. Their crossing is not visible on the scale of Fig. 9.11. Experimental tests of the behaviour predicted in Fig. 9.10 or 9.11 are being actively pursued in several laboratories, but are not yet definitive.[48]

We see from (9.123) that on the critical isotherm [locus (ii) in the two-phase region, Fig. 9.9], $\sigma^{\alpha,\beta\gamma} - \sigma^c$ vanishes proportionally to the fourth power of the deviation of the composition of the $\beta\gamma$ phase from the critical composition; and with a coefficient j^- for that branch of the isotherm on which $\sigma^{\alpha,\beta\gamma} > \sigma^c$ that is about three and a half times as great as the coefficient j^+ for the branch on which $\sigma^{\alpha,\beta\gamma} < \sigma^c$. In Fig. 9.12 we show, schematically, $\sigma^{\alpha,\beta\gamma}$ as a function of $\omega^{\beta\gamma}$ as given by (9.123).

The fourth power of $\omega^{\beta\gamma}$ in (9.123) would more generally be the μ/β power,[25] but since we here have $\delta = 5$ and $\mu = \gamma$, we necessarily have

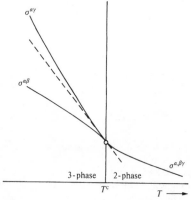

FIG. 9.11. Same as Fig. 9.10 but over a greater range of T.

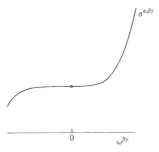

FIG. 9.12. Surface tension $\sigma^{\alpha,\beta\gamma}$ of the liquid phase $\beta\gamma$, as a function of composition $\omega^{\beta\gamma}$, at the critical solution temperature. The small circle marks the critical point.

$\mu/\beta = 4$ because of (9.52). This μ/β is the same as the exponent in the Macleod relation (§ 9.3), which we know to be roughly 4 but more accurately 3.88. In mean-field approximation $\mu/\beta = 3$; but in that case,[25,47] $\sigma^{\alpha,\beta\gamma} - \sigma^c \sim (\omega^{\beta\gamma})^3 \ln|\omega^{\beta\gamma}|$, the logarithmic factor appearing because in mean-field approximation $\delta = 3$, which is borderline for the behaviour of $\sigma^{\alpha,\beta\gamma}$ on locus (ii) in the two-phase region.[25] (Recall that, by contrast, it is $\delta = 5$ that is borderline for the behaviour of $\sigma^{\alpha\beta}$ and $\sigma^{\alpha\gamma}$ in the three-phase region and of $\sigma^{\alpha,\beta\gamma}$ on locus (i) in the two-phase region.)

We thus see that the critical-point exponent μ of (9.1), which is associated with the vanishing of a critical interfacial tension—or, equivalently, the exponent μ/β in the corresponding Macleod relation, the generalization of (9.9)—appears also in the non-critical tensions $\sigma^{\alpha\beta}$, $\sigma^{\alpha\gamma}$, and $\sigma^{\alpha,\beta\gamma}$ near a $\beta\gamma$ critical endpoint; but it appears then as the exponent in a singular correction term, not in the leading term, which is just σ^c, the limiting, non-vanishing tension of the non-critical interface. The critical point therefore manifests itself in the non-critical tensions, but more subtly than in the critical tension. The critical-point singularities in the non-critical tensions reflect the special structure of the non-critical interface at the critical endpoint (Fig. 9.7), at which there is infinite adsorption.

We may ask how the adsorptions at the non-critical interface diverge as the critical endpoint is approached. This question was first addressed by Beaglehole in two important papers, one experimental[49] and one theoretical.[50] We shall here calculate the adsorptions from the composition-dependence of the surface tensions, which we have already determined.

We again consider, for definiteness, a binary liquid mixture near its consolute point, in equilibrium with vapour. We shall calculate the relative adsorption $\Gamma_{2(1)}$, where 1 and 2 are the species labels. From

(2.47), this adsorption at the $\alpha, \beta\gamma$ interface in the two-phase region is

$$\Gamma_{2(1)}^{\alpha,\beta\gamma} = -(\partial\sigma^{\alpha,\beta\gamma}/\partial\mu_2)_T = -(\partial\sigma^{\alpha,\beta\gamma}/\partial x_2)_T(\partial x_2/\partial\mu_2)_T \qquad (9.128)$$

where x_2 is the fraction of species 2 in the near-critical liquid. It is $x_2 - x_2^c$ that is measured by $\omega^{\beta\gamma}$ in (9.123). We know that on the critical isotherm $\sigma^{\alpha,\beta\gamma} - \sigma^c$ vanishes as the μ/β power of $|\omega^{\beta\gamma}|$ [(9.123) being the special case $\mu/\beta = 4$], hence as $|x_2 - x_2^c|^{\mu/\beta}$; so on the critical isotherm $(\partial\sigma^{\alpha,\beta\gamma}/\partial x_2)_T \sim |x_2 - x_2^c|^{\mu/\beta - 1}$. Also on the critical isotherm $\mu_2 - \mu_2^c \sim |x_2 - x_2^c|^\delta$, by definition of δ as the exponent that defines the shape of the critical isotherm; so $(\partial x_2/\partial\mu_2)_T \sim |x_2 - x_2^c|^{-(\delta-1)}$ on that locus, or $(\partial x_2/\partial\mu_2)_T \sim |x_2 - x_2^c|^{-\gamma/\beta}$, by (9.52), and the exponent relation $\mu + \nu = \gamma + 2\beta$ in (9.62),

$$\Gamma_{2(1)}^{\alpha,\beta\gamma} \sim |x_2 - x_2^c|^{-\nu/\beta + 1} \qquad (9.129)$$

on the critical isotherm. Correspondingly, $\Gamma_{2(1)}^{\alpha\beta}$ at the $\alpha\beta$ interface in the three-phase region, or $\Gamma_{2(1)}^{\alpha,\beta\gamma}$ at the $\alpha, \beta\gamma$ interface on the locus (i) in the two-phase region, may be inferred from (9.129) by critical-point scaling; that is, by merely replacing $|x_2 - x_2^c|$ by $|T - T^c|^\beta$:

$$\Gamma_{2(1)}^{\alpha\beta} \quad \text{or} \quad \Gamma_{2(1)}^{\alpha,\beta\gamma} \sim |T - T^c|^{-\nu + \beta}. \qquad (9.130)$$

This is the same divergence found by Fisher and de Gennes in a different but related context.[51] In $d = 3$ this divergence is as $|T - T^c|^{-0.3}$. The general result (9.130) is of the suggestive form $\Gamma \sim \xi\Delta\rho$ with ξ the thickness of, and $\Delta\rho$ the difference in bulk densities across, a *critical* interface.

We remarked earlier that in the equivalent of (9.123) in mean-field approximation, $\sigma^{\alpha,\beta\gamma} - \sigma^c$ on the critical isotherm vanishes as $(\omega^{\beta\gamma})^3 \ln|\omega^{\beta\gamma}|$. Then from the same argument we have just given, now with $\mu = \frac{3}{2}$, $\gamma = 1$, and $\beta = \frac{1}{2}$, we would conclude that $\Gamma_{2(1)}^{\alpha\beta}$ or $\Gamma_{2(1)}^{\alpha,\beta\gamma} \sim \ln|T - T^c|^{-1}$. This is the same logarithmic divergence that Beaglehole found when he based his calculation on a free energy taken from the regular-solution theory.[50] In mean-field approximation $\nu = \beta$ (their common value being $\frac{1}{2}$), so it could already have been apparent from the general (9.130) that the adsorption would then diverge more slowly than any power. In reality, as we saw, we expect a divergence like $|T - T^c|^{-0.3}$, so the difference between classical and non-classical behaviour is in this instance considerable.

It is a different adsorption that we would find directly from $d\sigma^{\alpha\beta}/dT$ in the three-phase region or from $(\partial\sigma^{\alpha,\beta\gamma}/\partial T)_x$ on the locus (i) in the two-phase region. These derivatives, by (9.126) and (9.127), are finite at the critical endpoint; and that remains true when the terms $b_2 \ln(1/|b_2|)$ are more generally terms of order $|t|^\gamma$ and $|t|^\mu$, because $\gamma \geq 1$ and $\mu \geq 1$. From the adsorption equation in the form (2.46) we find for these derivatives

$$\frac{d\sigma^{\alpha\beta}}{dT} = -\Gamma^{\alpha\beta}_{2(1)}\frac{d\mu_2}{dT} - \Gamma^{\alpha\beta}_{3(1)}$$

$$\left(\frac{\partial\sigma^{\alpha,\beta\gamma}}{\partial T}\right)_x = -\Gamma^{\alpha,\beta\gamma}_{2(1)}\left(\frac{\partial\mu_2}{\partial T}\right)_x - \Gamma^{\alpha,\beta\gamma}_{3(1)} \qquad (9.131)$$

where now the subscript 3 refers to the entropy density. We know that $\Gamma^{\alpha\beta}_{2(1)}$ and $\Gamma^{\alpha,\beta\gamma}_{2(1)}$ diverge on approach to the critical endpoint; but then so must $\Gamma^{\alpha\beta}_{3(1)}$ and $\Gamma^{\alpha,\beta\gamma}_{3(1)}$, because thermodynamically the entropy density is just another density, not distinguished in its formal thermodynamic role from the densities of the chemical species. The divergent parts of the two terms on the right-hand side of each of the two equations (9.131) then cancel, leaving both left-hand sides finite at the critical endpoint.

The treatment of critical endpoints in this section has been entirely by the one-density van der Waals theory (as modified to yield a non-classical critical point). Then just from the structure of the theory, irrespective of its details, we have Antonow's rule, (9.121), and thus conclude that the $\alpha\gamma$ interface is perfectly wet by the β phase. We saw from Cahn's argument as outlined in § 8.5, and also from the Moldover–Cahn experiment referred to there, that we should expect and do indeed have such spreading when we are in the neighbourhood of a critical endpoint; so in that sense our one-density van der Waals theory based on the free-energy densities in Fig. 9.8 is realistic. Whether the functions $W(\omega)$ that we chose in (9.108), (9.113), (9.116), and (9.117) [depicted schematically in Fig. 9.8(a)–(d), respectively] are quantitatively accurate is another matter. They are probably realistic enough to have yielded correctly or nearly correctly the leading singular terms in the several interfacial tensions near the critical endpoint; that is, to have described those singularities with nearly correct exponents. The present description of that singular behaviour is probably not more literally correct than that, for our free-energy densities $-W(\omega)$ are otherwise artificial and, except for implying realistic values of the critical-point exponents, were chosen mainly for analytical convenience. A two- or three- [more generally a $(c+1)$-] density theory, with a realistic $W(\omega_1, \omega_2, \ldots)$, would yield a more nearly correct description of the phenomenon. If the critical endpoint in question were associated with and in close proximity to a tricritical point, as in § 9.5, then the many-density van der Waals theory would reduce to the one-density theory,[4] by the same argument with which we saw such a reduction near an ordinary critical point in § 9.1. But when the α phase is totally distinct from the near-critical β and γ phases, as in our example in which α is vapour while β and γ are liquids near their critical mixing point, the one-density theory may only be taken as figuratively true.

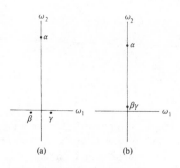

FIG. 9.13

In the rare circumstances in which, by symmetry, Antonow's rule cannot hold near the critical endpoint, the one-density theory for the non-critical interfaces is not even figuratively correct, and those interfaces must then be described by at least two independently varying densities. The hypothetical case of an enantiomeric liquid pair in equilibrium with vapour while near a critical point of liquid–liquid phase separation, already referred to in §§ 8.3 and 8.5, is an example. A real case is that of the liquid–vapour interface of helium (the ^{4}He isotope) near the liquid's superfluid transition. The lambda-point that marks that transition is thermodynamically equivalent to a critical point, although the superfluid 'phases' β and γ are not spatially separated so there is no physically realizable $\beta\gamma$ interface. The vapour phase, α, is symmetrically related to β and γ, so the phase equilibrium cannot be described by the one-dimensional free-energy densities $-W(\omega)$ of Fig. 9.8. Instead, a two-dimensional $-W(\omega_1, \omega_2)$ defined over an ω_1, ω_2 plane such as that in Fig. 9.13(a) ($T < T_\lambda$) and 9.13b ($T > T_\lambda$), would be required.[25,47] It is desirable that there be such a theory—one has not yet been given—because there already exist the very beautiful experiments of Magerlein and Sanders,[52] in which they observed and characterized the singularity in the liquid–vapour surface tension at the λ-point. It had already been suggested[53]—although the theoretical arguments that led to it were of a structure entirely different from those presented here—that the singular term in that surface tension would be $|t|^\mu$ [the exponent μ not recognized as such but expressed in the equivalent forms $2-\alpha-\nu$, $(d-1)\nu$, and $(2-\alpha)(d-1)/d$], just as we find here. The measurements agree well with that prediction.[52]

9.7 Renormalization-group theory; field-theoretical models

The renormalization-group theory[22,54] has been an important advance in the theory of critical points. It has provided new techniques for calculating equations of state, correlation functions, and their associated critical-point exponents. As was inevitable, it has been applied also to the determination of the surface tension[55,56] and the interfacial composition profile[56] near critical points and tricritical points.[57]

We begin with a brief description of the renormalization-group theory. It is illustrated most simply for a lattice spin system (Ising model), equivalent to the lattice-gas model in §5.3. First, the cells of the lattice gas are replaced by lattice sites, and the two possible states of occupancy of a cell—occupied or empty—are replaced by the two possible states, $\uparrow$ or $\downarrow$, of a discrete, two-state spin at the corresponding site. The deviation $\rho - \rho^c$ of the density of the lattice gas from its value at the critical point ($\rho^c = \frac{1}{2}$) translates into the excess density of $\uparrow$ spins over that of $\downarrow$ spins in the related Ising model, hence into the magnetization density M. Below the critical point two magnetic phases (domains) of magnetization density $\pm M^*$ (the spontaneous magnetization) may coexist, with a boundary between them in which the magnetization density is some $M(z)$, with $M(\pm\infty) = \pm M^*$. This $M(z)$ is then the same as the density profile $\rho(z) - \rho^c$ of the corresponding lattice gas. The excess free energy per unit area associated with the inhomogeneity in the spin system is analogous to and essentially the same as the surface tension of the lattice gas.

Next, for convenience, in one common form of the renormalization-group theory the discrete-spin model ($s_i = \pm 1$ for spins $\uparrow$ or $\downarrow$) is replaced by a continuous-spin model (each spin s_i such that $-\infty < s_i < \infty$), and, further, the spins s_i at the discrete lattice sites i are in turn replaced by a continuous spin density $s(\mathbf{r})$. The Hamiltonian density is taken to be of the so-called Landau–Ginzburg–Wilson form,

$$\tfrac{1}{2}m\,|\nabla s(\mathbf{r})|^2 - W[s(\mathbf{r})] \tag{9.132}$$

with

$$-W(s) = as^2 + bs^4 + \dots . \tag{9.133}$$

If $s(\mathbf{r})$ is allowed to have no Fourier components of wavelength less than some fixed constant, that constraint sets a scale of length analogous to the lattice spacing of the original Ising model. The integration of the Hamiltonian density (9.132) over all $\mathbf{r}$, to yield the Hamiltonian, is the analogue of a summation over lattice sites i. The square gradient is the analogue of the discrete $s_i^2 + s_j^2 - 2s_i s_j$ for a pair of neighbouring sites i and j; the term $s_i s_j$ here is the same as that arising from the interaction of neighbouring spins in the lattice model. The term as^2 in (9.133) is just another term like the continuum analogues of the s_i^2 or s_j^2 that we have just seen to be, in effect, contained in the square gradient. If there were no terms in $-W$ beyond the term as^2 in (9.133) the model would then be a continuous form of the so-called Gaussian model, with no external field. [To include the effect of such a field we would need an additional term in $-W(s)$ that was linear in s. It is often necessary to augment (9.133) in this way, for an external field in the spin model plays the role of a variable chemical potential in the lattice gas.] The term proportional to s^4 adds to the Boltzmann factor that weights each spin configuration $s(\mathbf{r})$ a factor that makes the model more like the Ising model and less like the Gaussian model: a purely Gaussian weighting is uniform on every sphere $\sum_i s_i^2 = constant$ in spin space, whereas, by contrast, $\sum_i s_i^4$ is a minimum on such a sphere (the associated Boltzmann factor is a maximum) when all the s_i^2 are equal, as they are in the Ising model, where $s_i = \pm 1$.

The Hamiltonian, we noted, is obtained by integrating the Hamiltonian density (9.132) over all $\mathbf{r}$; so the partition function $\exp(-\Omega/kT)$ (Ω is the corresponding free energy) is then

$$e^{-\Omega/kT} = \int \exp\left(-\frac{1}{kT} \int \{\tfrac{1}{2}m \, |\nabla s(\mathbf{r})|^2 - W[s(\mathbf{r})]\} \, d\mathbf{r}\right) d\{s(\mathbf{r})\} \quad (9.134)$$

where $\int \ldots d\{s(\mathbf{r})\}$ denotes functional integration over all $s(\mathbf{r})$ (subject to the exclusion of wavelengths less than some fixed constant, as already mentioned). The integrand is the Boltzmann factor that gives the relative weight of the spin distribution $s(\mathbf{r})$, or of the equivalent density distribution $\rho(\mathbf{r})$ in the related fluid model.

The free energy Ω is obtained from (9.134), and the various correlation functions and n-particle densities are obtained from related expressions. We may imagine evaluating such functional integrals stepwise: first integrating over some of the spin variables, which yields a modified Boltzmann factor that depends on fewer of the variables; and then repeating the process, thus producing at each stage an equivalent system of fewer degrees of freedom with its own Boltzmann factor. In the renormalization-group theory it is seen that by an appropriate choice of spin variables [the so-called block spins:[58] roughly the sums of the long-wavelength Fourier components of $s(\mathbf{r})$ down to those of a chosen wavelength L, or roughly the sums of the spins at the sites of the original model in blocks of length L on a side] this may be done systematically, at each stage preserving the form of the original problem; and that if the system is near its critical point, then the effective Boltzmann factor has at each stage its original functional form (or asymptotically approaches a limiting functional form), in which, however, the values of the parameters that play the roles of the original kT, m, a, b, ... in (9.132)–(9.134) change from one stage to the next. The recurrence relations (difference or differential equations) that determine the evolution of the parameters as the process unfolds are called the renormalization-group equations. Their fixed points are central to the theory. The trajectories that describe the evolution of the parameters in a space whose coordinates are the values of those parameters, are called the 'flows'. They flow towards a fixed point, and then either flow into it, if the system is at a critical point, or ultimately veer away from it, if the system is near but not at a critical point. The fixed point itself thus lies in the surface or on the locus of critical points in the parameter space, and is in that sense a representative one of the whole manifold of critical points. The behaviour of the system near a critical point depends on the nature of the trajectories near the associated fixed point. The renormalization-group equations may be linearized in the neighbourhood of the fixed point; and then the eigenvalues of the matrix of coefficients of the linearized equations are simply related to and determine the critical-point exponents. That, in briefest outline, is the structure of the theory.[22,54]

At any stage in the process, (9.132), with the values of the parameters that have been reached by that stage and as a function ot the spin variables then remaining, is the Hamiltonian density of a transformed but related system whose properties are identical with those of the original system. Alternatively, if in the spin variables that remain at that stage there are present only those Fourier components of the original $s(\mathbf{r})$ that are of wavelength greater than some L, then the effective Hamiltonian density (9.132) at that stage may equally well be interpreted as the free-energy density of a system that is constrained to be uniform over distances greater than L; that is, of a system in which fluctuations over distances greater than L have been suppressed. Further integration of the associated Boltzmann factor over the remaining block spin variables, as in (9.134), would then yield the partition function of the fully unconstrained system. This is as discussed in § 3.5; and we then see that, interpreted as a free-energy density, (9.132) is precisely that of the van der Waals theory. Indeed, (9.133), which may be truncated at the quartic term, is essentially the same as (9.87) in the two-phase region.

If the functional integrals that remain to be evaluated, such as that in (9.134), are approximated by their integrands evaluated with the spin density $s(\mathbf{r})$ that maximizes them, that is the mean-field approximation, and the result for an inhomogeneous fluid is clearly just the van der Waals theory. (That is *one* way of understanding the status of the van der Waals theory. An important alternative interpretation, due to Evans,[59] will be expounded presently.) An exact theory would take account of the fluctuations that are suppressed in the mean-field theory; that is, would include the contributions to the functional integrals that are made by spin densities $s(\mathbf{r})$ other than that which maximizes the Boltzmann factor in the integrand. That is what is in principle done in the renormalization-group theory,[56,57] and also in related theories of fluid interfaces,[60-62] which include the so-called field-theoretical models. In actual calculations, these typically take account only of small departures from the mean-field $s(\mathbf{r})$, usually by a perturbation expansion or an appropriate linearization. Among the fluctuations that are suppressed entirely in the mean-field approximation are the capillary waves (§ 4.9), which we have seen lead to a transverse correlation length and an interfacial thickness that both diverge in the limit of vanishing gravitational field. The capillary-wave fluctuations with their attendant divergences are present in the renormalization-group and field-theoretical models.[56,60–62]

We now digress briefly to describe the alternative interpretation of the van der Waals theory alluded to above, which was given by Evans[59] in the paper we already referred to in § 4.9 (ref. 77 of that chapter). We saw in § 4.5 that there is a unique functional $\Omega[\hat{\rho}(\mathbf{r})]$, or $\Omega[\hat{s}(\mathbf{r})]$ in our present language, which assumes a minimum value equal to the equilibrium free energy Ω when $\hat{\rho}(\mathbf{r})$ or $\hat{s}(\mathbf{r})$ is the equilibrium density. Such functionals are very difficult to construct for dense, inhomogeneous fluids, and remain largely unknown (except for the fluid of hard rods in one dimension).[63] The density functional may, however, be approximated by

$$\Omega[\hat{s}(\mathbf{r})] = \int \{\tfrac{1}{2}m \, |\nabla\hat{s}(\mathbf{r})|^2 - W[\hat{s}(\mathbf{r})]\} \, d\mathbf{r} \qquad (9.135)$$

with the function $-W(s)$ identified as the density of free energy Ω when the spin density in the system is uniform with the value s. If the free-energy density in the functional (9.135) is thought of as a truncated gradient expansion [cf. (4.161)], then (9.135) is an approximation to the true density functional Ω, valid for small departures of $\hat{s}(\mathbf{r})$ from uniformity. But Evans remarks that (9.135) may now be postulated to be the *exact* density functional for some system, thus defining what he calls the 'van der Waals model', the properties of which he then derives from (9.135) without further approximation. The $\hat{s}(\mathbf{r})$ that minimizes (9.135) is then the exact equilibrium $s(\mathbf{r})$ for this model, and the minimum in question is the exact free energy Ω; although the same $s(\mathbf{r})$ and the same free energy are merely approximations (the mean-field approximation) when (9.135) is viewed, not as the exact density functional $\Omega[\hat{s}(\mathbf{r})]$ for some model, but as the free energy Ω of a constrained system, with a functional integration yet to be done, as in (9.134). As we remarked in § 4.9, Evans shows that in the model for which (9.135) is imagined to be exact the transverse correlation length diverges as $[\sigma/mg(\rho^l - \rho^g)]^{\frac{1}{2}}$ for vanishing g (where m is the molecular mass); that is, as $1/\sqrt{2}$ times the capillary length, just as in the capillary-wave theory, and as predicted by Wertheim;[64] yet the interfacial thickness remains finite, as in the van der Waals theory. Further, Evans shows that the lowest order capillary-wave theory yields a poor approximation to the exactly known density profile $\rho(z)$ in this model. He remarks, however, that these results may not be representative, for (9.135) entails an artificial δ-function-containing direct correlation function, $c(\mathbf{r}_1, \mathbf{r}_2)$, obtained from (9.135) by (4.127) and

(4.134). Evans also illuminates the relation between the capillary-wave and field-theoretical models on the one hand, and the density-functional theory on the other. The mathematical calculations in all of them are similar; but what in the field-theoretical models, for example, appears as the calculation of corrections to the mean-field approximation, appears in the van der Waals model as an analysis of the stability of the planar interface, with its equilibrium profile $\rho(z)$, against density fluctuations, particularly capillary-wave fluctuations.

Returning now to the renormalization-group theory, after that digression, we recall the earlier remark (§ 9.3) that there prove to be two fixed points of the renormalization-group equations (only two, in the simplest cases): the Gaussian fixed point, leading to classical critical-point exponents, and the Ising fixed point, with which are associated the non-classical, d-dependent exponents.[22,54] At $d = 4$ these coincide, marking that as the borderline dimensionality, d^*. The problem is in a sense exactly soluble at $d = 4$; and the solution for $d < 4$ may be obtained by perturbation about $d = 4$, particularly by expansion in

$$\varepsilon = 4 - d, \tag{9.136}$$

treating it as a small parameter. This is the 'ε-expansion'.[22,54] To obtain the critical-point exponents correctly to second and higher order in ε has required Feynman-graph or other field-theoretical techniques.[65]

We may illustrate the results of the ε-expansion with that for the surface-tension exponent μ. From Wilson,[65] we have

$$\gamma = 1 + \tfrac{1}{6}\varepsilon + \tfrac{25}{324}\varepsilon^2 + \ldots, \qquad \eta = \tfrac{1}{54}\varepsilon^2 + \ldots. \tag{9.137}$$

But from (9.61) (which is confirmed in the renormalization-group theory[55]) and (9.57), we have $\mu = (d-1)\gamma/(2-\eta)$; whereupon, from (9.136) and (9.137),

$$\mu = \tfrac{3}{2} - \tfrac{1}{4}\varepsilon + \tfrac{5}{108}\varepsilon^2 + \ldots. \tag{9.138}$$

The first-order approximation (Ohta and Kawasaki[56]), $\mu = \tfrac{3}{2} - \tfrac{1}{4}\varepsilon$, is remarkable: it is exact, as it must be, in $d = 4$ ($\varepsilon = 0$), where it gives the classical $\mu = \tfrac{3}{2}$ of van der Waals; it is exact in $d = 2$ ($\varepsilon = 2$), where it gives $\mu = 1$; and it is virtually exact in $d = 3$ ($\varepsilon = 1$), where it gives $\mu = 1\cdot25$, while the best present estimate is $\mu \simeq 1\cdot26$ (Table 9.1 in § 9.3). The second-order approximation is not nearly so good, showing that the expansion (9.138) is probably asymptotic rather than convergent. The experience with such approximations for most of the other critical-point exponents that have been studied is that they improve in second order and worsen in third order.[66]

Ohta and Kawasaki also find (9.82), with

$$K = \frac{1}{6} \left[1 - \left(\frac{\sqrt{3}}{9} \pi - \frac{1}{2} \right) \varepsilon + \ldots \right]$$ (9.139)

to first order in ε. This gives the correct classical $K = \frac{1}{6}$ at $\varepsilon = 0$. At $\varepsilon = 1$ it gives $K = 0\cdot1492$, which is close to the $K = 0\cdot1426 \simeq \frac{1}{7}$ found in (9.92) and confirms that K in $d = 3$ is slightly less than the classical $K = \frac{1}{6} = 0\cdot1667$.

The density profile has also been found[56] to first order in ε. Since the ε-expansion is an expansion about $d = 4$, and since there are no capillary-wave divergences when $d > 3$ (§ 4.9), one may obtain in this way a determinate, limiting profile as a function of ε, in which one may then formally set $\varepsilon = 1$ to obtain an approximation appropriate to $d = 3$ yet free of capillary-wave divergences. Ohta and Kawasaki[56,67] cast their result for the profile in a form directly comparable with (9.93):

$$X(y) = \tanh(\tfrac{1}{2}y) \left[1 + \frac{2a}{3+a} \operatorname{sech}^2(\tfrac{1}{2}y) \right]^{-\frac{1}{2}}$$ (9.140)

where

$$a = \frac{\sqrt{3}}{6} \pi \varepsilon + \ldots$$ (9.141)

to first order in ε. This reduces properly to the classical $X(y) = \tanh(\tfrac{1}{2}y)$ when $\varepsilon = 0$. Rewriting (9.93) as

$$X(y) = \tanh(\tfrac{1}{2}y)[1 + \tfrac{1}{2} \operatorname{sech}^2(\tfrac{1}{2}y)]^{-\frac{1}{2}}$$ (9.142)

we see that it is very close to (9.140) with (9.141) if ε is set equal to unity in the latter, when $a = 0\cdot9069$.

The capillary-wave divergences at $d = 3$ may be recaptured in the renormalization-group calculations of the profile by setting $d = 3$ at an early enough stage:[56] all integrations, subtractions, etc., where it matters, are done explicitly in $d = 3$ rather than in $d = 4 - \varepsilon$ near 4. The resulting profile is then not entirely of the form (9.78), in which $[\rho(z) - \rho^c]/(\rho^l - \rho^g)$ is a universal function of z/ξ, but has some additional pieces that are not universal. Jasnow and Rudnick[56] express their final result as the q-dependent (angle- and wavelength-dependent) reflectivity

$$R(q) \propto \left| \int_{-\infty}^{\infty} \rho'(z) e^{2iqz} \, dz \right|^2$$ (9.143)

of an interface with density profile $\rho(z)$. They find $R(q)$ to be of the form

$$R(q) \propto r(q\xi) e^{-4q^2\langle \zeta^2 \rangle}$$ (9.144)

where $r(q\xi)$ is a universal function of $q\xi$ differing slightly from the corresponding function of $q\xi$ that would follow from the classical

$\tanh(|z|/2\xi)$ profile; while $\langle \zeta^2 \rangle$ is some function of ξ that is proportional to the logarithm of the gravitational acceleration g when g is small, and is roughly interpretable as the mean-square amplitude of interfacial capillary waves [cf. (4.243)]. The reflectivity in (9.144) is then the product of a factor that is a universal function of $q\xi$ and may be thought of as arising from the intrinsic interfacial structure, and a factor that is Gaussian in q but is an otherwise non-universal function of q and ξ arising from the capillary waves. Since $\langle \zeta^2 \rangle \to \infty$ as $g \to 0$, the reflectivity (9.144) vanishes in this limit; this is due to the erosion of the gradient $\rho'(z)$ by long-wavelength capillary waves of ever-increasing amplitude. Sullivan[62] derives results similar to those of Jasnow and Rudnick by other methods.

This development has gone some way towards giving us a theoretically well-founded expression for the density profile, that takes account of the full spectrum of fluctuations and thus incorporates both non-classical critical-point behaviour and the effects of long-wavelength capillary waves. We may expect that this problem will continue to excite interest, and that the pioneering work that has already been done will inspire further pursuit.

Notes and references

1. van der Waals, J. D. *Zeit. phys. Chem.* **13**, 657 (1894); English translation, *J. stat. Phys.* **20**, 197 (1979).
2. Cahn, J. W. and Hilliard, J. E. *J. chem. Phys.* **28**, 258 (1958).
3. Macleod, D. B. *Trans. Faraday Soc.* **19**, 38 (1923); Guggenheim, E. A. *Thermodynamics*, 5th ed., § 3.65, pp. 163–6, North-Holland, Amsterdam (1967).
4. Widom, B. in *Statistical Mechanics and Statistical Methods in Theory and Application* (ed. U. Landman), pp. 33–71. Plenum, New York (1977).
5. Watson, G. N. *Theory of Bessel Functions*, 2nd ed. §§ 3.7, 3.71, 7.23, and 13.21 (pp. 78, 80, 202, and 388). Cambridge University Press (1944).
6. Fisher, M. E. (privately communicated) observes that the same solution of the linear equation (9.31) gives both the large- and small-r/ξ behaviour of $h(r)$, and that this is in contrast to what is found from a roughly analogous, but non-linear, equation for the near-critical $h(r)$ implied by the Yvon–Born–Green equation: Jones, G. L., Kozak, J. J., Lee, E., Fishman, S. and Fisher, M. E. *Phys. Rev. Lett.* **46**, 795 (1981) and Fisher, M. E. and Fishman, S. *Phys. Rev. Lett.* **47**, 421 (1981). We are grateful to Professor Fisher for a discussion of this point.
7. van Hove, L. *Physica* **16**, 137 (1950).
8. Stillinger, F. H. and Frisch, H. L. *Physica* **27**, 751 (1961).
9. Landau, L. D. and Lifshitz, E. M. *Statistical Physics* (trans. E. Peierls and R. F. Peierls), § 116, pp. 366–9, Pergamon Press, Oxford (1958); Hart, E. W. *J. chem. Phys.* **34**, 1471 (1961); Fixman, M. *J. chem. Phys.* **33**, 1357 (1960); **36**, 1965 (1962); Lebowitz, J. L. and Percus, J. K. *J. math. Phys.* **4**, 248 (1963); Fisher, M. E. *J. math. Phys.* **5**, 944 (1964).
10. Fisher, ref. 9.

11. Kaufman, B. and Onsager, L. *Phys. Rev.* **76,** 1244 (1949).
12. From Fisher's relation $(2 - \eta)\nu = \gamma$ among the critical-point exponents ν, γ, and η (Fisher, ref. 9), which we shall discuss in § 9.3, and from the values $\gamma = 1\cdot237$ and $2\nu = 1\cdot259$ recently calculated by B. G. Nickel [to appear in *Proceedings of the Cargèse Summer Institute on Phase Transitions*, 1980, Plenum Press, New York] for a lattice model believed to be equivalent in its critical-point behaviour to a continuum fluid, we find $\eta = 0\cdot035$.
13. Among the many recent reviews of critical phenomena in fluids are those by Sengers, J. V. and Levelt Sengers, J. M. H. in *Progress in Liquid Physics* (ed. C. A. Croxton) Chapter 4, pp. 103–74, Wiley, Chichester (1978); Scott, R. L. in *Specialist Periodical Reports: Chemical Thermodynamics* (ed. M. L. McGlashan), Vol. 2, Chapter 8, pp. 238–74, Chemical Society, London (1978); Greer, S. C. *Accts. chem. Res.* **11,** 427 (1978); Stanley, H. E. *Introduction to Phase Transitions and Critical Phenomena*, 2nd ed., Oxford University Press (in preparation); Rowlinson, J. S. and Swinton, F. L. *Liquids and Liquid Mixtures*, 3rd ed., Chapters 3, 4, and 6, Butterworth, London (1982).
14. Keesom, W. H. and Keesom, A. P. *Commun. Physics Lab. Leiden* **20,** no. 221d (1932); Keesom, W. H. *Helium*, p. 216, Elsevier, Amsterdam (1942); Pippard, A. B. *Elements of Classical Thermodynamics*, p. 126, Cambridge University Press (1957).
15. Widom, B. *J. chem. Phys.* **43,** 3898 (1965); *Physica* **73,** 107 (1974); Griffiths, R. B. *Phys. Rev.* **158,** 176 (1967).
16. Onsager, L. *Phys. Rev.* **65,** 117 (1944).
17. Buff, F. P. and Lovett, R. A. in *Simple Dense Fluids* (ed. H. L. Frisch and Z. W. Salsburg), pp. 17–30, particularly p. 22, Academic Press, New York (1968); Lovett, R. A. Thesis, University of Rochester (1965), p. 13; Leng, C. A., Rowlinson, J. S. and Thompson, S. M. *Proc. Roy. Soc.* A**352,** 1 (1976).
18. Nickel, ref. 12; Zinn-Justin, J. *J. Physique* **42,** 783 (1981).
19. Fisher, M. E. *Rep. Prog. Phys.* **30,** 615 (1967), particularly p. 716; Tarko, H. B. and Fisher, M. E. *Phys. Rev. Lett.* **31,** 926 (1973); *Phys. Rev. B* **11,** 1217 (1975).
20. Fisher, ref. 19, particularly p. 713; Stell, G. *Phys. Rev. Lett.* **20,** 533 (1968).
21. Gunton, J. D. and Buckingham, M. J. *Phys. Rev.* **166,** 152 (1968); Hall, C. K. *J. stat. Phys.* **13,** 157 (1975).
22. Wilson, K. G. *Phys. Rev. B* **4,** 3174, 3184 (1971); Wilson, K. G. and Fisher, M. E. *Phys. Rev. Lett.* **28,** 240 (1972).
23. Riedel, E. K. and Wegner, F. J. *Phys. Rev. Lett.* **29,** 349 (1972); Stephen, M. J., Abrahams, E. and Straley, J. P. *Phys. Rev. B* **12,** 256 (1975); Stephen, M. J. *Phys. Rev. B* **12,** 1015 (1975).
24. Fisk, S. and Widom, B. *J. chem. Phys.* **50,** 3219 (1969).
25. Ramos Gómez, F. and Widom, B. *Physica* **104A,** 595 (1980).
26. See also Stauffer, D. in *Proceedings of 18th Conference on Magnetism and Magnetic Materials*, A. I. P. Conf. Proc. No. 10, Pt. 1, p. 828. Amer. Inst. Phys., New York (1973).
27. Warren, C. and Webb, W. W. *J. chem. Phys.*, **50,** 3694 (1969), in cyclohexane–methanol mixtures near the consolute point; Zollweg, J., Hawkins, G. and Benedek, G. B. *Phys. Rev. Lett.* **27,** 1182 (1971), and Zollweg, J., Hawkins, G., Smith, I. W., Giglio, M. and Benedek, G. B. *J. Physique* **33,** C1-135 (1972), in xenon; Bouchiat, M.- A. and Meunier, J. (1971, privately communicated), in carbon dioxide; and Wu, E. S. and Webb, W. W. *J.*

Physique **33**, C1–149 (1972) and *Phys. Rev. A* **8**, 2077 (1973), in sulphur hexafluoride.

28. Huang, J. S. and Webb, W. W. *J. chem. Phys.* **50**, 3677 (1969); Wu, E. S. and Webb, W. W. *Phys. Rev. A* **8**, 2065 (1973).
29. van der Waals, J. D. and Kohnstamm, Ph. *Lehrbuch der Thermodynamik*, Vol 2, pp. 39–40, Barth, Leipzig (1912); Kohnstamm, Ph. in *Handbuch der Physik* (ed. H. Geiger and K. Scheel) Vol. 10, § 45, Springer, Berlin (1926). Van der Waals and Kohnstamm used the name 'critical point of higher order' for what we now call the tricritical point. Schreinemakers, F. A. H., in H. W. Bakhuis Roozeboom, *Die heterogenen Gleichgewicht*, Part III, 2, pp. 24, 49, Vieweg, Braunschweig (1913), uses the name 'critical point of second order' for a critical point in a ternary system which is at a temperature maximum, but otherwise unremarkable.
30. Radyshevskaya, G. S., Nikurashina, N. I. and Mertslin, R. V. *J. gen. Chem. USSR* [or *Zh. Obshch. Khim.*] **32**, 673 (1962).
31. Krichevskii, I. R., Efremova, G. D., Pryanikova, R. O. and Serebryakova, A. V. *Russ. J. phys. Chem.* **37**, 1046 (1963) [*Zh. Fiz. Khim.* **37**, 1924 (1963)].
32. Griffiths, R. B. *J. chem. Phys.* **60**, 195 (1974).
33. Widom, B. *J. phys. Chem.* **77**, 2196 (1973).
34. Knickerbocker, B. M., Pesheck, C. V., Scriven, L. E. and Davis, H. T. *J. phys. Chem.* **83**, 1984 (1979), and Knickerbocker, B. M., Pesheck, C. V., Davis, H. T. and Scriven, L. E., preprint, 1981; Fleming, P. D., Vinatieri, J. E. and Glinsmann, G. R. *J. phys. Chem.* **84**, 1526 (1980); Fleming, P. D. and Vinatieri, J. E. *J. Coll. Interf. Sci.* **81**, 319 (1981); Herrmann, C.-U., Klar, G. and Kahlweit, M. *J. Coll. Interf. Sci.* **82**, 6 (1981).
35. Cash, L., Cayias, J. L., Fournier, G., MacAllister, D., Schares, T., Schechter, R. S. and Wade, W. H. *J. Coll. Interf. Sci.* **59**, 39 (1977); Bellocq, A. M., Bourbon, D. and Lemanceau, B. *J. Coll. Interf. Sci.* **79**, 419 (1981).
36. Widom, B. *Phys. Rev. Lett.* **34**, 999 (1975).
37. Fox, J. R. *J. stat. Phys.* **21**, 243 (1979). See also Fisher, M. E. and Sarbach, S. *Phys. Rev. Lett.* **41**, 1127 (1978); Sarbach, S. and Fisher, M. E. *Phys. Rev. B* **20**, 2797 (1979); Kaufman, M., Griffiths, R. B., Yeomans, J. M. and Fisher, M. E. *Phys. Rev. B* **23**, 3448 (1981); and Yeomans, J. M. and Fisher, M. E. *Phys. Rev. B* **24**, 2825 (1981).
38. Griffiths, R. B. *Phys. Rev. Lett.* **24**, 715 (1970); *Phys. Rev. B* **7**, 545 (1973).
39. Leiderer, P., Poisel, H. and Wanner, M. *J. low Temp. Phys.* **28**, 167 (1977); Leiderer, P. in *Quantum Fluids and Solids* (ed. S. B. Trickey, E. D. Adams, and J. W. Dufty), pp. 351–60, Plenum, New York (1977).
40. Papoular, M. *Phys. Fluids* **17**, 1038 (1974).
41. Widom, B. *J. chem. Phys.* **62**, 1332 (1975).
42. Ramos Gómez, F. *J. chem. Phys.* **74**, 4737 (1981).
43. Kim, M. W., Goldburg, W. I., Esfandiari, P., Levelt Sengers, J. M. H. and Wu, E.-S. *Phys. Rev. Lett.* **44**, 80 (1980).
44. Lang, J. C. and Widom, B. *Physica* **81A**, 190 (1975); Creek, J. L., Knobler, C. M. and Scott, R. L. *J. chem. Phys.* **67**, 366 (1977); Bocko, P. *Physica* **103A**, 140 (1980); Lovellette, M. *J. phys. Chem.* **85**, 1266 (1981).
45. Lang, J. C., Lim, P. K. and Widom, B. *J. phys. Chem.* **80**, 1719 (1976).
46. Rusanov, A. I. in *Progress in Surface and Membrane Science* (ed. J. F. Danielli, M. D. Rosenberg, and D. A. Cadenhead) Vol. 4, pp. 57–114, particularly the concluding paragraph. Academic Press, New York (1971).
47. Widom, B. *J. chem. Phys.* **67**, 872 (1977).

48. Khosla, M. P. and Widom, B. *J. Coll. Interf. Sci.* **76**, 375 (1980).
49. Beaglehole, D. *J. chem. Phys.* **73**, 3366 (1980). See also Ruderisch, B., Mang, H. and Findenegg, G. H. abstract of paper D38 read at the meeting of the Deutsche Bunsengesellschaft in Marburg, 29 May 1981.
50. Beaglehole, D. *J. chem. Phys.* **75**, 1544 (1981).
51. Fisher, M. E. and de Gennes, P. G. *C.R. Acad. Sc. Paris* **287**, B207 (1978).
52. Magerlein, J. H. and Sanders, T. M. *Phys. Rev. Lett.* **36**, 258 (1976).
53. Sobyanin, A. A. *Soviet Phys. JETP* **34**, 229 (1972); Hohenberg, P. C. *J. low Temp. Phys.* **13**, 433 (1973).
54. An elementary account of the theory is given by B. Widom in *Fundamental Problems in Statistical Mechanics III* (ed. E. G. D. Cohen) pp. 1–45, North-Holland, Amsterdam (1975). See also Domb, C. and Green, M. S. (eds) *Phase Transitions and Critical Phenomena*, Vol. 6, Academic Press, London (1976); Ma, S.-K. *Modern Theory of Critical Phenomena*, Benjamin, Reading, Mass. (1976); Pfeuty, P. and Toulouse, G. *Introduction to the Renormalization Group and to Critical Phenomena* (trans. G. Barton), Wiley, London (1977); and Stanley, ref. 13, Part V.
55. de Oliveira, M. J., Furman, D. and Griffiths, R. B. *Phys. Rev. Lett.* **40**, 977 (1978); Curado, E. M. F., Tsallis, C., Levy, S. V. F. and de Oliveira, M. J. *Phys. Rev. B* **23**, 1419 (1981).
56. Ohta, T. and Kawasaki, K. *Prog. Theor. Phys.* **58**, 467 (1977); Rudnick, J. and Jasnow, D. *Phys. Rev. B* **17**, 1351 (1978); Jasnow, D. and Rudnick, J. *Phys. Rev. Lett.* **41**, 698 (1978).
57. Jasnow, D., Ohta, T. and Rudnick, J. *Phys. Rev. B* **20**, 2774 (1979).
58. The idea of block spins goes back to Kadanoff, L. P. *Physics* **2**, 263 (1966). It was then elaborated and incorporated into the renormalization-group theory by Wilson, ref. 22.
59. Evans, R. *Mol. Phys.* **42**, 1169 (1981).
60. Zittartz, J. *Phys. Rev.* **154**, 529 (1967).
61. Wallace, D. J. and Zia, R. K. P. *Phys. Rev. Lett.* **43**, 808 (1979); Wallace, D. J., to appear in *Proceedings of the Cargèse Summer Institute on Phase Transitions*, 1980, Plenum, New York.
62. Sullivan, D. E., preprint, 'Statistical Mechanics of a Nonuniform Fluid with Long-Range Attractions', 1981.
63. Robledo, A. and Varea, C. *J. stat. Phys.* **26**, 513 (1981). The integral equation for the density $\rho(\mathbf{r})$ that is implied by this functional was found also, by other methods, by Percus, J. K. *J. stat. Phys.* **15**, 505 (1976), and then by Robledo, A. *J. chem. Phys.* **72**, 1701 (1980).
64. Wertheim, M. S. *J. chem. Phys.* **65**, 2377 (1976).
65. Wilson, K. G. *Phys. Rev. Lett.* **28**, 548 (1972); see also Domb and Green, ref. 54, Vol. 6. The exponents were first obtained correctly to first order in ε by Wilson and Fisher, ref. 22, by other techniques.
66. Brézin, E., Le Guillou, J. C., Zinn-Justin, J. and Nickel, B. G. *Phys. Lett. A* **44**, 227 (1973).
67. We assume in (9.140) that Ohta and Kawasaki, ref. 56, intended the quantity printed as '$2a/(3+a)\mathrm{sech}^2(r/(2L))$' in their equation (4.17) to be read as $[2a/(3+a)]\mathrm{sech}^2(r/2L)$, for only then is it consistent with their statement that for $\varepsilon = 1$ their profile is numerically close to that of Fisk and Widom [ref. 24; the present (9.93)].

For further references, see p. 313

APPENDIX 1

THERMODYNAMICS

In this appendix we define some of the thermodynamic terms we have used and we derive and collect the formulae to which we have referred.

In any infinitesimal change in the equilibrium state of an isotropic fluid, with no or neglible external fields, the changes in energy U, entropy S, volume V, and masses (or numbers of moles or molecules) n_i of the independent components $i = 1, \ldots, c$, are related by

$$dU = T \, dS - p \, dV + \boldsymbol{\mu} \cdot \mathbf{dn} \qquad (A1.1)$$

where T is the absolute temperature, p the pressure, and μ_i the chemical potential (mass, molar, or molecular, according to the meaning of n_i) of the ith species. We have used the vector notation introduced in (2.11) and (2.12), but not yet the more compressed form introduced after (2.34). In (A1.1) the only reversible-work term is $-p \, dV$. If the system is not an isotropic fluid, or if it is subject to variable external fields, there are additional or more general infinitesimal reversible-work terms: stress $\times$ strain (tensors), gravitational field $\times$ mass, electric field $\times$ electric polarization, magnetic field $\times$ magnetization, etc.; but $-p \, dV$ is representative, and for the purpose at hand (A1.1) is general enough. It relates only the bulk properties of a macroscopic system; the generalization to include surface properties is the subject of Chapter 2.

The function $U(S, V, \mathbf{n})$ is an extensive function of extensive arguments, and so is homogeneous of degree 1 in those arguments: $U(\lambda S, \lambda V, \lambda \mathbf{n}) \equiv \lambda U(S, V, \mathbf{n})$ for all λ. (Cf. footnote in § 9.4.) Hence, from Euler's theorem,[1]

$$U = TS - pV + \boldsymbol{\mu} \cdot \mathbf{n}. \qquad (A1.2)$$

The Helmholtz and Gibbs free energies F and G are

$$F = U - TS, \qquad G = F + pV = \boldsymbol{\mu} \cdot \mathbf{n}. \qquad (A1.3)$$

The energy density $\phi = U/V$, entropy density $\eta = S/V$, Helmholtz-free-energy density $\psi = F/V$, and mass (or molar or molecular) densities $\rho_i = n_i/V$ then satisfy

$$d\phi = T \, d\eta + \boldsymbol{\mu} \cdot \mathbf{d\rho} \qquad (A1.4)$$

$$\psi = \phi - T\eta = -p + \boldsymbol{\mu} \cdot \boldsymbol{\rho} \qquad (A1.5)$$

$$d\psi = -\eta \, dT + \boldsymbol{\mu} \cdot \mathbf{d\rho} \qquad (A1.6)$$

$$dp = \eta \, dT + \boldsymbol{\rho} \cdot \mathbf{d\mu}. \qquad (A1.7)$$

The last of these is the Gibbs–Duhem equation.

A function of appropriately chosen independent variables, from which all thermodynamic properties are derivable by differentiations alone, with no integrations required, is said to be a thermodynamic potential. Examples are $U(S, V, \mathbf{n})$, $S(U, V, \mathbf{n})$, $F(T, V, \mathbf{n})$, $G(T, p, \mathbf{n})$, and the grand-canonical free energy $-pV = \Omega(T, V, \boldsymbol{\mu})$. Intensive functions of intensive arguments, from which all the intensive properties of the system are derivable, are also called thermodynamic potentials; examples are $\phi(\eta, \boldsymbol{\rho})$, $\eta(\phi, \boldsymbol{\rho})$, $\psi(T, \boldsymbol{\rho})$, and $-p(T, \boldsymbol{\mu})$. These are not essentially different from the corresponding extensive potentials, here U, S, F, and Ω, respectively.

To illustrate the derivation of thermodynamic properties from a potential, we may consider $F(T, V, \mathbf{n})$, which, from (A1.1) and the first of (A1.3), satisfies the differential identity

$$dF = -S\,dT - p\,dV + \boldsymbol{\mu}\cdot d\mathbf{n}. \tag{A1.8}$$

The system's entropy, pressure, and chemical potentials are thus obtained as functions of T, V, $\mathbf{n}$ by

$$S = -(\partial F/\partial T)_{V,\mathbf{n}}, \qquad p = -(\partial F/\partial V)_{T,\mathbf{n}}, \qquad \mu_i = (\partial F/\partial n_i)_{T,V,\,\text{all}\,n_{j(\neq i)}}; \tag{A1.9}$$

and then the energy by $U = F + TS$; the Gibbs free energy by $G = F + pV$ or $\boldsymbol{\mu}\cdot\mathbf{n}$; the coefficient of thermal expansion by $\alpha = V^{-1}(\partial V/\partial T)_{p,\mathbf{n}} = -(\partial p/\partial T)_{V,\mathbf{n}}/V(\partial p/\partial V)_{T,\mathbf{n}}$; the isothermal compressibility by $\kappa = -[V(\partial p/\partial V)_{T,\mathbf{n}}]^{-1}$; the constant-volume heat capacity by $C_V = (\partial U/\partial T)_{V,\mathbf{n}}$ or $T(\partial S/\partial T)_{V,\mathbf{n}}$; the constant-pressure heat capacity by $C_p = C_V + TV\alpha^2/\kappa$; etc.

The extensive thermodynamic potentials arise in statistical mechanics as the logarithms (or k or $-kT$ times the logarithms) of partition functions that are themselves functions of the appropriate independent variables. Thus,

$$\begin{aligned}
S(U, V, \mathbf{n}) &= k\ln Q(U, V, \mathbf{n}) \\
F(T, V, \mathbf{n}) &= -kT\ln Z(T, V, \mathbf{n}) \\
G(T, p, \mathbf{n}) &= -kT\ln \check{Z}(T, p, \mathbf{n}) \\
\Omega(T, V, \boldsymbol{\mu}) &= -kT\ln \Xi(T, V, \boldsymbol{\mu})
\end{aligned} \tag{A1.10}$$

where Q, Z, $\check{Z}$, and Ξ are the partition functions of the microcanonical, (petit-) canonical, 'isothermal-isobaric', and grand(-canonical) ensembles, respectively.

We consider the differential identities (A1.4), (A1.6), and (A1.7) satisfied by the intensive potentials ϕ, ψ, and p, and the equivalent identities

$$d\eta = \frac{1}{T} d\phi - \frac{\boldsymbol{\mu}}{T} \cdot d\boldsymbol{\rho} \qquad (A1.11)$$

$$d\frac{\psi}{T} = \phi \, d\frac{1}{T} + \frac{\boldsymbol{\mu}}{T} \cdot d\boldsymbol{\rho} \qquad (A1.12)$$

$$d\frac{p}{T} = -\phi \, d\frac{1}{T} + \boldsymbol{\rho} \cdot d\frac{\boldsymbol{\mu}}{T} \qquad (A1.13)$$

satisfied by the potentials η, ψ/T, and p/T. [That (A1.12) and (A1.13) are equivalent to their earlier counterparts follows from (A1.5).] Identifying the differential coefficients in those six relations, we obtain important thermodynamic identities, of which the most used are

$$\left(\frac{\partial \phi}{\partial \eta}\right)_{\boldsymbol{\rho}} = \left(\frac{\partial \eta}{\partial \phi}\right)_{\boldsymbol{\rho}}^{-1} = T \qquad (A1.14)$$

$$\left(\frac{\partial \psi}{\partial T}\right)_{\boldsymbol{\rho}} = -\left(\frac{\partial p}{\partial T}\right)_{\boldsymbol{\mu}} = -\eta \qquad (A1.15)$$

$$\left(\frac{\partial \psi/T}{\partial 1/T}\right)_{\boldsymbol{\rho}} = -\left(\frac{\partial p/T}{\partial 1/T}\right)_{\boldsymbol{\mu}/T} = \phi \qquad (A1.16)$$

$$\left(\frac{\partial \psi}{\partial \rho_i}\right)_{T, \text{all } \rho_{j(\neq i)}} = \mu_i, \qquad \left(\frac{\partial \psi}{\partial \rho}\right)_T = \mu \quad \text{(one component)} \qquad (A1.17)$$

$$\left(\frac{\partial p}{\partial \mu_i}\right)_{T, \text{all } \mu_{j(\neq i)}} = \left(\frac{\partial p/T}{\partial \mu_i/T}\right)_{T, \text{all } \mu_{j(\neq i)}} = \rho_i \qquad (A1.18)$$

$$\left(\frac{\partial p}{\partial \mu}\right)_T = \left(\frac{\partial p/T}{\partial \mu/T}\right)_T = \rho \quad \text{(one component).} \qquad (A1.19)$$

Further, from (A1.16) and (A1.17),

$$\left(\frac{\partial \mu_i/T}{\partial 1/T}\right)_{\boldsymbol{\rho}} = \left(\frac{\partial \phi}{\partial \rho_i}\right)_{T, \text{all } \rho_{j(\neq i)}}$$

$$\left(\frac{\partial \mu/T}{\partial 1/T}\right)_{\boldsymbol{\rho}} = \left(\frac{\partial \phi}{\partial \rho}\right)_T \quad \text{(one component);} \qquad (A1.20)$$

while from (A1.17) alone,

$$\left(\frac{\partial \mu_i}{\partial \rho_j}\right)_{T, \text{all } \rho_{k(\neq j)}} = \left(\frac{\partial \mu_j}{\partial \rho_i}\right)_{T, \text{all } \rho_{k(\neq i)}}$$

$$\left(\frac{\partial \mu_a}{\partial \rho_b}\right)_{T, \rho_a} = \left(\frac{\partial \mu_b}{\partial \rho_a}\right)_{T, \rho_b} \quad \text{(two components).} \qquad (A1.21)$$

In the text, we have often replaced μ_i by $kT \ln(\Lambda_i^3 \zeta_i)$ in these formulae, or sometimes taken only the configurational (non-kinetic) parts of the thermodynamic functions, which satisfy the same relations. In (5.11), for example, U is potential energy rather than total energy; but $kT \ln \zeta_i$ is the configurational part of μ_i, and (5.11) is then the one-component form of (A1.20).

Considering again the differential identities (A1.4), (A1.6), (A1.7), and (A1.11)–(A1.13), we observe that in each of the partial differentials $x \, dy$ of which the right-hand sides are composed, the thermodynamic function y and its differential coefficient x are a conjugate pair: one is a field—an intensive function that takes equal values in coexisting phases, and, more generally, has a uniform value even in an inhomogeneous system; and the other is a density—an intensive function that need not, and in general does not, have equal values in coexisting phases, and thus, more generally, need not be spatially uniform. Indeed, it is when its densities are not uniform on a macroscopic scale that a system is said to be inhomogeneous. In thus distinguishing fields and densities we follow Griffiths and Wheeler,[2] to whom we also owe those names. The distinction is more important than that between intensive and extensive functions, which is often superficial. 'Densities' may be general ratios of extensive quantities (as we shall see shortly) and so may include mole ratios or other measures of chemical composition. Mistura[3] has shown what choices of variables are allowable if the partial differentials that make up the differential of the potential are still to be composed properly of conjugate fields and densities: the fields may be chosen to be more or less arbitrary functions of T and μ, and then the Gibbs–Duhem relation (A1.7) or an equivalent determines the densities that are conjugate to those fields.

We return to the Gibbs–Duhem equation, and for the moment rewrite it in the extensive form

$$0 = -V \, dp + S \, dT + \mathbf{n} \cdot \mathbf{d\mu}. \qquad (A1.22)$$

Of the $c+2$ fields $-p, T, \mu_1, \ldots, \mu_c$, we choose any $c+1$ to be the components of a new, augmented vector, which for convenience we continue to call μ. [This is the more compressed notation we introduced after (2.34).] Of the $c+2$ fields in the list above, let the one that is not chosen to be one of the $c+1$ components of the new vector μ, be called μ_{c+2}; let the extensive coefficient of $d\mu_i$ in (A1.22) be called X_i; and let X_i/X_{c+2} for all $i = 1, \ldots, c+1$ be called the 'density' ρ_i, the components of a new, augmented vector ρ. If μ_{c+2} has been chosen to be the field $-p$, then X_{c+2} is V and ρ_i is literally a density, as were the components of the original vector ρ in (A1.7). Alternatively, if μ_{c+2} is the chemical potential of one of the constituents, then $c-1$ of the $c+1$ 'densities' ρ_i are the mole or mass ratios n_i/n_{c+2}, which determine the chemical composition.

In this new notation, (A1.22) is

$$-d\mu_{c+2} = \boldsymbol{\rho} \cdot \mathbf{d\mu}. \tag{A1.23}$$

Thus, the field $-\mu_{c+2}$, which is now a potential, is a function $-\mu_{c+2}(\boldsymbol{\mu})$ of the $c+1$ fields $\boldsymbol{\mu}$, with differential coefficients $\boldsymbol{\rho}$. In a one-phase system, the fields $\boldsymbol{\mu}$ may vary independently, and (A1.23) then determines the concomitant variation of the dependent potential $-\mu_{c+2}$. But when two phases, α and β, are in equilibrium, the $c+1$ fields $\boldsymbol{\mu}$ are no longer independent, for their possible variations are constrained by the equilibrium condition $\mu_{c+2}^{\alpha}(\boldsymbol{\mu}) = \mu_{c+2}^{\beta}(\boldsymbol{\mu})$; or equivalently, from (A1.23), by

$$(\boldsymbol{\rho}^{\alpha} - \boldsymbol{\rho}^{\beta}) \cdot \mathbf{d\mu} = 0. \tag{A1.24}$$

This is a general form of the Clapeyron equation. If the system is of one component, and if we choose $\mu_{c+2}(=\mu_3)$ to be its chemical potential μ, so that μ_1 and μ_2 are $-p$ and T and ρ_1 and ρ_2 are V/n and S/n, then (A1.24) becomes $-[(V/n)^{\alpha} - (V/n)^{\beta}] \, dp + [(S/n)^{\alpha} - (S/n)^{\beta}] \, dT = 0$, the Clapeyron equation in its most familiar form.

Although a density does not generally have the same value in two coexisting phases, it may have. We call that event an azeotropy, thus extending the meaning of the word, which originally meant the circumstance of two equilibrium phases' having the same chemical composition. At an azeotropic point of the phase equilibrium, then, one or more of the components of $\boldsymbol{\rho}^{\alpha} - \boldsymbol{\rho}^{\beta}$ in (A1.24) vanishes. Now consider those fields μ_i that are conjugate to densities ρ_i *not* involved in the azeotropy; i.e. fields μ_i for which $\rho_i^{\alpha} - \rho_i^{\beta} \neq 0$. Their possible variations are restricted by

$$\sum_i (\rho_i^{\alpha} - \rho_i^{\beta}) \, d\mu_i = 0, \tag{A1.25}$$

where now the summation is taken to extend over only those i for which $\rho_i^{\alpha} - \rho_i^{\beta} \neq 0$. At an azeotropic point, then, fields fewer in number than $c+1$—say, $c'+1$ of them, with $c' < c$—satisfy a Clapeyron equation among themselves, the fields in question being those whose conjugate densities are not party to the azeotropy. The two-phase equilibrium at such a point is like that in a system of $c' < c$ components, the difference $c - c'$ being the number of densities that take part in the azeotropy.

Of the $c'+1$ fields in (A1.25) suppose we fix all but one—some chosen μ_j, say. Then at the azeotropic point, in any variation of the state of the system that preserves the two-phase equilibrium, $d\mu_j = 0$; that is, at fixed values of all the other fields $\mu_i (i \neq j)$ in (A1.25) the field μ_j is at an extremum. This is the most familiar property of an azeotrope. Suppose, for example, we have an azeotropy of the kind that gives the phenomenon its name; viz. one in which the chemical compositions of the two phases are identical. Choose μ_{c+2} in (A1.23) to be the chemical potential of one of the constituents, so that, as we saw there, the $c-1$ chemical-composition variables are then among the $c+1$ densities $\boldsymbol{\rho}$, while $-p$ and T are the fields conjugate to the remaining two densities. Then at the azeotrope in which the chemical compositions of the two phases are identical, $-p$ and T satisfy (A1.25); from which we conclude that p is extremal at fixed T and T is extremal at fixed p, the familiar property of ordinary azeotropy. But as a thermodynamic concept azeotropy is more general than that, and we say it occurs whenever $\rho_i^\alpha - \rho_i^\beta = 0$ for one or more densities ρ_i.

Azeotropy is equivalent, as we saw, to an extremal property of the fields that are conjugate to the remaining densities. In a one-component system, as we know from the ordinary form of the Clapeyron equation, if $(V/n)^\alpha = (V/n)^\beta$ at some point, the tangent to the p, T equilibrium curve at that point is parallel to the p-axis, and the temperature on the equilibrium curve is then a minimum or a maximum there; while if at some point $(S/n)^\alpha = (S/n)^\beta$, the tangent to the equilibrium curve at that point is parallel to the T-axis, so that p on the curve is at an extremum there. These are azeotropies, if the word is taken in its general sense.

The phenomenon of azeotropy is most readily understood geometrically in the thermodynamic space of the $c+1$ independent fields, like the p, T space we just considered for a one-component system. If the coexistence surface in this space is at some point parallel to one of the coordinate axes; that is, if the (hyper-) plane that is tangent to the surface at that point is parallel to one of the axes, there the density conjugate to the field that is measured in the direction of that axis has equal values in the two phases; and the values, on the surface, of the remaining fields (if the surface is not there parallel also to one or more of their coordinate axes) are then extremal at that point. Where the coexistence surface is parallel simultaneously to two or more of the coordinate axes, each of two or more densities has equal values in the two phases, and the remaining fields are extremal. There may in general be a $(c-1)$-dimensional manifold of points of the c-dimensional coexistence surface at which the surface is parallel to a selected one of the coordinate axes and hence at which the density conjugate to that field has equal values in the coexisting phases; a $(c-2)$-dimensional manifold of states that are azeotropic in two selected densities; etc.; and, finally, there can be at most single, isolated states that are azeotropic in c densities at once.

Azeotropies do not necessarily occur; the foregoing are merely limitations on their occurrence. On the other hand, they can always be made to occur by a suitable choice of fields. With each point of the two-phase coexistence surface we may associate a new coordinate system in which any number up to c of the coordinate axes is chosen parallel to the surface at that point. In a description based on the new fields defined by such coordinates, that arbitrary two-phase state is azeotropic in up to c densities—the new densities that are conjugate to those new fields. Thus, to some extent azeotropy is only a matter of representation; but it is noteworthy when it occurs with ordinary laboratory densities, and not when it is made to occur by special choice of variables.

Notes and references

1. If $f(x, y, ...)$ is homogeneous of degree n in $x, y, ..., ;$ i.e., if $f(\lambda x, \lambda y, ...) \equiv \lambda^n f(x, y, ...)$ for all λ, then $x \, \partial f/\partial x + y \, \partial f/\partial y + ... = nf$.
2. Griffiths, R. B. and Wheeler, J. C. *Phys. Rev.* **A2,** 1047 (1970).
3. Mistura, L. *Physica* **104A,** 181 (1980).

Additions to References

Chapter 8

§ 8.3 On the thickness of gravity-thinned layers, see Kayser, R. F., Moldover, M. R. and Schmidt, J. W. *J. chem. Soc. Faraday Trans.* 2 **82,** 1701 (1986).
§ 8.5 The name 'Cahn transition' has now generally been replaced by 'wetting transition'. See further work and reviews by de Gennes, P. G. *Rev. mod. Phys.* **57,** 827 (1985); Sullivan, D. E. and Telo da Gama, M. M., in *Fluid Interfacial Phenomena* (ed. C. A. Croxton) Wiley, Chichester (1986); Indekeu, J. O. *Phys. Rev.* B **36,** 7296 (1987); Dietrich, S., in *Phase Transitions and Critical Phenomena,* Vol. 12 (ed. C. Domb and J. L. Lebowitz) Academic Press, New York (1988); Kahlweit, M., Busse, G., Haase, D., and Jen, J. *Phys. Rev.* A **38,** 1395 (1988); R. Lipowsky, *Critical Behavior of Interfaces: Wetting, Surface Melting and Related Phenomena,* Habilitations-Schrift, Ludwig-Maximilians-Universität, München (1987) (Spez. Ber. der Kernforschungsanlage, Jülich–Nr. 438); Schick, M., in *Liquides aux Interfaces/Liquids at Interfaces* (ed. J. Charvolin, J. F. Joanny, and J. Zinn-Justin) Les Houches, Session XLVIII, 1988, Elsevier, Amsterdam (1989).

Chapter 9

New reviews on critical phenomena at interfaces include: Binder, K., in *Phase Transitions and Critical Phenomena,* Vol. 8 (ed. C. Domb and J. L. Lebowitz) Academic Press, New York (1983); Jasnow, D. *Rep. Prog. Phys.* **47,** 1059 (1984).
§ 9.4 Extensions and tests of this theory include Binder, K. *Phys. Rev.* A **25,** 1699 (1982); Gielen, H. L., Verbeke, O. B., and Thoen, J. *J. chem. Phys.* **81,** 6154 (1984); Chaar, H., Moldover, M. R., and Schmidt, J. W. *J. chem. Phys.* **85,** 418 (1986); Mon, K. K. and Jasnow, D. *Phys. Rev.* A **31,** 4008 (1985); *J. stat. Phys.* **41,** 273 (1985); Mon, K. K. *Phys. Rev. Lett.* **60,** 2749 (1988).
§ 9.5 For experimental measurements of interfacial tensions near tricritical points, see Pegg, I. L., Goh, M. C., Scott, R. L., and Knobler, C. M. *Phys. Rev. Lett.* **55,** 2320 (1985); Sundar, G. and Widom, B. *J. phys. Chem.* **91,** 4802 (1987).

APPENDIX 2

DIRAC'S DELTA-FUNCTION

Dirac called his delta-function an 'improper function' since it is not a function as that term is defined in analysis. It has a meaning only when used within an integral, or when such use is implied. The function $\delta(x-a)$ can then be described by saying that it is infinitely sharply peaked at $x=a$, and zero elsewhere, and that it is normalized to unity. Hence

$$\int \delta(x-a)\,dx = 1, \qquad \int f(x)\delta(x-a)\,dx = f(a), \qquad (A2.1)$$

where $f(x)$ is any smooth function of x. The limits of integration are unimportant as long as they include the point $x=a$; generally they are $\pm\infty$. The function can be given a more concrete form by expressing it as the limit of a sequence of proper functions, for example,

$$\delta(x-a) = \lim_{\alpha \to 0} [\alpha^{-1}\pi^{-\frac{1}{2}} \exp(-(x-a)^2/\alpha^2)]. \qquad (A2.2)$$

It is the derivative of a unit step-function, which is zero for $x < a$, and unity for $x > a$.

The following properties follow from the definition (A2.1).

$$\delta(b(x-a)) = |b|^{-1}\delta(x-a), \qquad (A2.3)$$

of which a special case $(b=-1)$ is

$$\delta(a-x) = \delta(x-a). \qquad (A2.4)$$

These two results follow at once from (A2.1) by changing the variable of integration to $y = bx$. Similarly,

$$\int \delta(x-a)\delta(x-b)\,dx = \delta(a-b), \qquad (A2.5)$$

$$\frac{\partial}{\partial x}\delta(x-a) = -\frac{\partial}{\partial a}\delta(x-a), \qquad (A2.6)$$

$$(x-a)\frac{\partial}{\partial x}\delta(x-a) = -\delta(x-a). \qquad (A2.7)$$

The dimensions of $\delta(x-a)$ are those of a^{-1}.

In statistical mechanics we usually use three-dimensional delta-functions whose arguments are vector separations,

$$\delta(\mathbf{r}_1 - \mathbf{r}_2) = \delta(x_1 - x_2)\delta(y_1 - y_2)\delta(z_1 - z_2) \qquad (A2.8)$$

and whose dimensions are V^{-1}. From the results above,

$$\int f(\mathbf{r}_2)\delta(\mathbf{r}_1 - \mathbf{r}_2) \, d\mathbf{r}_2 = f(\mathbf{r}_1), \qquad (A2.9)$$

$$\int \delta(\mathbf{r} - \mathbf{r}_1)\delta(\mathbf{r} - \mathbf{r}_2) \, d\mathbf{r} = \delta(\mathbf{r}_1 - \mathbf{r}_2), \qquad (A2.10)$$

$$\delta(b(\mathbf{r}_1 - \mathbf{r}_2)) = |b|^{-3}\delta(\mathbf{r}_1 - \mathbf{r}_2), \qquad (A2.11)$$

$$\nabla_1 \delta(\mathbf{r}_1 - \mathbf{r}_2) = -\nabla_2 \delta(\mathbf{r}_1 - \mathbf{r}_2). \qquad (A2.12)$$

These results, and particularly (A2.9), (A2.10), and (A2.12), are used in many places in Chapter 4.

Finally, we quote but do not use the Fourier transform

$$\delta(\mathbf{q}) = \frac{1}{(2\pi)^3} \int e^{i\mathbf{q}\cdot\mathbf{r}} \, d\mathbf{r}. \qquad (A2.13)$$

NAME INDEX

Abraham, D. B. 115, 120
Abraham, F. F. 115, 176–7, 187, 196, 198–203
Abrahams, E. 270
Adam, N. K. 212, 226–7
Adams, D. J. 82, 178–9
Adamson, A. W. 11, 37
Alexander, A. E. 11
Andersen, H. C. 192, 194, 201, 203
Andrews, T. 15, 93
Antonow, G. N. 212–13
Ashcroft, N. W. 192
Atkins, K. R. 115
Ažman, A. 187, 191

Bach, J. 11
Bacon, R. C. 174
Bailey, A. I. 8
Bakhuis Roozeboom, H. W. 276
Bakker, G. v, 3, 16, 33, 47, 112, 211
Barker, J. A. 92, 157, 176–9, 182–5, 187, 192, 196, 200–3
Barker, R. 187
Bartell, F. E. 213
Baxter, R. J. 196
Beaglehole, D. 174, 217, 293–4
Bellemans, A. 8, 33, 37, 38, 42–3
Bellman, R. 119
Bellocq, A. M. 278
Benedek, G. B. 276
Bennett, C. H. 182
Berne, B. J. 175–6, 182–3, 187
Bernoulli, D. 3
Bikerman, J. J. 3, 8
Billups, R. 8
Binder, K. 176
Bishop, A. R. 231
Blum, L. 196
Bocko, P. 284
Bogoliubov, N. N. 75, 77, 85
Boltzmann, L. 20, 69
Bongiorno, V. 101, 124, 197, 199

Bonissent, A. 187
Bonnet, J. C. 11
Born, M. 75, 115
Borštnik, B. 187, 191
Boruvka, L. 38, 232, 235
Boscovich, R. J. 2
Bottomley, G. A. 11
Boucher, E. A. 9, 10
Bouchiat, M.-A. 174, 276
Bourbon, D. 278
Bowditch, N. 2, 18, 19
Boys, C. V. 237, 240
Brézin, E. 301
Briant, C. L. 176
Bricmont, J. 120
Brossel, J. 174
Broughton, J. Q. 187
Brout, R. 131
Bruce, H. D. 174
Brush, S. G. 3, 14, 18
Buckingham, M. J. 57, 268–9
Buff, F. P. 37, 38, 79, 86, 88–90, 93, 94, 98, 101–3, 110, 112–13, 115–16, 124, 174, 195, 211, 231–2, 239–40, 264
Burton, J. J. 176
Byckling, E. 82
Byrne, D. 174

Cahn, J. W. 50, 56, 62, 208, 213–19, 224, 228–31, 244, 249, 295
Cape, J. N. 187
Carey, B. S. 45, 91
Cash, L. 278
Castle, P. J. 174, 191, 195
Cavendish, H. 2
Cayias, J. L. 278
Challis, J. 3, 19
Chandler, D. 192, 194, 201, 203
Chapela, G. A. 177–9, 182–3, 185, 192, 199
Chester, G. V. 187
Clark, P. 25
Clarke, J. H. R. 187
Clausius, R. 25, 86
Co, K. U. 191–2, 194

Cole, M. W. 115
Cotter, M. A. 132
Cramér, H. 99
Creek, J. L. 284
Crosland, M. 10
Crowell, A. D. 227
Croxton, C. A. 177, 190, 192, 194
Cruchon, D. 174
Curado, E. M. F. 297, 301
Currie, J. F. 231

Dahler, J. S. 69, 105
Dash, J. G. 227
Davis, H. T. 45, 91, 101, 115, 124, 197–9, 213, 215, 224, 230–1, 278
de Boer, J. 14, 131, 137
De Dominicis, C. 94
Defay, R. 8, 33, 37, 38, 42–3
de Feijter, J. A. 237–8, 240
De Feraudy, M. F. 176
de Gennes, P. G. 202, 217, 224, 294
DeHaven, P. W. 101–3, 113
de Oliveira, M. J. 82, 297, 301
Desai, R. C. 69, 91, 105
Dickinson, E. 187
Domb, C. 141, 296, 299, 301
Donahue, D. J. 213
Dupré, A. 5, 13–15, 19

Earnshaw, J. C. 174
Ebner, C. 94, 97–8, 196–9, 202–3
Edmonds, B. 37
Efremova, G. D. 276–7, 284
Erpenbeck, J. J. 175
Esfandiari, P. 284
Etters, R. D. 176
Evans, R. 57, 69, 73, 78, 79, 89, 94–6, 98, 120–1, 197, 199, 202–3, 299–301
Everett, D. H. 42

Faraday, M. 3
Farges, J. 176
Ferrier, R. P. 177, 192

SUBJECT INDEX

action (in dynamical analogy) 53, 56, 58, 150–1, 218, 220, 227–8, 231
activity or chemical potential 31, 33, 71, 75, 81–4, 99, 124, 129–48, 152, 154, 163, 168, 178, 197, 244–6, 250, 262, 271, 307–12
  intrinsic 73, 96
  local 82, 124, 178, 190, 194
adhesion 43
adsorption 32, 34–7, 40, 60–4, 113, 161, 173–4, 185–7, 225–7, 232–3, 236, 264, 285–6, 293–5
  linear 233, 236
  see also Gibbs adsorption equation
$\alpha\gamma$ interface 208, 214–25, 227–32, 237, 244, 280–1, 295
amphiphile 277
aneotropy 48
angle of contact, *see* contact angle(s)
Antonow's rule 208, 212–20, 230, 280, 284, 295–6
argon
  capillary constant 116
  interatomic potential 176, 184, 201
  internal pressure 19
  separation of dividing surfaces 37–8
  surface tension 11–12, 176, 184, 201
  thickness of surface 117, 174, 187, 201
azeotropy 37, 57, 225, 311–13
  surface 37
  see also aneotropy

background potential, *see* mean field
barometric law 74–5
BBGKY equation, *see* Yvon–Born–Green equation
Bethe–Guggenheim approximation 141–5, 149
black film, Newton's 240
block spins 298–9, 306
borderline dimensionality 122, 268–70, 282–3, 291, 301
Bose gas 268–9
bubble 27, 38, 42, 167–8
  see also curvature, surface; spherical surface

Cahn transition 208–9, 214, 220, 225–32, 247–8, 295
  critical point 227–32

caloric theory 2–3, 21
capillarity, defined v, 22
capillary
  constant or length 10–11, 116, 300
  phenomena v, 1–3, 9–10, 15, 17, 20–4
  rise 10
  tube xii, 1, 9–11
  waves 11, 115–22, 180–1, 198, 300–3
capillary-wave divergences, *see* thickness of interface, divergence
carbon dioxide 304
carbon monoxide 173
chemical potential, *see* activity or chemical potential
chlorine 185, 187
Clapeyron equation 35, 225, 236, 311–12
  surface 226
coexistence curve and densities 59, 143–4, 147–8, 152–3, 155, 178–9, 196, 230, 250, 252–3, 261, 265, 288–9, 310, 312–13
coherence length, *see* fluctuations, coherence length
cohesion 1, 3, 8, 25
  see also potential, intermolecular
common-tangent construction 51–8, 64–5
compressibility
  equation 74, 101, 104, 157, 196, 256
  isothermal 72, 79, 85, 101, 127, 168, 250, 256, 258–9, 261–2, 265–6, 273–5, 288, 308
  local 200–1
  osmotic 262, 282–4, 288
  of water 11
computer simulation v, 37, 82, 121, 126, 173–87, 190–5, 198–9, 201–4
contact angle(s) 8–9, 61, 207, 209–12, 231, 237, 239, 241, 245
contact line, *see* three-phase line
convex-envelope construction 57, 64–5, 68
coordination number 131, 136–49
correlation function(s) 20, 69–85, 105–6, 297–8
  density-density 77, 91, 108, 118, 121
  direct, *see* direct correlation function
  near a wall 196
  time-dependent 105–7
  total 72–4, 77, 85, 117–20, 127, 157, 195–6, 255–61, 263, 266, 273–4, 303

parallel plates 1, 11
partition functions 70–1, 85, 95, 152, 298–9, 308
penetrable-sphere (and related) models 129, 151–71, 173, 180, 195–6, 199, 202, 246
    primitive or two-component 149, 152, 163–4, 167–9, 244, 246
    three-component 244–6
    see also surface tension of penetrable-sphere models
Percus–Yevick approximation 101, 127, 196–9
perfect gas 74, 85–6, 95–6, 290
perturbation theories 98, 185, 191–2, 194, 200–4, 299, 301
phase
    non-autonomous 33
    surface, see surface phase
    transition in interfaces 225–32
    see also Cahn transition
Plateau border 237
Poisson brackets 105–7
polymers 204
potassium chloride, molten 187
potential-distribution theorem 82, 84, 94, 124, 126, 129–32, 154–6, 178, 190, 194
    see also test particle
potential-energy surface, see free-energy surface
potential, external, see field, external
potential, intermolecular 1–9, 13–14, 16–21, 70, 79–82, 86, 89, 101, 130–3, 151, 157, 173, 175, 200, 271
    moments of 6, 17–19, 101, 135
    orientational 76, 79, 91, 105, 109, 173, 201, 204
    pair or two-body 75, 77, 79, 83, 85–7, 91, 98, 104–5, 123, 136, 140–1, 190, 201–2
    range of 3, 18–20, 25, 28, 47, 59, 65–6, 82, 90, 92, 98, 123–4, 129–31, 152, 176, 178, 180, 202, 217, 224, 256–60
    square-well 192
    sticky-sphere 196
    three-body, etc. 83, 86, 91, 105, 109, 152, 182, 184–5, 201
    virial of, see virial, intermolecular
    see also argon, interatomic potential; Lennard-Jones potential
potential, thermodynamic 33–4, 57, 62, 94, 225–6, 308–11
pressure
    equation, see virial equation
    gradient 79, 86, 105, 108
    hydrostatic 10, 79, 110, 113
    internal 3–7, 9–10, 14–15, 17, 19

moments of 28, 45, 90–1, 110, 124
tensor 43–6, 69, 85–93, 105–6, 109, 112–13, 124, 171, 187, 195, 204, 239, 307
    arbitrariness of 86, 91, 93, 113, 124, 187
profile
    bare 115, 121, 198
    density or composition 16–17, 24, 47, 50, 53–6, 60, 64, 66, 74, 76, 79–84, 120–3, 131–5, 139, 145–7, 151, 153, 156, 161–5, 168–9, 173–4, 177–82, 185–7, 190–204, 217–18, 222–5, 240–7, 250–5, 260, 270–6, 280–1, 285, 287, 297, 300–3.
    hyperbolic tangent 156, 180, 201, 244, 251–2, 276, 281, 302–3, 306
    of liquid near a solid 187, 196–8, 205
    of orientation, see orientation of molecules at an interface
    oscillations 177–9, 191–2, 194
    of refractive index 174
    'wings' of 180, 195, 202, 224, 273–4

quantal liquids 94, 204
    see also helium; mixtures, ^{3}He + ^{4}He; superfluid
quasi-chemical approximation, see Bethe–Guggenheim approximation

Rayleigh's theory of surface tension 13, 52, 55, 57, 89, 93, 101, 199
reflectivity 302–3
regular-solution theory 294
renormalization-group theory 66–7, 255, 269, 275, 296–303, 306
roughening of an interface 120

scaling and hyperscaling 255, 267–9, 271, 280, 283, 294
soap films 237, 240
spherical surface 38–43, 91, 109–14, 167–71
    see also bubble; curvature, surface; drop
spreading, see wetting or spreading
stability, mechanical or hydrostatic 1, 15, 43–4, 79, 86, 88, 110, 113, 124
statistical mechanics of liquid-gas surface 18, 33–4, 43, 50, 52, 69–128, 190–204, 255–61, 296–303
stress tensor, see pressure tensor
sulphur hexafluoride 305
sum rules 105, 191, 195
superfluid 262, 283, 296
surface curvature, see curvature, surface
surface density, see adsorption
surface, dividing, see dividing surface
surface energy 29, 123, 155–9, 173, 183
surface, equimolar, see dividing surface
surface excess, see adsorption

with non-classical exponents 270–6
see also two- (or more-) density van der
  Waals theory
theory of liquids and phase transitions
  15–16, 51, 54–5, 132–3
vapour pressure 37, 42–3
virial coefficient, second 184
virial
  equation 86, 104, 157
  intermolecular 18, 86, 91, 124, 182
    three-body 86
  theorem 86, 126
vortex theory of atoms 25

water
  capillary constant 10
  compressibility 11
  in contact with glass 9
  density profile 204
  internal pressure 19
  surface tension 11–12, 36
  vapour pressure of drops 42
waves
  capillary, *see* capillary waves

gravitational, *see* gravity waves
Weiss approximation 131, 137
  see also mean-field approximation
wetting or spreading 3, 9, 22, 208, 212–25,
  227–31, 247, 280, 295
  coefficient 9, 208, 214–16, 231
  film or layer 208, 216–19, 223–5, 228, 230
  transition on a solid 247
  see also Cahn transition

xenon 304
X-rays
  group and phase velocities 174
  scattering and reflection 173–4

YBG equation, *see* Yvon–Born–Green
  equation
Young's equation 9–10, 211
Yvon–Born–Green equation 75–6, 79, 84,
  88, 91, 125, 190–5, 198, 202–4, 303
  orientational 76
Yvon equation
  first 77–8, 118
  second 77–8, 102, 108, 118